Forestry and Biodiversity

Edited by Fred L. Bunnell and Glen B. Dunsworth

Forestry and Biodiversity

Learning How to Sustain Biodiversity in Managed Forests

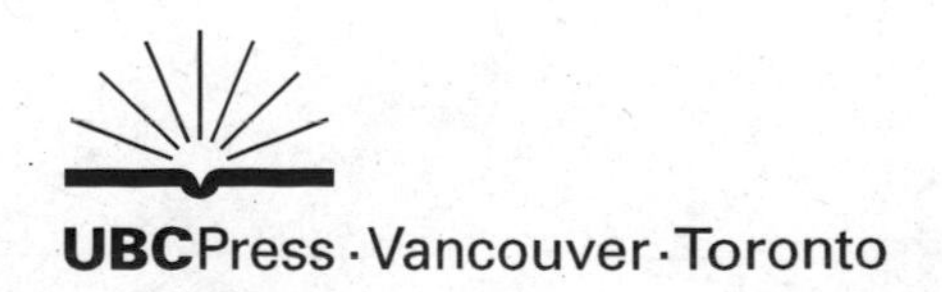

UBCPress · Vancouver · Toronto

20 19 18 17 16 15 14 13 12 11 10 09   5 4 3 2 1

Printed in Canada with vegetable-based inks on FSC-certified ancient-forest-free paper (100 percent post-consumer recycled) that is processed chlorine- and acid-free.

Library and Archives Canada Cataloguing in Publication

Forestry and biodiversity : learning how to sustain biodiversity in managed forests / edited by Fred L. Bunnell and Glen B. Dunsworth.

Includes bibliographical references and index.
ISBN 978-0-7748-1529-1 (bound); ISBN 978-0-7748-1530-7 (pbk.);
ISBN 978-0-7748-1531-4 (e-book)

1. Biodiversity conservation – Case studies. 2. Forest management – Case studies. 3. Adaptive natural resource management – Case studies. 4. Sustainable forestry – Case studies. 5. Forest ecology. 6. Conservation biology. 7. Coastal forests – Management. 8. Rain forests – Management. I. Bunnell, Fred L., 1942- II. Dunsworth, Glen B., 1952-

QH75.F67 2009   333.95′16   C2008-906491-7

Canada

UBC Press gratefully acknowledges the financial support for our publishing program of the Government of Canada through the Book Publishing Industry Development Program (BPIDP), and of the Canada Council for the Arts, and the British Columbia Arts Council.

This book has been published with the help of Cascadia Forest Products and the K.D. Srivastava Fund.

UBC Press
The University of British Columbia
2029 West Mall
Vancouver, BC V6T 1Z2
604-822-5959 / Fax: 604-822-6083
www.ubcpress.ca

Contents

Illustrations

Figures

Tables

Preface

This book arose from an attempt to solve a complex, real-world problem – how do we sustain biological diversity in managed forests? MacMillan Bloedel harvested natural forests in coastal temperate rainforest, largely on public land. It had the largest forest holdings in coastal British Columbia and faced effective market campaigns to stop clear-cutting of old growth and reduce its harvest. In 1998, the company adopted a completely new form of forestry that it believed could be shown to sustain native species richness in coastal temperate rainforests. Shortly after making this commitment, MacMillan Bloedel was acquired by Weyerhaeuser Company. Weyerhaeuser maintained the commitment to evaluating ecological consequences of forest practices, so efforts to learn how to sustain biodiversity in managed forests also continued. The forest planning and practices introduced (zoning and variable retention) were novel, as was the large-scale effort to evaluate the effectiveness of these practices within an adaptive management program.

As part of its quality control in developing and implementing its adaptive management program, Weyerhaeuser hosted meetings of an International Scientific Advisory Panel. Panel members were recognized experts drawn from Australia, Europe, Canada, and the United States. These members praised the efforts and accomplishments of the adaptive management program and were instrumental in Weyerhaeuser receiving the Ecological Society of America's Corporate Award in 2001. The panel also urged the authors to publish the approach, yielding this book.

The book is divided into three parts. Part 1, "Introduction," introduces the generic nature of the problem, including elements of wicked problems and complex challenges faced by forest managers, plus a potential solution to the problem, an effective adaptive management program. Part 2, "The Indicators," treats the major indicators of success in sustaining biological diversity and learning acquired from evaluating these indicators. Part 3,

"Summary," summarizes the monitoring design for the adaptive management program, noting elements that must be considered and summarizing lessons learned.

To present this process of learning how to sustain biodiversity in managed forests, we draw on literature from six continents but present it in the context of a real-world case study. Employing a specific case study has the benefit of linking abstract concepts to the real world of unyielding topography, nature's frequent caprice, and economic constraints. It does this for a coastal temperate rainforest – historically the most controversial forest type in the northern hemisphere. Conflicts between the extraction of wood and other forest values in this region are often magnified, frequently to national and international levels. The area is also uncommonly rich in species among northern temperate regions, thus presenting a significant challenge to efforts to sustain biological diversity.

Material summarized in this book is the product of a great many individuals – both researchers and practitioners of forestry. The authors are those who had responsibility for distilling those contributions into an accessible story.

Bill Beese chaired the Variable Retention Working Group that developed guidelines for implementation of retention systems and conducted implementation monitoring. He is trained as a forest ecologist and silviculturist and worked in ecological classification, reclamation, prescribed fire, and silvicultural systems research for twenty-five years with MacMillan Bloedel and Weyerhaeuser. He has great respect for the foresters and loggers who put the theory of retention into practice and managed to improve safety performance at the same time.

Fred Bunnell held the senior forest renewal chair in conservation biology at the University of British Columbia for fifteen years, where he is now professor emeritus. During that period, he was for ten years director of the Centre for Applied Conservation Biology. He was trained as a forester and received six provincial, national, and international awards for applied research. He has also served on more than eighty resource management committees and panels, many of which have dealt with contentious issues. He admits that the problem addressed in this book is the most complex and most frustrating one on which he has worked.

Glen Dunsworth chaired the Adaptive Management Working Group that conducted the field studies and helped to design the program. He was trained as a genecologist and worked in regeneration research, genetics, and modelling and conservation biology for twenty-five years with MacMillan Bloedel and Weyerhaeuser. He believes that this case study reflects the realities of design and implementation of adaptive management in an industrial setting, a truly challenging learning experience.

Dave Huggard is a research associate at the University of British Columbia and an independent consultant in forest ecology. He has an MSc in animal ecology and a PhD in forestry. He is a lapsed naturalist and field biologist, an aspiring statistician, and currently most interested in how we can obtain and use reliable knowledge in natural sciences. These interests have led him to roles in a variety of integrated forest ecology projects, including as a member of the Adaptive Management Working Group.

Laurie Kremsater is a research associate at the University of British Columbia and a consultant in forest wildlife ecology. Trained in forestry and wildlife ecology, she has worked in conservation biology and forestry for twenty years, primarily with vertebrates and their forest habitats. She has served on a number of land use planning teams, including the Clayoquot Scientific Panel, she has co-ordinated a variety of integrated forest ecology projects, and she has been a member of the Adaptive Management Working Group that addressed the challenges described in this book.

Acknowledgments

This book and the journey of the Forest Project comprise an example of corporate leadership responding to concerns of social licence – social expectations beyond the law that a corporation must meet. The project would not have been possible without the determination and vision of CEO Tom Stephens, Chief Forester Bill Cafferata, and the MacMillan Bloedel Board of Directors. The corporate critical thinking and Forest Project strategy were the product of intense and focused teamwork facilitated by Dennis Fitzgerald and led by Glen Dunsworth and Fred Bunnell (ecology); Bill Beese, Ken Zielke, and Bryce Bancroft (silviculture); Nick Smith and Steven Northway (growth and yield); Tom Holmes, Lorne High, Walt Cowlard, Marv Clark, Ray Krag, and Lorne Pelto (Harvesting); Robert Prins, Bill Stanbury, Ian Vertinsky, and Casey Van Kooten (economics); and Linda Coady, David McPhee, Bill Shireman, Terry Morely (social aspects).

Heightened awareness of social licence and partnership in applying the science behind the project was provided by ENGOs including the Natural Resources Defense Council, Greenpeace Canada, Western Canada Wilderness Committee, Sierra Legal Defence Fund, Ecotrust Canada, World Wildlife Fund, and Sierra Club of British Columbia. We are specifically grateful for the advice and intellectual support provided by Jody Holmes, Rachel Holt, and Matt Price.

Putting the project on the ground could not have happened without the dedicated work of MacMillan Bloedel and Weyerhaeuser field foresters and engineers and the members of the Variable Retention Working Group, including the assistance and guidance of Ken Zielke and Bryce Bancroft.

The adaptive management framework was the product of the collaborative efforts of the scientists in the Adaptive Management Working Group (which the authors helped coordinate). We thank Lynn Baldwin, Mark Boyland, Wayne Campbell, Emina Krcmar, Ann Chan-McLeod, Renata Outerbridge, Kristiina Ovaska, Isobel Pearsall, Mike Preston, Kella Sadler,

Lennart Sopuck, Nick Stanger, Tony Trofymow, Pierre Vernier, and Elke Wind. We are especially indebted to Jeff Sandford for his intrepid work and gentle hand in coordinating this disparate group, providing for their field needs and safety, and carefully integrating all their data into a geo-referenced database.

The annual review and advice of the science panel were two of the most helpful aspects of the project. The panel reviews provided a third-party validation of the work and an opportunity to share in the development of the science and aid in its implementation. More importantly, given the stature of the scientists chosen, the panel provided global recognition of the novelty of the work. The support and leadership of Jerry Franklin, David Lindenmayer, Bruce Marcot, Reed Noss, and M.A. Sanjayan were instrumental in sustaining the effort and guiding the science.

We appreciate the corporate and government funding without which much of the science could not have been developed. We received corporate funding from MacMillan Bloedel, Weyerhaeuser, and Cascadia Forest Products and grants from the BC Enhanced Forest Management Pilot Project, BC Forest Science Program, BC Forest Investment Account, Canadian Forest Service, Forest Renewal BC, National Science and Engineering Research Council, and Social Sciences and Humanities Research Council.

Comments of anonymous reviewers improved the manuscript.

Finally, we thank our partners and families for their love, patience and sacrifice for those innumerable times when we were away from home and in front of the computer late at night. Without their support, the project and this book would not have been possible.

Part 1: Introduction

Part 1 provides the context for much of this book and the book's focus on learning how to sustain biological diversity in managed forests. It describes the problem, the major case study, and the general approach employed. It also reviews the implementation-monitoring program that was necessary because a new approach to forestry had been undertaken.

As noted in Chapter 1, the problem addressed is that of managing to sustain biodiversity in managed forests. It is a wicked problem with interrelated, interdependent parts encompassing diverse values. We cannot maximize our efforts to produce wood products or to sustain biodiversity without harming the other major objective. The single objective of sustaining biodiversity in forested systems confronts special difficulties. Among them are international agreements that have encouraged policies and practices well ahead of direct experience, thus placing forest practitioners in a situation of playing "catch-up" and having to learn as quickly as possible. The most effective way of learning while implementing new planning and practices is adaptive management. The broad goal of this book is to illustrate an approach to learning how to sustain biodiversity in managed forests. It describes a process for learning, not a tidy recipe for success.

Chapter 2 presents the case study used to illustrate the process. The area is about 1.1 million hectares in rugged terrain, primarily within the coastal temperate rainforest, hosts some of the largest, longest-lived trees in North America, and has attracted market campaigns to limit forest practices. The turbulent social environment led the forest company to adopt novel approaches to forest planning and practice. Although these approaches were well reasoned, their consequences were unknown. The company implemented an adaptive management program to evaluate the effectiveness of practices, new structures to help the program work, and an extensive monitoring program to provide feedback on its actions to practitioners and policy makers. This book draws on experiences from six continents but expresses much of it around a single, complex case study. Among the advantages of employing a case study is linking abstract concepts to the real world.

Chapter 3 describes the approach taken to the problem in terms of four questions that managers confront when addressing any complex problem. Where do we want to go? How do we get there? Are we going in the right direction? How do we change if the direction is wrong? These four questions reflect the four steps critical to creating an effective adaptive management program: (1) clearly defined objectives, (2) planning and practices to attain those objectives, (3) ways to assess proximity to those objectives, and (4) ways to modify practices if those objectives are not attained (links to management action). In Chapter 3, we provide an overview of how these steps were pursued in the major case study from coastal British Columbia.

The bulk of this book is devoted to the rationale, implementation, and interpretation of the indicators used to assess success and the overall design and implementation of the adaptive management program. That is, much of the book is about effectiveness monitoring. However, the approach to both forest planning and practice is novel. The company also had to employ implementation monitoring that recorded the rate at which it was attaining corporate goals for the implementation of new practices and the degree to which the practices were adhering to principles outlined by the company. Chapter 4 describes the implementation-monitoring program.

1
The Problem

Fred L. Bunnell, Glen B. Dunsworth, David J. Huggard, and Laurie L. Kremsater

1.1 "Wicked" Problems

When the term "ecosystem" was coined, the "system" part was invoked to represent the remarkable number of interactions that occur among components of any chunk of nature (Tansley 1935). The practice of forestry modifies ecosystems and changes the nature of interactions within an ecosystem and its ability to provide different values that we desire. Society desires a variety of values from the forest, some of them competing. Management of forests becomes a challenge of dealing with organized complexity and directing clusters of interrelated parts toward specific ends. Problems that cannot be resolved by treating these clusters in relative isolation Ackoff (1974) termed "messes." He argued that we fail more often because we solve the wrong problem than because we get the wrong solution to the right problem. Ackoff (1974), King (1993), and others[1] noted that using techniques of systems analysis, we could solve many seemingly intractable problems by examining interactions among parts rather than breaking problems into separate parts and fixing individual components. Others argue that some problems defy any form of analysis and are insoluble in terms of a simple "yes" or "no" answer. They term such problems "wicked."[2] There is no moral sense to this use of the word but a recognition that some problems defy tidy solutions no matter the efforts and good intentions directed toward them.

Wicked problems arise when the boundaries of messes expand to include sociopolitical and moral-spiritual issues (King 1993). They have been defined as a "class of social systems problems which are ill-formulated, where the information is confusing, where there are many clients and decision-makers with conflicting values, and where the ramifications in the whole system are thoroughly confusing" (Buchanan 1992, 15).

Many of the challenges facing natural resource management fit the definition of wicked problems (Allen and Gould 1986; Barrie, McCool, and Stankey 1988; Shindler and Cramer 1999; Wang 2002). That is certainly true of efforts to design an approach to sustaining biological diversity in managed forests. When designing an approach to management, we typically engage in a kind of linear thinking that entails two major phases: problem definition and problem solution. In the problem definition phase, all elements of the problem are systematically analyzed, while in the solution phase, these elements are synthesized into a solution. We use this approach believing that it offers the best hope of a logical solution to a problem. That is especially true in many optimizing models used to create harvest schedules in a managed forest. With wicked problems, it is not possible to clearly define the problem, let alone the solution.

We illustrate more completely in the following sections why sustaining biodiversity in managed forests is a wicked problem. Here we consider generic elements of such problems and their implications. Among their features are that

- they can be described in different ways that have different solutions – there is no one way to formulate the problem;
- the problem is unique;
- there is always more than one plausible explanation for outcomes;
- there is no single right or true test for a solution; and
- the solutions cannot be true or false, although they can be more or less effective (see Rittel 1972 and Rittel and Webber 1973, 1984).

Few argue that they have found *the* approach to sustaining biological diversity in managed forests. The first feature of wicked problems indicates that we should not become wedded to any single approach. It appears equally true that every specific example of the generic problem is unique. Over the past decade, we have addressed sustaining biodiversity in managed forests in several regions. Factors ensuring that each case was unique included the social and economic environment, the size of the area, tree species present, other organisms present, typography, harvest methods, and the disturbance history. This feature of wicked problems implies that what we offer in this book is an approach to learning to solve the problem in ways that can be no more than broadly effective – details will vary among areas. The third feature (more than one plausible explanation or mechanism for observed outcomes) clearly applies to the problem of sustaining biological diversity, in part because interconnections in nature encourage alternative mechanisms and because many spatial, temporal, and cumulative effects are poorly understood. That is why we have subtitled this book *Learning How to Sustain Biodiversity in Managed Forests*. It is the process of learning that we describe, not a tidy solution. The absence of a single best test of approaches is frustrating to a researcher and severely challenges some practitioners, policy makers, and other decision makers. We can take heart from the last feature noted –

providing that we focus on what we can agree is more or less effective rather than what is right or wrong. Combined, the features suggest that we are likely to agree only on a direction rather than a prescription. That argues for the use of adaptive management and suggests that we generally learn more from comparisons among alternative practices than from comparisons of specific targets (§5.3).

1.2 Expanding and Competing Values

The values that we ascribe to biodiversity or economic returns are often strongly held and frequently in conflict. They help to make the problem we address a wicked one. In some instances, these values have deep historical roots but have evolved in ways that make the practitioner's life more complex and confused. In his classic silvicultural textbook, D.M. Smith (1962, 369) wrote that "there is hardly any objective more ancient and honorable ... [than sustainable yield]." Indeed, foresters have pursued sustainable yield since the late Middle Ages, when coppice forests were managed to guarantee a perpetual supply of fuel (Knuchel 1953). Only one tradition in forestry is older.[3] Joint management of forests and wildlife began at least 500 years before the first efforts at sustainable yield in coppice systems. Our word *forest* (*forestis*) first appears in 556 CE to describe a tree-covered area in which hunting and fishing rights were preserved for King Childebert I (Glacken 1967, 325). The concept of sustainability for more than one resource is well entrenched in forest practices. What has expanded rapidly is the number of values to be sustained.

Values desired from forests have continually changed but never so rapidly as in the past three decades. Rapid expansion began with the environmental movements of the 1970s and gained momentum in 1987 when the Brundtland Commission introduced the concept of "sustainable development." That concept considered *all* resources and came to mean not just the uninterrupted flow of a particular product but also retaining resources for *future alternative* uses. These uses were unknown and thus unspecified. Biodiversity was mentioned only twice in the commission's report (Brundtland 1987, 148, 157). Just five years later, the Earth Summit in Rio de Janeiro produced three international documents directly addressing biological diversity in forests.[4] The scope of responsibility for land managers, particularly forest managers, had broadened enormously to include things that had not yet been named or identified. We often fail to appreciate the relatively slow rates at which some forests change. At the time of the Earth Summit in 1992, not all consequences of forest practices initiated fifty years earlier were evident. Compared to the common pace of events in forestry, or to common progress surrounding international agreements, the pace was extremely rapid. For some newly emphasized values, there was almost no experience in developing forest planning and practices to sustain them. Sustaining biological diversity has proven to be the greatest challenge, and "policy created to sustain biological diversity is well ahead of research and understanding" (Bunnell and Chan-McLeod 1998, 7). The fact that enacting policy has preceded research or experience on how to enact it is a serious challenge. The challenge is amplified because some desired values are competing.

Competition is most evident between biological diversity and wood products. In both temperate and tropical regions, species richness is greater in tall, well-stratified forests than in vegetation of lower stature (Bunnell and Kremsater 1990; Wilson 1988). In many areas, forests also contribute significantly to regional economies and help to support amenities. Managers thus confront two potentially competing goals: (1) to extract renewable wood products from the richest terrestrial ecosystems on Earth, and (2) to maintain biological diversity.

While addressing this competition, we should be mindful that wood is a renewable resource and that the creation of wood products is more environmentally friendly than that of many substitute products. Lumber, for example, is highly energy efficient compared with other construction products, requiring only 580 kilowatts per tonne to produce. In contrast, aluminum requires about 73,000 kilowatts per tonne, plastics 3,480, and cement 2,900.[5] In most instances, the production of forest products produces a smaller "ecological footprint" than does the production of alternatives. Likewise, there is little evidence suggesting that production of wood products and associated economic contributions cannot be sustained indefinitely when appropriately managed.[6] However, there is increasing evidence that some forest practices will not permit the long-term maintenance of forest-dwelling species (Angelstam 1997; Berg et al. 1994; Morrison and Raphael 1993; O'Hara et al. 1994; Spies, Franklin, and Thomas 1988). Because the major cause of species' decline appears to be habitat loss (Diamond 1984; Hayes 1991; Wilcove et al. 1998), it is important to understand how forest practices modify habitat and associated species. The challenge is to sustain biological diversity while extracting wood products.

Describing this challenge, Bunnell observed that "[foresters] are proceeding at great speed, over very difficult terrain, towards an unknown goal" (1998a, 822). We have acknowledged the speed. We next examine why the terrain is so difficult, and we present our efforts to clarify the goal in §3.2.1.

1.3 Special Difficulties in Forests

Of all terrestrial systems, forests contain the largest and most long-lived components. The long lives of many tree species, the spatial extent of forests, and the social perceptions of forests combine to create special difficulties in managing to sustain biological diversity. Bunnell and others noted five (Bunnell and Chan-McLeod 1998; Bunnell, Kremsater, and Wind 1999).

1 For historical and social reasons, forest practitioners bear an uncommon responsibility for sustaining biological diversity. In western North America, for example, both forestry and agriculture provide highly valued crops, but only forestry is consistently charged with sustaining biological diversity. This condition appears to result from history (some species evicted by agriculture have survived in forests: e.g., grizzly bears and wolves) and from greater social value being placed on food crops than on wood products. Both species richness and genetic richness are under greater threat from agriculture than from forest practices (Askins 1993; Dobson et al. 1997; Hamrick, Mitton, and Linhart 1979; McCracken 2005).
2 More species reside in forests than in other plant communities. Monitoring species richness in forests requires surrogates, such as habitat elements, that represent groups of species, including those difficult to census.
3 Forestry is usually planned over large areas. In any area large enough to permit sustainable harvest, both natural disturbances and forest practices create stands of different ages and structures. Moreover, large areas are rarely homogeneous and contain stands of varying species composition and distinctive communities. That variation encourages species richness but complicates setting objectives and planning.
4 Forestry must be planned over long periods. Other than for some broadleaved species, rotation lengths in Canadian forests are rarely less than sixty years. Natural disturbance frequencies in forests can be highly variable in extent, frequency, duration, location, and intensity (Agee 1993; Cumming, Burton, and Klinkenberg 1996). In any forest, especially unmanaged forests, there will be a diverse array of stands with a variety of structures that change over time. Long periods of time are required to account for all components contributing to biological diversity. Many would argue that sixty years is much too short (see, e.g., Goward 1994 and Norse 1990).
5 Public goals and perceptions change quickly, but practices are evaluated slowly. Only recently have concerns about forest-dwelling organisms expanded to include all species. Well-documented information is available for only a very small portion of species, even among vascular plants and vertebrates. Moreover, because trees can grow large (thereby inducing considerable environmental change), but grow slowly, there has been little time for practitioners to evaluate the reliability of newly developed tactics to sustain forest-dwelling species (Binkley 1992; Bunnell 1998c; Namkoong 1993).

These difficulties consistently challenge the larger public and forest managers. Difficulties are especially challenging when a commitment has been made to sustain both forest-dwelling biodiversity and economic production of wood fibre. They are sufficiently challenging to elude tidy solution and thus yield a wicked problem. Because practices to sustain biodiversity have preceded experience with those practices, land managers must play "catch-up." Fortunately, there is an effective way of learning while practices are being implemented – adaptive management.

1.4 Adaptive Management

Adaptive management[7] is a formal process for continually improving management practices by learning from the outcomes of operational and experimental approaches. Four elements of this definition are key to its utility. First, it is *adaptive* and intended to be self-improving. Second, it is a well-designed, *formal* approach that connects the power of science to the practicality of management. Third, it is an ongoing process for *continually improving management,* so the design must connect directly to the actions that it is intended to improve. Fourth, although experimental approaches can be incorporated into adaptive management effectively, *operational* approaches and scales are emphasized to permit direct connection to the efforts of managers.

There is nothing easy about implementing adaptive management well. The design itself is difficult to conceive and implement, particularly in forested systems (see Chapter 12). Two of the largest barriers have been effectively linking the somewhat different worlds of science and management and developing concepts and structures that permit feedback from findings to changes in action. At its best, adaptive management is a conscious choice to link science with management so that each responds to the needs and information of the other (Halbert 1993; Lee and Lawrence 1986; Marzluff, Raphael, and Sallabanks 2000; Meretsky, Wegner, and Stevens 2000; Salafsky et al. 2002). Neither science nor management is consistently subordinate; rather, the priorities of each are reconciled through the design of the process. Managers help to establish the program's direction so that it connects to management issues and commit to using the results of the monitoring. Researchers understand management issues well enough to design sampling and experiments that connect directly to management questions and help to interpret findings and their implications for management. Even when this ideal is attained, the process may fail and not be adaptive because institutional structures or philosophies become barriers to implementing changes (Lee 1993, 1999; Ludwig, Hilborn, and Walters 1993; Stankey and Shindler 1997).

Despite these barriers, efforts to attain and evaluate success proceed best through the framework of adaptive management. For adaptive management to attain its theoretical promise, it must contain four broad elements (see Figure 1.1):

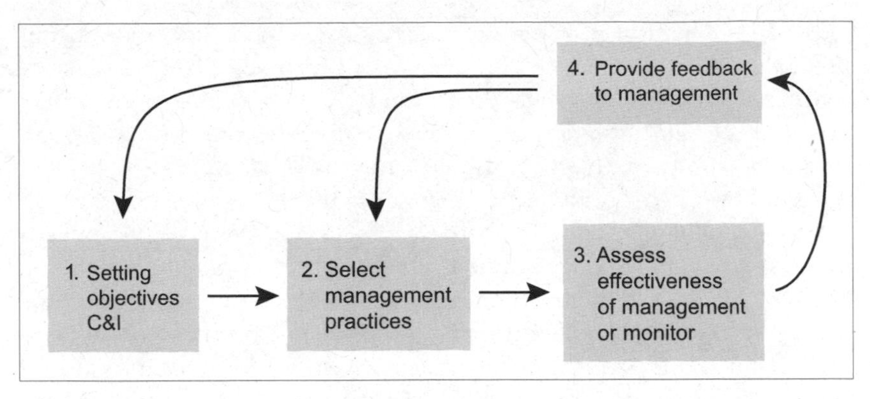

Figure 1.1 Major elements of a successful adaptive management process.
Note: C & I represent Criteria and Indicators used in some forest certification systems.

1 clearly defined objectives;
2 planning and practices to attain those objectives;
3 ways to assess proximity to the objectives; and
4 ways to modify practices if objectives are not attained or are changed (links to management action).

Specify clear objectives: Objectives must be sufficiently clear that means of assessing proximity to them are likewise clear. Objectives cannot be rigid and must respond to changes in both values and information. In sustainable forest management, major objectives are increasingly being described as criteria for success, while indicators are used to measure outcomes that should be evident when a criterion is successfully attained. Criteria and indicators for sustaining biodiversity in managed forests are described in §3.2.2.

Select planning and practices: Once objectives are clear, the planning and practices likely to attain those objectives can be chosen. In forests, effects of practices at the stand level both affect and are affected by forest planning over broader areas and vice versa. That is particularly true for effects on forest-dwelling biodiversity (Bunnell and Huggard 1999). Evaluation of success in sustaining biological diversity must occur at a variety of scales.

Assess proximity to objectives: The objective of sustaining biological diversity cannot apply to specific cutblocks but must be evaluated for broad areas (Bunnell and Chan-McLeod 1998; Bunnell, Kremsater, and Wind 1999). Assessments everywhere are too costly, so they must be designed in a fashion that can be "scaled up" to reflect larger areas. Scaling up often relies on computer projections of habitat and organisms' responses to habitat. We learn fastest, or reduce uncertainty fastest, when we organize our current information in a way that can be directly challenged by new information – that is, by making predictions and evaluating their accuracy. Some elements of the task of managing to sustain biodiversity are too ill known to permit prediction. A key element of assessing proximity to objectives is providing a structure for learning. Where prediction is difficult, comparisons remain useful (§5.2). It is important to recognize that monitoring, or a way of assessing proximity to objectives, is only one part of adaptive management. To make it work, the fourth broad element of the process must be present.

Ways to modify practices: The fourth element of effective adaptive management is a set of known ways to modify practices if objectives are not attained or if they change in response to changing values. New information is of little use to practitioners if it does not link to management practice. Linkage is provided by both the evaluation system developed for each indicator of success and descriptions of potential management actions that are expected to help correct any failures or strengthen areas of weakness. The evaluation system should assess the success of current management activities and track improvements in management over time. The ability to change may require a formal mechanism for accepting results and associated management or policy changes.

1.5 Bounding the Book: What It Is and Is Not

Of all objectives sought by sustainable forest management, sustaining biodiversity is the most challenging and fraught with uncertainty. This book addresses the problem of learning how to sustain biodiversity in managed forests in a scientifically credible and cost-effective fashion. The focus on learning emphasizes the importance of monitoring consequences of actions. Because this learning is occurring while management actions are occurring, the appropriate framework for the monitoring is adaptive management. That is, the findings of monitoring are intended to guide future management actions. The book *is* an illustration of how we can learn to sustain biological diversity in the midst of continuing management actions.

The book *is not* a review of the wide range of strategies and tactics that have been proposed or attempted to sustain biological diversity. The range in approaches reflects the complexity of the problem and the fact that there is no single correct solution to the problem (§1.1). The approach to sustaining biological diversity discussed in this book borrows heavily from other efforts but usually differs somewhat to meet local goals or conditions (see Aubry et al. 1999; Aubry, Halpern, and Maguire 2004; Hollstedt and Vyse 1997; Lovejoy 1985; Margules 1993; Schmiegelow, Machtans, and Hannon 1997; and Vyse, Hollstedt, and Huggard 1998). For example, designed experiments are included but only as one part of a broader program that includes many issues not amenable to simple experimental resolution. The design of experiments also incorporated a social decision to involve all the regional divisions of the company in the adaptive management effort, so the questions and geographic layout are more complicated than are those involved in more focused research efforts. Moreover, the variability in forest practices addressed is wider than that normally encountered in commercial forestry.

Principles of various regional approaches to zoning (Kneeshaw et al. 2000; Noss 1993; Oliver 1992) are incorporated into the management approach (§2.3), as are coarse filter approaches such as gap analysis and landscape trend analysis (see Chapter 6), but again, as one part of the overall monitoring approach. Nor was our approach focused on particular species of *a priori* interest. We did not set out to monitor particular species at risk or focal species, nor did we attempt to monitor entire communities. Our interest was species that could help to answer specific forest management questions. Elsewhere, initial emphasis on species has been driven primarily by regulations, such as the US Endangered Species Act or the survey and manage approach of the Northwest Forest Plan (Molina, Marcot, and Lesher 2006), or simply by relative ease of monitoring.[8] We also include, but do not emphasize, a broad assessment of trends in multiple species developed in programs such as the Alberta Biodiversity Monitoring Program (ABMP) and the Finnish Wildlife Triangles (ABMP 2005; Pellikka, Rita, and Lindén 2005). We focus on species that are sensitive to the forestry practices and management issues of interest.

Incorporating useful elements of a range of approaches reflects an effort to enact a cost-effective approach to effectiveness monitoring in an adaptive management framework. The emphasis is consistently focused on how we learn to do better rather than on championing a particular approach. A criteria and indicators approach was adopted because it is now so much a part of forest companies' efforts at certification globally.[9] We present a rationale for the broad indicators and specific measurements selected (§3.2).

The practice of forestry is complex. Developing the planning and practice of forestry to sustain biodiversity is still more complex. To cope with this complexity, applied researchers often invoke simplified concepts such as natural disturbance regime and range of natural variability (Haeussler and Kneeshaw 2003; Hessburg, Smith, and Salter 1999; Perera, Buse, and Weber 2004; Swanson et al. 1994). This book does not evaluate competing concepts invoked to guide management. The planning and practices evaluated here focus on consequences of the most dramatic departures from the natural disturbance regime rather than on mimicking the regime (see Chapter 2). The most dramatic departures are the reduction of old-forest elements and intact old-growth forests.

There are both general and specific reasons for not attempting to mimic the natural disturbance regime. Generally, the extremes of natural variation are simply unacceptable to the vast majority of the public, whether induced by logging or more naturally (neither the current mountain pine beetle outbreak nor a larger forest fire is considered desirable). The specific nature of the natural disturbance regime in the example area would create additional problems if logging attempted to mimic natural patterns. The case study is an area of 1.1 million hectares in coastal British Columbia (see Chapter 2). The driest portion of the area is now well populated by people who do not want a return to the frequent fires that occurred naturally. The small openings created by gap dynamics in the wetter portions of the study area are impractical to re-create widely from economic, safety, and ecological concerns (they would interject roads and their associated disadvantages throughout the region).

Once a social decision has been made to truncate natural disturbance regimes, the issue becomes one of examining consequences of various forms of truncation (Bergeron and Harvey 1997; Bergeron et al. 2001; Bergeron et al. 2002; Hansen et al. 1991; Hunter 1993). Such an examination is challenging. Natural disturbance regimes do not exist as simple, measurable, and fixed "things." Quantifying the abstraction of natural disturbance regimes for specific locations has proven very elusive and varies with the scale and period over which it is defined (Boychuck et al. 1997; Cumming, Burton, and Klinkenberg 1996; Wallenius 2002). Moreover, mimicking natural disturbance closely is simply not achievable because logging and natural disturbance are inherently different. Logging, for example, does not produce charcoal or create the upturned mounds of earth following from windthrow that help to perpetuate one forest type in the study area (McRae et al. 2001; Zackrisson, Nilsson, and Wardle 1996).[10]

The case study used for illustrative purposes differs from other forest types in features beyond the disturbance regime. For example, variable retention (VR) is employed. VR was invoked to mitigate the main detrimental effects of industrial logging when compared to the major regional disturbance regime – potential loss of large, long-lived trees and abundant coarse woody debris, creation of large openings. That is not the primary objective for management in many forest types, simply because they do not have trees nearly so large or long lived. Other elements of the physical and ecological setting had to be acknowledged in the approach to forest planning and practice. For example, there is a major variation in ecosystem types and elevation across the management area, which is not true of much of the boreal forest in North America and Europe. The history of logging is relatively short compared with those of most of Europe and elsewhere in North America, which means that the presence of unlogged, old-growth stands must be considered and can be exploited as opportunities for conservation. Features of the case study, such as rugged terrain and natural forests of exceptionally large, long-lived trees, must be accommodated in the approach to forestry and monitoring.

There are also features absent in the area that must be accommodated elsewhere. For example, forests and the practice of forestry dominate most of the landscape. Other land use practices such as agriculture, mining, or oil and gas development are uncommon, so there is less need for cumulative effects analysis (Davies 1995; Grant and Swanson 1991). Likewise, where the practice of forestry dominates the landscape, the potential negative impacts of fragmentation are less well expressed than where agricultural and urban areas are interspersed (Bayne and Hobson 1997; Bunnell 1999; Rudnicky and Hunter 1993). Although there is private land in the management area, there is nowhere near the interspersion of public and private land that complicates the practice and monitoring of forestry in the United States and parts of eastern Canada. The tree species potentially available for harvest are few relative to temperate Australia, eastern North America, and parts of eastern Canada, requiring less silvicultural attention to maintaining tree species diversity, particularly for hardwoods (Bergeron and Harvey 1997; Greene et al. 2002).

As always, there are the differences in the organisms' responses to forest practices that require localized approaches. The American marten (*Martes americana*), for example, is considered an indicator of old-forest conditions in some areas (Sturtevant, Bissonette, and Long 1996; Thompson and Curran 1995) but exploits young forests as well in British Columbia (Lofroth 1993; Poole et al. 2004). The inherent dangers in extrapolating findings from elsewhere require breadth in the monitoring system and are exacerbated by the relative lack of regional knowledge for some organism groups compared with other regions (e.g., lichens, bryophytes, and invertebrates in Fennoscandia).

Such ever-present differences in the ecological and physical setting, and thus the practice of forestry, ensure that specifics of the forest planning and practices and of the monitoring system described here are limited primarily to the Pacific Northwest of North America. This book necessarily addresses a specific area and asks questions about effects of specific management practices, but our purpose is to illustrate generic concepts that will help others to design adaptive management and monitoring programs in a variety of situations. More specifically, our purpose is to illustrate generic concepts applicable for forests anywhere by providing concrete examples of these concepts within a real corporate enterprise.

1.6 Summary

The problem that this book addresses is that of managing to sustain both biodiversity and production of wood products in managed forests. It is apparent that

1. sustaining biological diversity in managed forests is encumbered by difficulties, many specific to forest management;
2. sustaining biodiversity is a "wicked" problem with no single correct solution; and
3. adaptive management is well suited to exposing solutions that are most effective in a range of alternatives.

Our goal is to illustrate an approach for learning how to sustain biodiversity in managed forests. In this chapter, we introduced broad elements of the problem. In Chapter 2, we present the major example that we use to illustrate the process.

2
The Example

Fred L. Bunnell, William J. Beese, and Glen B. Dunsworth

2.1 Physical and Ecological Setting

The setting described here is similar to most coastal forests in the Pacific Northwest. Specific details are provided for the case study, a large coastal tenure in British Columbia. MacMillan Bloedel managed the tenure when the study began; it has since experienced changes in ownership.[1]

2.1.1 Physical Landscape

When the program began, the tenure covered about 1.1 million hectares.[2] It is made up of separate pieces that are scattered from the southern Gulf Islands and southeastern Vancouver Island, over sizable portions of Vancouver Island, a portion of the central mainland coast, and north to the Queen Charlotte Islands or Haida Gwaii (Figure 2.1).

About 75 percent of the tenure was public or crown land. Private lands (25 percent of the tenure) were restricted to sites on southeastern Vancouver Island. One year after the provincial government announced that it would follow all recommendations made by the Clayoquot Science Panel (CSP),[3] the company (then MacMillan Bloedel) entered into negotiations with the Nuu-Chah-Nulth to form a joint venture for the area of Clayoquot Sound, on western Vancouver Island. These negotiations were completed successfully in 1998, and Iisaak Forest Resources Limited now administers that area. Although the Nuu-Chah-Nulth had a controlling interest in Iisaak (51 percent), the area is recognized as part of the larger tenure.

The tenure shares predominating features of coastal British Columbia and the major islands. Watercourses of various sizes and gradients are abundant, yielding different kinds of riparian habitat. Much of the terrain is rugged and mountainous. Valleys are usually deep, glaciated troughs, with gentle slopes restricted to valley floors. Ridge tops defining valleys commonly rise over 1,000 metres, with peaks considerably higher. Gently sloping uplands are relatively uncommon, and valley sides are commonly steeper than 30 or 60 percent. Bedrock outcrops, usually of crystalline metamorphic and

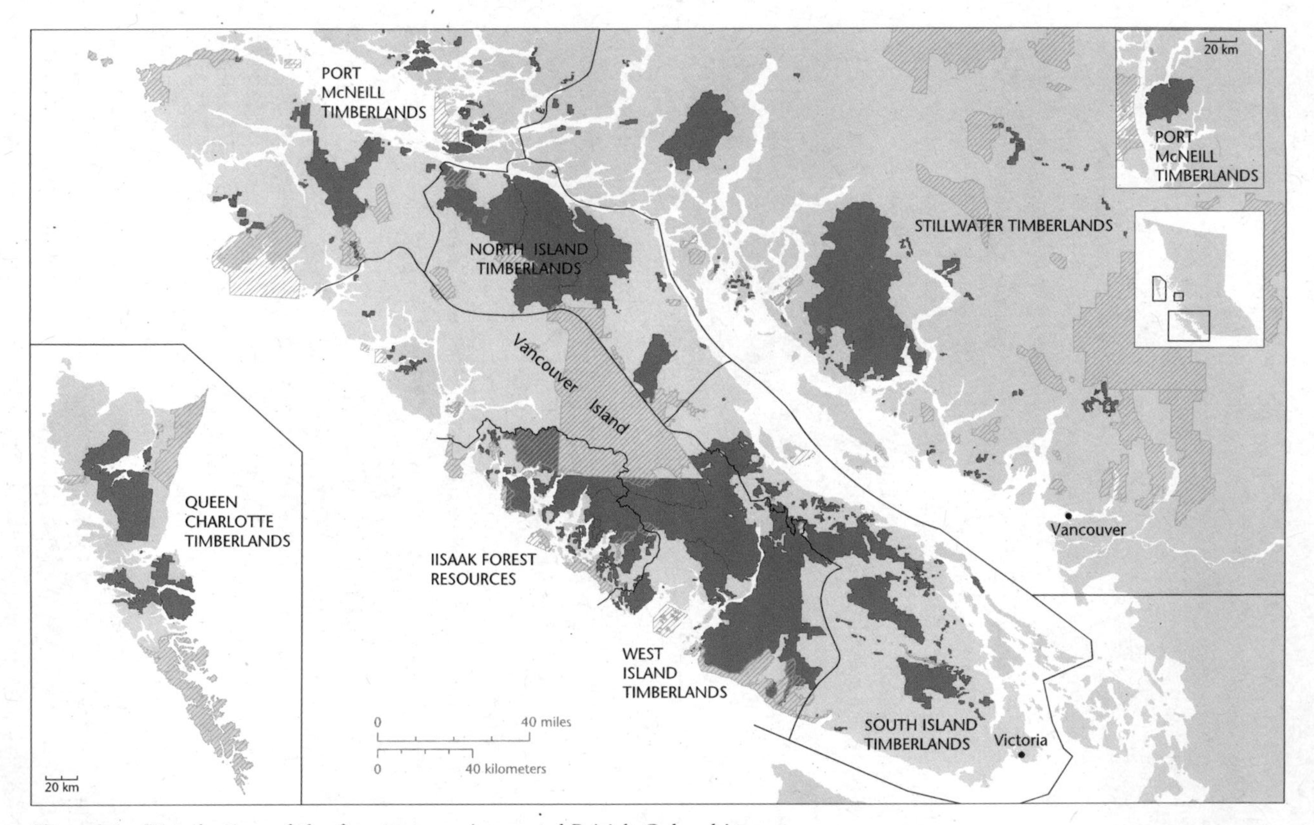

Figure 2.1 Distribution of the forest tenure in coastal British Columbia.

Note: The tenure in British Columbia (dark grey) occurs as separate pieces on the Gulf Islands, Vancouver Island, the central mainland coast, and the Queen Charlotte Islands (Haida Gwaii). Parks are in grey.

Source: MacMillan Bloedel Limited.

intrusive rocks, are common on steep slopes and at high elevations. On gentle slopes and in the coastal lowlands, glacial and post-glacial deposits generally bury bedrock. On steeper slopes, weathered till deposited by melting glaciers is naturally susceptible to debris slides and debris flows. On Vancouver Island, where much of the tenure is located, the mountains form a broad spine down the centre of the island. Areas of more gentle terrain and deeper soils occur in the southeast, east of Strathcona Park (Figure 2.1), and around the northern tip.

2.1.2 Climate, Vegetation, and Fauna

Most relief is perpendicular to prevailing weather systems, so annual precipitation differs greatly from windward to leeward sides. Windward areas on the mountainous coast are among the wettest temperate regions of the world. Measurable precipitation usually occurs 200 or more days of the year, and annual totals in most of the region range from about 1,750 to 4,000 millimetres. Orographic effects and topographic steering of winds create local pockets of higher precipitation. Afternoon relative humidity rarely falls below 70 percent. Throughout the year, the ocean modifies temperatures, so that winters are relatively mild and summers are relatively cool other than in the driest regions, where temperatures above 30°C are not uncommon (Figure 2.2).

The climate of the region encourages a landscape dominated by coniferous trees. Nonetheless, there are recognizable, sometimes pronounced, differences in climate and vegetation. Practitioners and researchers in British Columbia use a well-developed ecological classification system – the Biogeoclimatic Ecological Classification (BEC) system. The BEC system is hierarchical and uses climate, soil, and characteristic vegetation to group ecosystems regionally and more locally.[4] The system has proven useful for a variety of research and management tasks and is employed by all provincial ministries. Levels in the hierarchy used in this text range from zones at the broadest level to site series at the finest level (Table 2.1).

On the coast of British Columbia, the zones, other than Alpine Tundra, are broad forest types reflecting similar regional climates. These are divided into subzones that reflect major differences in climate in the zone. Subzones are divided into variants, which have the same general climatic conditions but different plant communities because they are in different geographic areas or at different elevations. Site series within variants have similar soil moisture and nutrient regimes and would produce similar climax vegetation.

Biogeoclimatic zones are usually named after one or more of the dominant climax species (e.g., Coastal Western Hemlock) and are referred to by a two- to four-letter acronym (e.g., CWH). On the coast, subzone names are derived from classes of relative precipitation and continentality. The first part of the

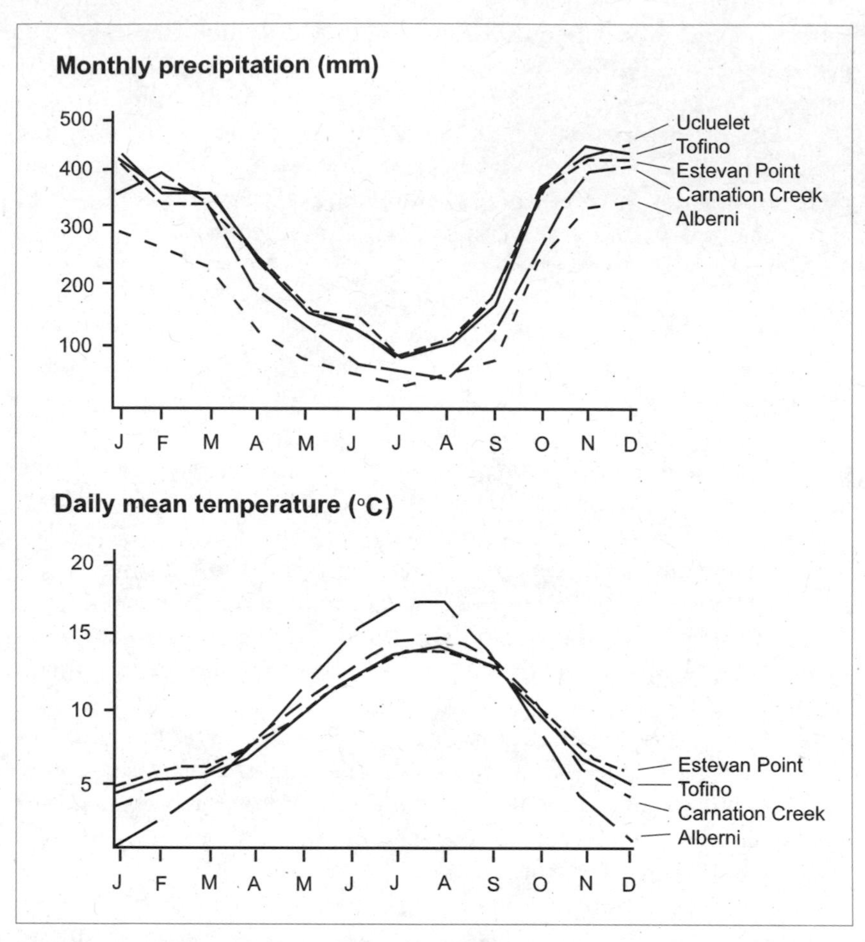

Figure 2.2 Mean monthly temperature and precipitation from selected sites in Weyerhaeuser's coastal tenure in British Columbia.
Source: CSP (1995b: Figure 2.1).

subzone name describes the relative precipitation, and the second part describes relative continentality. For example, the CWHwh represents wet, hypermaritime Coastal Western Hemlock. See the Glossary and Chapter 7 for more details on variants and site series.

The study tenure includes four BEC zones. Alpine Tundra is not forested. The other three can be considered broad forest types: Coastal Western Hemlock (CWH), Mountain Hemlock (MH), and Coastal Douglas-fir (CDF).

Coastal Western Hemlock: This zone occurs throughout the tenure outside the rainshadow at elevations below about 900 metres above sea level in the south and about 400 metres in the north. It represents about 87 percent of the tenure. The most common tree species is usually western hemlock (*Tsuga*

Table 2.1

The Biogeoclimatic Ecological Classification (BEC) system used in British Columbia

Unit	Determining features
Zones	Climatic features; dominant climax of zonal ecosystems[1]
Subzones	Climax or near-climax plant association on zonal sites; modified by precipitation and continentality on the coast[2]
Variants	Geographical divisions of a subzone; often reflect elevation on the coast
Site series	Areas with similar site moisture and nutrient regimes within a variant; characterized by assemblages of indicator vegetation

1 A zonal ecosystem or site is that which best reflects the mesoclimate or regional climate of an area.

2 In coastal zones, continentality is classified as hypermaritime (h), maritime (m), and submaritime (s).

heterophylla). Western redcedar (*Thuja plicata*) and Douglas-fir (*Pseudotsuga menziesii*) are common in southern parts of the zone, where Douglas-fir is more common on drier sites, particularly south-facing slopes. Amabilis fir (*Abies amabilis*) and yellow cedar (*Chamaecyparis nootkatensis*) occur commonly on wetter, usually higher elevation sites. Sitka spruce (*Picea sitchensis*) is common in the northern portions and on selected sites farther south, while shore pine (*Pinus contorta* ssp. *contorta*) occurs on very dry and very wet sites. Other relatively common tree species include grand fir (*Abies grandis*), western white pine (*Pinus monticola*), and bigleaf maple (*Acer macrophyllum*) in warmer, drier, southern portions; black cottonwood (*Populus balsamifera* ssp. *trichocarpa*) in flood plains; and red alder (*Alnus rubra*) on disturbed sites. The zone is characterized by its frequent precipitation, cool (though sometimes hot and dry) summers, and mild winters. It receives more rain than any other zone in the province, typically 1,000 to 4,400 millimetres of precipitation (more in some areas), with snowfall contributing about 15 percent in the south and 40-50 percent in the north. Mean annual temperature ranges from about 5.2 to 10.5°C among subzones. The CWH zone is broadly equivalent to coastal temperate rainforest and hosts some of the largest, longest-lived trees in North America. Productivity is often high but variable, with a site index of about 28 ± 6 m at 50 years (mean ± SD).

Mountain Hemlock: This zone occurs above the CWH, about 900 to 1,800 metres in the south and 400 to 1,000 metres in the north. Alpine Tundra sometimes exists above it. It represents about 12 percent of the tenure. The most common tree species are mountain hemlock (*Tsuga mertensiana*), amabilis fir, and yellow cedar. Species more common in the CWH make incursions at lower elevations. The zone is characterized by short, cool summers and long, cool, wet winters, with heavy snowfall for several months. The MH zone receives more total precipitation than the CWH zone (about 1,700 to 5,000 millimetres annually), but 20 to 70 percent falls as snow. Mean annual temperatures vary from 0 to 5°C, and average monthly temperature remains below 0°C for one to five months, depending on subzone. The zone generally is not productive: commercial forest sites and tree growth become progressively poorer with increasing elevation, cooler temperatures, and increasing duration of snow cover. Forests are not continuous in the upper MH and are largely confined to lower elevations with parkland of isolated clumps of trees above. Relatively little harvest occurs in the MH, and the best site indices are about 24 ± 5 m at 50 years (mean ± SD).

Coastal Douglas-Fir: This zone occurs on the southern, rainshadow portion of the tenure. It represents about 1 percent of the tenure and is present primarily on private land. The most common tree species is Douglas-fir. Western redcedar, grand fir, arbutus (*Arbutus menziesii*), and red alder are frequently present, depending on disturbance, site moisture, and nutrient regime. The zone is characterized by warm, dry summers and mild, wet winters. This is the driest zone, receiving about 650 to 1,250 millimetres of precipitation annually, about 5 percent as snow. Mean annual temperatures range from about 9.2 to 10.5°C. Forested sites in this zone typically are productive, with a site index of about 29 ± 5 m at 50 years (mean ± SD).

Natural disturbance regimes differ markedly among the three zones. Fire was the dominant disturbance regime in the CDF zone. Most fires were small, <30 ha, and the average fire return internal was about 100 to 300 years (British Columbia Forest Service 1990; Parminter 2004). Much of the CDF is now well populated, and fire suppression efforts are intense. Fire adapted species, such as Lewis' Woodpecker (*Melanerpes lewis*), that were formerly common are now uncommon. The dry and very dry maritime (dm, xm) and submaritime (ds, ms) subzones of the CWH zone, occupying rainshadow areas of the southern BC coast, have a similar climate and fire frequency to the CDF zone. Conversely, the montane and windward subzones of the CWH are noted for the near absence of recent fire and the dominance of late-seral forest communities that experience a disturbance regime of small-scale, tree-fall gaps (Lertzman et al. 1996; Lertzman, Fall, and Dorner 1998; Veblen and Alaback 1996). Along the windward portions of the coastal mountains, evidence of fire (charcoal and Douglas-fir) is largely restricted to specific topographic features such as south-facing slopes (Gavin, Brubakér, and Lertzman 2003; Schmidt 1960, 1970; Veblen and Alaback 1996). The pattern thus differs from similar forest farther south where fire and Douglas-fir are more common and widespread (Hemstrom and Franklin 1982). Median time since fire on terraces of the Clayoquot River is significantly greater (4,410 years) than on adjacent hill slopes (740 years). On hill slopes, all south- and southwest-facing sites had burned in the past 1,000 years, compared with only 27 percent of north- and east-facing sites (Gavin, Brubaker, and Lertzman 2003). The small openings that permit forest regeneration in the CWH are created primarily by mortality from root rot, mistletoe, snow break, and wind. Wind is the most common agent and produces two broad, self-perpetuating patterns of disturbance (CSP 1995b, 20).

Where western hemlock and amabilis fir are dominant, patches ranging from a few trees to several hundred hectares are episodically blown down by wind. The low mounds of soil produced by upturned root wads of windthrown trees provide seedbeds for western hemlock and amabilis fir. The subsequent stand structure (dense canopies with high "sail" area and shallow-rooted trees often on the tops of hummocks) makes the forest susceptible to further windfall and subsequent recruitment of the same species. Where western redcedar and western hemlock mixes dominate, crowns are less dense with less "sail" area. That, along with redcedar's firmer rooting habit, yields more wind-firm stands. Wind is still the major natural disturbance factor, but trees tend to be blown down as isolated individuals.

Less is known of the natural disturbance regimes in the MH zone of British Columbia other than that there is relatively little disturbance from fire, insects, and disease (Brink 1959; Brooke, Peterson, and Krajina 1970; Lertzman 1992). The stands are far less affected by large-scale disturbances, particularly fire, than are mountain hemlock stands farther south in the United States (Agee 1993). At lower elevations, the pattern is one of small, dispersed gaps often resulting from the fall of individual trees, whereas at higher elevations in subalpine parkland the pattern is that of tree islands and clumps of regeneration (Brett and Klinka 1998; Brooke, Peterson, and Krajina 1970; Lertzman 1992; Lertzman and Krebs 1991). Regeneration is not well understood under either disturbance regime.

Recently, logging has changed the pattern of forest renewal. Logging began where the terrain was most level, accessibility easiest, and forest productivity highest – the warmer, drier sites. As a result, the most extensively harvested areas are on the southeast of Vancouver Island and other less rugged areas east of Strathcona Park and on northern Vancouver Island (Figure 2.1)

The flora in the management area is rich, particularly in the Coastal Western Hemlock zone. Forests there are characterized by an extraordinary richness of bryophytes (mosses and liverworts) and lichens. Many of the bryophytes are endemic to the rainforest (McCune 1993; McCune and Geiser 1997; Newmaster et al. 2003; Schofield 1994). Natural forests in the area contain an abundance of dead trees and rotting wood. Lichens, bryophytes, and vascular plants of the region all include species with close affinities to dying and dead trees or rotting wood and the organic layers that it produces. Similarly, all three groups include species that require, or are more abundant in, older forests. Among species endemic to British Columbia, five mosses and eleven vascular plants occur in the management area. The potential of

Figure 2.3 Structural complexity encourages rich flora and fauna adapted to its features. | *Photograph by W. Beese.*

forest practices to assist the spread of exotics is a concern. Some exotics, such as Scotch broom (*Cytisus scoparius*), are already well established, particularly in drier areas. Chapter 10 provides more detail on the local flora.

Much more is known about the vertebrate fauna than the invertebrates. Richness of the forest vertebrate fauna is strongly influenced by the structural diversity of the vegetation. The natural vegetation of coastal British Columbia is dominated by large trees, structurally complex forests, and naturally hosts one of the richest terrestrial vertebrate faunas of northern temperate regions. The non-forest vegetation of bogs, rock outcrops, subalpine parkland, and alpine tundra contributes to the richness. Excluding very rare occurrences and accidentals, 380 terrestrial vertebrate species occur along the narrow band of coastal temperate rainforest in the Pacific Northwest (Bunnell and Chan-McLeod 1997).[5] Not all of these are forest-dwelling, and the occurrence of some on rocky islets may confer immunity from forest practices, though species using lakes, streams, or estuaries could be impacted. Data are sparse, but this richness likely extends to the invertebrate fauna.

The nature of the disturbance regime and climate over most of the area appears to influence the native fauna (Bunnell and Chan-McLeod 1997). First, because of the longevity of the trees and the long intervals between stand-initiating disturbances, most natural stands are dominated by large, old trees. In unmanaged areas, even-aged young stands are small and rare. Second, the natural growth rates, longevity, and low frequency of disturbance allow large amounts of biomass to occur. Much of this biomass exists as large live trees, large-diameter snags, and large logs on the ground. Downed wood on the lower slopes of the CWH zone, for example, can exceed 400 cubic metres or 100 metric tonnes per hectare. Third, the shade-tolerant nature of tree species present and the generally small-scale effects of common agents of disturbance ensure a complex structure in the stands. The canopy is usually uneven, with gaps where old trees have died or been windthrown. In these gaps, the understorey is often well developed and includes young conifers. Although dominated by old trees, natural stands contain trees of a wide range of ages. Fourth, the climate is wet, and streams of various sizes are abundant.

The vertebrate fauna responds to each of these broad features. Across the broad forest types in the province (forested BEC zones), the proportion of the vertebrate fauna exploiting early-seral stages is positively correlated with mean fire size and burn rate (mean fire size divided by mean fire-return interval; hectares/year), while species preferring late-seral stages are negatively correlated (Bunnell 1995). Forests in the management area host a smaller portion of early-seral species and a larger portion of late-seral species than most forest types in the province. Greater numbers of species using cavities and downed wood are found in these forests than in other broad forest types of the province (Bunnell 1995). The uncommonly high species richness is likely a product of the naturally multi-aged, complex stand structures. About 180 forest-dwelling, terrestrial vertebrate species breed in the management area. As well, more species show strong affinities for riparian areas than in any other of the broad forest types of the province. Implications to forest practices are that the native fauna is adapted to a complex forest structure that includes older age classes as well as abundant dead trees and downed wood. Riparian habitat likewise contributes significantly to sustaining biodiversity.

Riparian habitat is important to more than terrestrial vertebrates. Streams and lakes in the tenure host twenty species of fish. Conditions in lakes and, especially, streams can be affected by adjacent forest practices (Brown and Krygier 1970; Budd et al. 1987; Castelle, Johnson, and Conolly 1994; Christie and Regier 1988). Traditionally, we make a distinction between terrestrial or land-dwelling and aquatic fauna. That tradition does not reflect reality. Evidence of their interconnection is especially clear for invertebrates in subterranean water, where flowing water hosts immature stages of insect species and more stable ground water hosts subterranean crustaceans (Gibert, Danielopol, and Stanford 1998; Stanford and Ward 1988, 1993). It is also evident among vertebrates, such as stream-dwelling amphibians and fish, that are influenced by the input of forage litter and nutrients from terrestrial systems or adjacent shading effects on in-stream organisms (Hawkins, Murphy, and Anderson 1982; Hawkins et al. 1983; Murphy, Hawkins, and Anderson 1981). The fauna is often a product of interactions between terrestrial and aquatic systems rather than of separate, unconnected systems. That is especially true in the management area, where so many species seek out riparian habitat. More detail on the fauna is presented in Chapter 10.

2.2 Social and Historical Contexts

Developing forest planning and practices in coastal forests of the Pacific Northwest has been challenging for several reasons. Foremost among them in British Columbia is the fact that 95 percent of the forestland is publicly owned or crown land. Interests of the public must therefore be addressed far more rapidly and directly than is the case on private land. Forestry is also difficult to pursue with little impact because slopes are steep, the soil is often wet, and the equipment is large. Fish-bearing streams are common and vulnerable to errors in forest practice, especially road construction. Much of the region is coastal temperate rainforest, which may be the most controversial forest type in northern temperate regions. Logging original rainforest has attracted effective environmental campaigns, including boycotts of some forest products. In the United States, this forest type stimulated controversies about the spotted owl and old growth. In British Columbia, controversies and campaigns have included Carmanah, Tsitika, Klanawa, Kitlope, Kutzhemateen, Clayoquot Sound, and the "spirit bear."[6] In the study area, any slope failures around clearcuts were clearly visible, some from the highway leading to a national park.

Because it was the largest company working in coastal rainforest, MacMillan Bloedel had long been at the centre of forest controversies and environmental campaigns in British Columbia. They were costly to the forest industry and tended to overshadow efforts to improve practices. Clayoquot Sound, part of the tenure, illustrates how demanding of time and resources environmental controversy can be.

In 1984, Meares Island in Clayoquot Sound was blockaded, and the Tla-O-Qui-Aht and Ahousaht Nations declared the island a tribal park. A year later, a court injunction was granted against logging on Meares Island pending treaty settlement. In 1989, the provincial government established the Clayoquot Sound Sustainable Development Task Force to create a development strategy for the entire sound. The task force failed due to disagreements over interim logging and representation at the table. In 1990, the government again attempted to produce a sustainable development strategy by creating the Clayoquot Sound Sustainable Development Strategy Steering Committee. Environmental groups withdrew from the committee following a decision to approve interim logging. In 1991, the provincial government instructed the Commission on Resources and Environment to develop a land use plan for Vancouver Island excluding Clayoquot Sound. The area was becoming too hot to handle, and planning was required for adjacent areas. One year later, 1992, a blockade protesting logging in Clayoquot Sound led to the arrest of sixty-five individuals. Given the failure of various tables to reach any decision on areas to be protected, the government announced the Clayoquot Land Use Decision in April 1993. Within months and despite court injunctions against disruption, opposition to logging in Clayoquot Sound led to widespread civil disobedience and arrests for blockading logging operations. At the time, the 12,000 participants made it Canada's largest exercise in civil disobedience; the record of 865 arrests of individuals demonstrating against forest practices is unsurpassed anywhere.

In the fall of 1993, the government of British Columbia enacted a new strategy to resolve the issues in Clayoquot Sound. An independent panel of scientists and First Nations representatives was charged with making recommendations on forest practices appropriate to Clayoquot Sound.[7] That panel reported its recommendations to the public in 1995. Among them was the introduction of variable retention as the appropriate harvest practice. The term "variable retention" was deliberately chosen to describe the new system: "retention" to emphasize focus on what was left to sustain other forest values rather than what was removed during harvest; "variable" because amounts retained would necessarily vary with topography and other site conditions and would have to vary if the entire range of biodiversity was to be sustained. The provincial government agreed to adopt all the panel's recommendations for the area of Clayoquot Sound.

The panel included recommendations about planning and more active engagement of the Nuu-Chah-Nulth in decision making for the area. MacMillan Bloedel engaged in negotiations with the Nuu-Chah-Nulth that culminated with the creation of a joint venture, Iisaak Forest Resources, in 1998. In 1997, MacMillan Bloedel had acquired a new chief executive officer who was determined to explore ways in which the company could move from the vortex of controversy and perhaps increase market share while still harvesting coastal rainforest. Among his first undertakings was a team of researchers and practitioners to address the question of whether MacMillan Bloedel could stop clear-cutting and still regenerate the forest and make a profit. The team was distributed over six topic areas. It collated and analyzed relevant data to determine that it would cost more but probably could be done effectively. The CEO and Board of Directors concluded that the additional cost would be offset by regaining market share and increasing shareholder value. MacMillan Bloedel subsequently announced, in June 1998, that it would no longer clear-cut coastal forests anywhere in British Columbia. The company recognized that actions for sustaining biological diversity and other forest values include both long-term forest planning and short-term practices. To sustain the entire mix of desired values, it chose to reserve more old growth from harvest through a system of stewardship zones and to implement variable retention (VR) instead of clear-cutting.

The zoning system and variable retention are described in the following section. Although well reasoned,[8] the approach had never been attempted before. MacMillan Bloedel met the challenge of assessing consequences of the new approach by creating "the Forest Project." The primary objectives of this project were to make variable retention operational and create an effective approach to adaptive management, including a cost-effective monitoring program. When Weyerhaeuser acquired MacMillan Bloedel, it embraced those objectives. Although data and experience from six continents are included in this book, it is the Forest Project that we describe as a working example. In 2001, the project was awarded the Ecological Society of America's corporate award for incorporating sound ecological concepts and practices into planning and operational procedures. Many of the project's efforts, intended to help design the adaptive management program, were considered pilot studies. In July 2003, Weyerhaeuser BC announced that the five-year implementation phase was complete, and the project was now the approach to doing business on 1.1 million hectares of coastal tenure. The approach would be called the Coast Forest Strategy.[9]

2.3 New Planning and Practices

To sustain the values requested by the public, the company made changes to both planning and practices.

2.3.1 Planning

A common goal of planning is to separate competing values in space to the degree possible. The company attempted this by designating zones having different intensities of harvest. In order of decreasing amounts of fibre extraction, there are three stewardship zones: Timber, Habitat, and Old-Growth. Modifications to zoning subsequently pursued within the sustainable forest management network were termed the "TRIAD approach." Key attributes of each zone, as they were initially conceived, are summarized in Table 2.2.

The Timber zone is the primary source of economic values, and its primary goal is commercial timber production. It comprises most of the productive forest in the tenure (about 65 percent). Within the zone, up to 80 percent of the productive forest area can be considered available for harvest throughout the harvest cycle. The silvicultural system for the harvestable area is predominantly the "retention system," a departure from traditional even-aged management with retention ranging upward from a minimum of either 5 percent dispersed retention or 10 percent group retention (Table 2.2). Although some harvest sites may receive other silvicultural systems (e.g., shelterwood), all treatments include some permanent retention of structural attributes. The provision of late-seral features through retention is intended

Table 2.2

Description of stewardship zones

	Timber	Zone Habitat	Old Growth
Management emphasis	Timber production	Habitat conservation	Late-seral forest conditions
Portion of managed land base in each zone (%)	65	25	10
Average proportion of productive forest area in reserves (%)	28	40	70
Minimum long-term retention in each cutblock (%)	Dispersed: 5 Group: 10	Dispersed or group: 15	Dispersed or group: 20
Primary silvicultural systems	Retention, shelterwood	Retention, shelterwood, selection	Selection, irregular, shelterwood

Source: Adapted from Beese et al. (2003).

to sustain many species that would not be present under clear-cutting. The degree of fibre removal is highest in this zone, but retention adds to elements of older forest already present in riparian buffer strips, deer winter ranges, and other areas reserved from harvest. Because of the large extent of the zone and uncertainty about the outcomes from variable retention, most monitoring has occurred in this zone.

The Habitat zone has as its primary goal the conservation of wildlife and other organisms comprising biological diversity. It is viewed as an area where refinements to practice can be implemented as guided by results of monitoring. It occupies about 25 percent of the tenure. Minimum retention in this zone is 15 percent, and no more than 70 percent of the forest area in the zone is considered available for harvest (Beese et al. 2003). Whereas VR in the Timber zone primarily uses group or mixed retention, a full range of silvicultural systems with retention is anticipated in the Habitat zone (Table 2.2). The name "Habitat zone" reflects the opportunity to designate specific practices (e.g., retention type and amount) to meet needs of specific species in specific places. For example, monitoring in the Timber zone may suggest helpful gains to cavity-nesters, from higher levels of retention of trees and snags, that could be achieved by suitably located Habitat zones. The Habitat zone also provides an opportunity for "mini-zoning." In the Habitat zone, specific areas ("mini-zones") of high amounts of retention could be located so as to reduce any gaps apparent in the representation of ecosystem types in non-harvestable areas. Although currently little different from the Timber zone, the Habitat zone could become the zone where early refinements documented by the adaptive management program were applied.

The Old-Growth zone is intended primarily to maintain late-seral forest conditions, so relatively little wood is removed from the zone. It comprises about 10 percent of the tenure. In the Old-Growth zone, two-thirds of the forest area is reserved from harvest, and harvest in the remaining third is by uneven-aged systems (e.g., group selection or irregular shelterwood with retention using <1 ha openings and return intervals of <33 percent of the rotation age). The minimum level of retention in harvested areas is 20 percent. By improving ecological representation in the non-harvestable land base, the Old-Growth zone is intended to contribute to sustaining poorly known ecological functions, meeting the needs of late-seral organisms, particularly those requiring larger areas of older seral stages, and providing a significant area of ecological benchmarks for monitoring. Late-seral species may decline significantly or even disappear from the Timber zone. To attain the broad objective of sustaining biological diversity, late-seral species must remain productive in the Old-Growth zone if they cannot be accommodated by late-seral inclusions in the Timber or Habitat zone (e.g., deer winter ranges, ecologically sensitive areas, or late-seral features sustained by variable retention).

After consultation with government and stakeholders, the tenure was divided into these three stewardship zones, using large landscape units (5,000 to 50,000 hectares) and based on the development history and ecological attributes of each area. For example, watersheds with more than 70 percent old-growth cover, high social values, or First Nation priorities were candidates for the Old-Growth zone. Watersheds with a long history of logging, occupying private land, and having a high proportion of younger age classes, were candidates for the Timber zone. Initial matches for the Habitat zone were sought among areas designated for "low-intensity" harvesting or medium to high biodiversity emphases in government land use plans. The proportion of land initially allocated to each zone was arbitrary and considered subject to change dependent on feedback from monitoring.

The potential for restoration silviculture to hasten acquisition of old-growth attributes was considered at the outset, and pilot efforts to examine consequences were initiated (Krcmar 2002; Perry and Muller 2002; see also Carey, Thysell, and Brodie 1999). If assessment of ecological representation found that late-seral attributes were lacking in particular areas or ecological types, then it might prove possible to restore them more rapidly than would occur without intervention. Such practices would employ variable density thinning or wide spacing to hasten development of late-seral stand attributes. Where successful, restoration would create small Old-Growth zones in Timber and Habitat zones. If restoration activities were applied, then they would become part of the adaptive management program. The need for restoration and the desirable extent were to be assessed through Indicator 1, representation, and Indicator 2, habitat structure (§3.2.2).

2.3.2 Practices

Zoning is a landscape-level allocation tool that helps to distribute intensity and timing of disturbance. Aggregating the most conflicting values into separate zones reduces local conflict but was believed unlikely to maintain all forest-dwelling organisms. To retain the kinds of habitat structures required by many forest-dwelling organisms, the company also implemented variable retention in all zones (Table 2.2). Variable retention (VR) was adopted because it preserves, in managed stands, far more of the characteristics of the natural forest that many of the public strongly value and to which organisms in the area are adapted (§2.1.2). VR was first defined and proposed by the Clayoquot Scientific Panel for application in the Clayoquot Sound region of the tenure. The panel noted that "the variable-retention system provides for the permanent retention after harvest of various forest 'structures' or habitat elements such as large decadent trees or groups of trees, snags, logs, and downed wood from the original stand that provide habitat for forest biota" (CSP 1995b, 83).

Others have since refined the approach.[10] The company adopted the term "variable retention" to refer to any silvicultural system that retains structural elements of the existing stand throughout the harvested area for the long term to attain specific objectives. These objectives include:

- retaining late-seral forest structures to increase species richness in managed stands, increase habitat connectivity across the landscape, and provide "lifeboats" for survival and dispersal of organisms after harvesting;
- creating opportunities to match changing market demands and harvesting profiles without "high-grading" or compromising forest health, vigour, genetic composition, or timber quality;
- meeting diverse social expectations of stewardship and visual aesthetics; and
- matching the range of possible retention and silvicultural systems to meet site-specific needs for regeneration and habitat (Beese et al. 2003; Bunnell, Kremsater, and Boyland 1998; CSP 1995b; Franklin et al. 1997).

Retention can vary in amount, type, and spatial pattern of structure depending on objectives. Because they have different management emphases, the three stewardship zones specify different amounts of retention at the stand and landscape levels (Table 2.2). The new "retention silvicultural system" was defined as the primary system used in the VR approach. Two broad types are recognized: *dispersed* retention (individual trees or small groups of trees are retained), and *aggregated* or *group* retention (larger groups or patches of trees are retained). These two types of retention can be used on the same cut block (and are then referred to as *mixed* retention). The retention system can be applied to all ages of forests and a wide range of tree species to maintain and recruit greater structural diversity than under short rotation clearcut systems.

The ecological rationale of the retention system is implicit in the objectives above. This rationale distinguishes it from other silvicultural systems, including "clearcut with reserves," which is primarily reservation of a patch of trees adjacent to a cutblock. The rationale is based on two points. First, the system retains enough tree canopy to have forest or residual tree influences over the majority of the harvested area. Clearcuts have been defined ecologically as "an area in which the above-ground and below-ground influence of the trees has been removed in more than half the area" (Kimmins 1997, 76), and "the minimum size of opening that constitutes a clearcut varies with the height of the surrounding forest, and is roughly equal to an area greater than about four tree heights in diameter" (Keenan and Kimmins 1993, 122). Some silvicultural systems (e.g., selection systems, shelterwood) meet the objective of increasing forest influence but do not fulfill the second part of the rationale for variable retention – that the live and dead trees retained, plus the amount of canopy cover, are sufficient to meet the needs of a broad array of organisms.[11]

To fulfill its rationale, the retention system must maintain greater than half the original forest area within the influence of surrounding trees, or trees retained in the harvested area, and create openings that generally are less than four tree heights across. Moreover, the retained trees must include a range of attributes appropriate to sustaining diverse organisms (e.g., Figure 2.4).

From its conception, the VR approach was intended to be variable for two reasons. The first was to provide flexibility in response to many site-specific features (safety, silviculture, harvesting feasibility, economics, and visual aesthetics). The second was to ensure that a variety of patterns of forest structure were retained to meet the needs of a variety of organisms. Figure 2.4 illustrates that a wide variety of forms of retention have been applied. The choice of approach (amounts to retain, type or pattern of retention, and structures to retain) is first influenced by the strategic direction set by zoning (Table 2.2) and then modified by more local landscape features. For that reason, the nature of retention varies widely. It was unclear at the outset how VR would be applied, particularly which structures would be retained. For that reason, the monitoring initially had to address features of implementation (see Chapter 4).

2.4 Structures to Make It Work

With the adoption of stewardship zones and VR, the company began a grand

Figure 2.4 Groups of trees are no more than four tree lengths apart; individual trees are no more than two lengths apart. | *Photographs by W. Beese and from Bunnell (2005: Figure 2).*

experiment in forestry. It now had to make it work. Adaptive management programs most commonly fail through lack of will or because institutional structures fail to connect scientific findings to changes in activities (Lee 1993, 1999; Lee and Lawrence 1986; Ludwig, Hilborn, and Walters 1993; Stankey and Shindler 1997). The company recognized this problem and created three new structures to make the program work: the Adaptive Management Working Group, the Variable Retention Working Group, and the International Scientific Advisory Panel.

The *Adaptive Management Working Group* was comprised primarily of researchers, including academics, consultants, employees of the company, representatives of companies with adjacent tenures, and government representatives. Initially, the group had two purposes. Its first purpose was to create an approach to adaptive management for biodiversity, including the objective, indicators of success in attaining that objective, and efficient means of assessing the indicators. Its second purpose was to design and implement pilot studies that would determine an efficient design for the adaptive management program. It included those who described the initial ecological rationale for the new planning and practice (see Bunnell, Kremsater, and Boyland 1998).

The *Variable Retention Working Group* was composed of practitioners – those who had to implement the new planning and practices. Early discussions in the Variable Retention Working Group focused on how to make the new practices work. With time and demonstrated ability to make new practices work, discussions focused increasingly on conflicts between economic return and sustaining biodiversity – illustrating that a wicked problem remains wicked. While broad elements of the new approach and specifics of the monitoring design were developed by the Adaptive Management Working Group, the innovation and the practicality of making new practices work operationally resided primarily in the Variable Retention Working Group. Company personnel who attended both groups helped to maintain communication between them. Such linkage is necessary to ensure that scientific findings do feed back to management action and close the loop in the adaptive management process (Figure 1.1).

The *International Scientific Panel* was created to provide credibility through quality control and to provide guidance from a breadth of experience. The panel also helped to evaluate the credibility of any findings of the program that were contrary to existing policy or regulations. The issue of managing to sustain biodiversity is value laden and must connect with (and perhaps influence) government regulation and policy. Those features determined the composition and means of selecting the panel. Panel members included both internationally recognized scientists and personnel of provincial and federal government agencies. There were twelve members – four chosen by the company, four chosen by environmental NGOs, and four selected by government agencies (two provincial and two federal). The panel met and reviewed progress every one or two years. Each year panel membership was altered somewhat to ensure a breadth of viewpoints, but a core group was retained. Scientists on the panel have been drawn from Australia, Europe, Canada, and the United States.[12]

Each group helped to facilitate the process. The diversity of viewpoints and talents within the Adaptive Management Working Group encouraged debate and helped to ensure that much of the breadth of biodiversity was represented. The Variable Retention Working Group was necessary to translate the concepts into actions and to introduce the innovation necessary to make the process work. The International Scientific Panel provided quality control and guidance from a broad base of experience, generally from outside the province.

2.5 Monitoring

Successive ownerships have committed to implementing adaptive management as part of the change to new planning and practices. The widespread use of variable retention and the approach to zoning employed were novel. Consequences of the new practices, including the degree to which such practices sustain components of biological diversity, were uncertain. Credible monitoring had to be implemented if the adaptive management program was to work and learning and refinement of practice were to occur.

Designing the monitoring program was the most intellectually challenging part of implementing adaptive management. Most of this book describes how the monitoring program was designed so that learning to sustain biodiversity could occur. Chapter 4 reviews the monitoring program employed to evaluate whether the changes were being made on time and as planned. Chapters 5 through 11 describe elements of the monitoring program designed to assess specific indicators of success. Chapter 12 summarizes the monitoring design, and Chapter 13 summarizes the lessons learned from attempting the design.

2.6 Summary

From this overview of the physical, ecological, and social settings of the case study, we draw six points.

1 The geology, weather, and terrain create conditions naturally susceptible to slope failures and debris slides, and are thus operational challenges to forest practices.
2 The rich flora of the region includes species with close affinities to dying and dead trees or rotting wood and the organic layers that it produces.
3 The rich fauna of the region appears to be adapted to the natural disturbance regime and includes a high proportion of species responding positively to a complex forest structure that includes older age classes as well as abundant dead trees and downed wood.
4 The forest of large trees hosting high species richness attracted international concern about forest practices.
5 The novel approaches to forest planning and practice proposed to address that concern had not been assessed. For both economic and ecological reasons, it was imperative that they be assessed.
6 The most effective assessment procedure was adaptive management, which required new structures to facilitate it and a strong emphasis on monitoring.

The book focuses on effectiveness monitoring for sustaining biological diversity in managed forests. Generic concepts applicable for forests anywhere are addressed but illustrated by concrete examples of these concepts as applied to a real corporate enterprise in British Columbia. This chapter provides ecological, historical, and social contexts for the case study and describes the new planning and practices. Chapter 3 outlines the generic approach taken to implement adaptive management.

3
The Approach

Fred L. Bunnell and Glen B. Dunsworth

3.1 Managers' Questions

Our interest is the problem of sustaining biological diversity in managed forests. Managers faced with any problem confront four general questions.

1 Where do we want to go?
2 How do we get there?
3 Are we going in the right direction?
4 How do we change if the direction is wrong?

These questions represent the management tasks of establishing objectives, deciding on actions, monitoring success, and linking findings to actions. They equate directly with the four broad elements of an effective adaptive management program: setting clear objectives, deciding on actions, monitoring success, and providing feedback to management (§1.4). Each question is treated briefly in the following sections.

This book focuses primarily on questions 3 and 4 as ways of learning how to sustain biological diversity in managed forests. Our answer to question 1 appears to be widely applicable, but regional answers to question 2 will vary greatly in response to local conditions. The scope of the adaptive management program presented here is intended to encompass all forest-dwelling organisms. These and other features of the forest will differ regionally. Little is known of the needs of most species, whatever the region. For that reason, the program uses informed surrogates for poorly known organisms and functions, such as ecosystem types and habitat structures. These surrogates have more general applicability. The case study used to illustrate practical issues covers a large forest tenure in British Columbia, both private and public lands (about 1.1 million hectares; see Chapter 2).

3.2 Establishing Objectives and Measures of Success

The managers' initial question is "where do we want to go?" We must answer that question to get anywhere specific. We need to know both where we are going and how we intend to get there to provide context for monitoring success. Monitoring success necessarily is related to objectives. Conversely, an objective is not helpful without some practical way of assessing whether or not it has been attained. Thus, monitoring must be linked to the features being manipulated to attain objectives. These features are manipulated by the planning and practices intended to sustain biodiversity – the "hows" of getting there. It is important to recognize that there is no specific "where," no static spot that we can define in time and place. That is a feature of any wicked problem (§1.1). "Where" is a direction, and determining success is limited to discerning what is more or less effective rather than a tidy recipe of planning and practices.

Explicitly describing the direction as an operational goal and developing measures of success in attaining that goal are complex and difficult tasks. Lack of agreement on the scientific content of the term "biological diversity" impedes the creation of operational definitions of biodiversity (e.g., Bunnell 1998c; Delong 1996; Wood 1997, 2000). The extensive and long-term nature of forest practice creates special difficulties that hinder the establishment of objectives and the development of monitoring protocols for forested systems. Major issues obscuring the definition of biological diversity and the special difficulties encountered in forested systems were treated in §1.3 and elsewhere (Bunnell and Chan-McLeod 1998; Bunnell, Kremsater, and Wind 1999). Below we summarize earlier efforts to define biodiversity, offer an operational definition for biodiversity, and describe broad indicators of success in sustaining biodiversity.

3.2.1 Defining Biological Diversity

To develop tactics for sustaining biological diversity, there must be agreement on what is to be sustained and what is meant by sustenance – where do we want to go? That has not proven easy. The scope of biological diversity yields a remarkably complex, largely unknown, and ill-defined objective. By 1998, there were at least ninety definitions of biological diversity. A review (Bunnell 1998a; Delong 1996) of their features suggested four key points.

1 The term "biodiversity" suffers from reification, the treatment of an abstract concept as if it were a thing. West (1994) termed biodiversity a "concept cluster" (see also Bowman 1993 and Merriam 1998). Practitioners attempting to manage for "the thing" risk failing to achieve all imagined benefits ascribed to the concept.
2 Most definitions include a phrase such as "the full range of life in all its natural forms, including genes, species, and ecosystems – and the ecological processes that link them" (review of Bunnell 1998a). This apparent agreement on the content of biological diversity is illusory. It belies the

enormous difficulties of translating such definitions into an operational statement that can be used by land managers to guide actions.

3 The most frequently copied definition and the only legally binding, international definition emphasize variability (OTA 1987; UN 1992).[1] Unlike earlier definitions that stressed entities such as genes or species (see, e.g., IUCN, UNEP, and WWF 1990 and WRI et al. 1992), the Convention on Biological Diversity makes a clear and specific distinction between "biological diversity" ("variability among living organisms") and "entities" or biological resources. Emphasis on variation acknowledges natural conditions but hinders explicit definition of a management objective by departing from traditional approaches to classification and inventory. It also implies an acceptable range rather than a static target.

4 When discussing biological diversity, the public, politicians, and courts do not mean what scientists mean, and managers pursue publicly defined goals. Five dominant public concerns were addressed by international agreements emerging from UNCED '92:[2] rates of extinction, future options, productive ecosystems, economic opportunities, and local participation in decision making (Bunnell 1997, 1998a). These concerns are easily understood. Major underlying mechanisms, such as genetic variation, are less well understood and of less concern to most of the public. The challenge is to create an operational surrogate for the term "biological diversity" that reflects both public concerns and scientific content of the term.

The scientific basis for sustaining biological diversity is not the wide range of elements noted in definitions but the variety or diversity itself (Bunnell 1998b; Wood 1997). Definitions of the United Nations and other major international organizations clarify this point (McNeeley et al. 1990; UN 1992). These definitions do not claim that biodiversity consists solely of genes, species, ecosystems, or ecological processes; rather, it encompasses them and their variability. Biological diversity is most simply an attribute of life, the differences among entities (Bunnell 1997; Wood 1997). These genetically based differences among organisms and taxa permit continued adaptability and continued creation of biological diversity (Frankel and Soulé 1981; Mayr 1997; Norton 1986). The public concerns noted thus connect directly with the scientific basis for sustaining biological diversity (Figure 3.1). Retaining a variety of individuals and species permits the adaptability that sustains ecosystem productivity in changing environments (Bailey et al. 2004; Mittelbach et al. 2001; Naeem et al. 1994; Tilman and Downing 1994) and begets further diversity (future adaptability and options), thereby potentially sustaining desirable economic opportunities. The only major public concern evident in the documents emerging from UNCED '92 that is not directly connected to genetic variability is local participation.

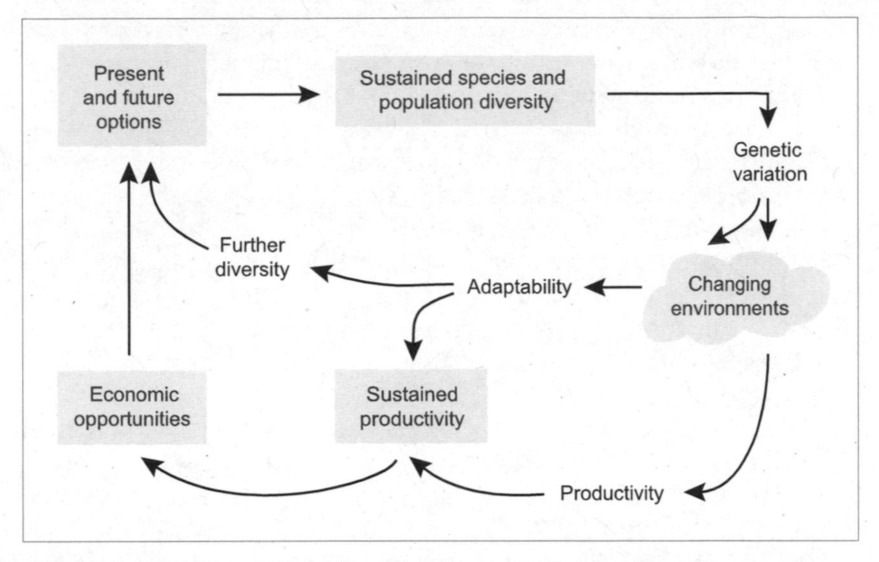

Figure 3.1 Relationships among sustained species and population diversity and other desired outcomes of sustained productivity, economic opportunities, and present and future opportunities.
Source: Bunnell, Kremsater, and Boyland (1998: Figure 1).

Although genetic diversity is a fundamental incentive for sustaining biological diversity, it presents intractable problems as an operational definition, including the fact that genes are not self-replicating units. For this reason, various workers have argued that native species richness (the number of native species) is currently the most tractable surrogate for biological diversity (see Bunnell 1998c and Namkoong 1998). It is, however, an incomplete surrogate. For example, it ignores the question "are all species equal?" There is increasing evidence for keystone species (e.g., woodpeckers, beavers) whose contributions to ecosystem function and integrity are greater than those of other species, indicating that species or genetic richness is not the only guiding principle for sustaining biological diversity.[3] Nonetheless, species richness can help to specify the broad objective, especially when combined with geographical distribution, to capture the range of genetic variability. Other indicators of biological diversity, however, must augment species richness. For example, maintaining unharvested areas of all distinct ecosystem types helps to maintain poorly known ecological functions (see the following section and Chapter 6).

3.2.2 A Criterion and Indicators of Success

One outcome of UNCED '92 was the system of criteria and indicators used to describe sustainable forest management for purposes of certification. Most simply, criteria represent sets of values that are to be sustained by the forest management plan. Indicators represent things that should be evident and capable of measurement if those values are being sustained. Forest certification inspired various systems of criteria and indicators but is not the primary reason for their utility. Specifying values clearly as criteria or objectives is necessary if we are to answer "where do we want to go?" Specifying measures or indicators of success is equally necessary if we intend to evaluate the success of our efforts.

From the preceding observations on the challenge of establishing an objective for sustaining biodiversity and the special difficulties in forested systems, we derive four characteristics of a criterion that reflect success in sustaining biological diversity.

1 *Objectives must be established over large areas and long time periods.* Although this characteristic follows from ecological principles, there are no guidelines beyond reasoned guesses of at least 200,000 hectares and one rotation (Bunnell 1997).[4]
2 *Operational objectives must be defined clearly as a measurable and achievable portion of the entire scientific content of "biological diversity."* Difficulties of creating operational definitions of biodiversity have been exposed.[5] Several workers have argued that sustaining species and populations of species is a measurable, credible objective (Figure 3.1). No operational definition can include explicitly all elements of all definitions of biological diversity. The indicators of success, however, can attempt to embrace the full meaning of the term "biological diversity."
3 *Surrogates for monitoring must use representative ecosystems, species, or groups of species, because all individual species are impossible to monitor.* This condition follows from the richness of forested systems and the practical need to monitor efficiently. The challenge remains to ensure that the surrogates correlate well with, or are causally linked to, the impacts of forestry on the totality of biological diversity.
4 *To sustain biological diversity in managed forests, we need to exploit relationships between groups of species and habitat structure and how that structure is changed by forest practices.* Appropriate tactics for attaining the objective must relate directly to forest practices and be readily measurable. That is, tactics and measurements must be related to features of the forest modified by forest practices. Monitoring surrogates for species such as ecosystem representation and habitat structure does not eliminate the need to monitor species or other measurable components of biological diversity.

Our approach to assessing success in sustaining biological diversity acknowledges these points. Indicators of success are needed before all underlying mechanisms generating and sustaining biological diversity are clarified. A critical part of the criterion for success must be to maintain native species richness over some large area and long time period. Four of the public concerns noted above relate to this component of biological diversity (Figure 3.1). Indicators beyond species richness are needed to encompass the content of "biological diversity" and to connect directly with forest planning and practice. The term "biological diversity" is often equated, implicitly or explicitly, with ecosystem productivity, health, resistance to pathogens, or even forest structure. The three indicators presented below are intended to encompass much of that content.

Criterion: Biological diversity (native species richness and its associated values) is sustained within the tenure.
The criterion defines one set of values provided by the sustainable forest management plan. The coastal tenure, more than 1 million hectares, is likely large enough to encompass the requisite variety of habitat needed to sustain biodiversity. The influences of time can be addressed only through projections by simulation models. Success in attaining this criterion of effective forest management is assessed by three indicators representing much of the implied content of the term "biological diversity." For example, poorly known species and functions are addressed through Indicator 1 and diverse habitat structure through Indicator 2. Other features, sometimes associated with the term "biological diversity," are assessed by the other parts of the monitoring program (e.g., site productivity, windthrow, resistance to pathogens). The three indicators are as follows.

Indicators of Success
Indicator 1: Ecologically distinct ecosystem types are represented in the non-harvestable land base of the tenure to maintain lesser known species and ecological functions.

Indicator 2: The amount, distribution, and heterogeneity of stand and forest structures important to sustain native species richness are maintained over time.

Indicator 3: The abundance, distribution, and reproductive success of native species are not substantially reduced by forest practices.

Each indicator is intended to assess different aspects of success in attaining the criterion.

- Indicator 1 assesses ecosystem representation in areas that will not be harvested and is intended primarily to ensure that little-known species and functions that may not be assessed by Indicators 2 and 3 are sustained. It also serves to identify unmanaged "benchmarks."
- Indicator 2 complements Indicator 1 by evaluating habitat elements and structures that we know are required by many species and projecting consequences of changes in those habitat features through time.

- Indicator 3 is intended to assess whether species naturally present on the tenure are likely to continue as well-distributed, productive populations. It serves to test the broader approaches of Indicators 1 and 2.

The three major indicators have associated subindicators. They encompass the complexity in the criterion and necessarily interact. For example, the distribution of ecologically distinct habitat types within the non-harvestable land base of the tenure determines the kinds of habitat provision required in harvestable areas to meet success. Assessing each of the three major indicators must proceed in parallel, because findings for each indicator inform the others. There is thus no order of primacy of effect implied in their number. Instead, the order represents increasing refinement, with increasing emphasis on individual forest attributes or species.

Objectives, or criteria of success, must be sufficiently clear that means of assessing proximity to objectives are likewise clear. Broad objectives for sustainable forest management for the case study are expressed in the framework of criteria and indicators. The broad criterion may appear clear but encompasses complexity that can be embraced only if different features contributing to it are recognized and assessed by different indicators. The criterion and three broad indicators of success were established early in the program, providing the necessary context for developing the adaptive management and monitoring program. Each indicator is discussed separately in following chapters. The company had created similar criteria and indicators for other forest values through certification processes.[6] The criterion and its indicators constitute an answer to the question "Where are we going?"

3.3 Deciding on Actions

Two broad kinds of actions help to answer the question "How do we get there?" They are long-term planning and short-term practices. Deciding on specific actions begins by determining issues that the actions are intended to resolve. Our issues were reserving older forests from harvest and sustaining biological diversity adapted to older forests with complex structures. Both planning and practice help to sustain biological diversity. Because biodiversity requires a large area to achieve its full expression, sustaining it must be planned over large areas. All organisms of a region are not found naturally on every hectare, nor can they be sustained on every hectare by management or preservation. Reserving more older forest from harvest is addressed primarily by planning and implementing stewardship zones (§2.3.1). The Old-Growth zone adds to areas of older forest already reserved by other means (e.g., riparian reserves, deer winter ranges, protected areas). Without very large tracts, relying on zones to provide late-seral habitat would prove to be insufficient for some species. Potential limitations of habitat structure are addressed by implementing variable retention (§2.3.2). Late-seral

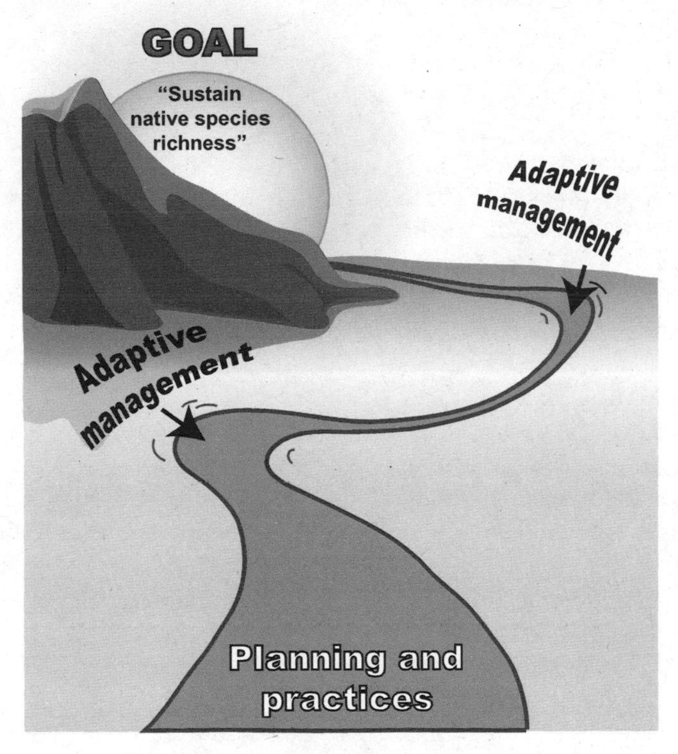

Figure 3.2 Adaptive management provides the course corrections to the road map provided by planning and practice.
Source: Prepared by I. Houde.

attributes are distributed across the tenure by retention systems. Public concern inspired the actions, but they were selected through review of scientific literature and practical experience.

Together, planning and practices provide a "road map for getting there." Each informs the other. If sufficient structure is retained to maintain species at local levels, then they do not have to be accommodated in broad-scale planning. Adaptive management helps to provide course corrections to the road map (see Figure 3.2). Although the choice of actions was well reasoned (Beese and Zielke 1998; Bunnell, Kremsater, and Boyland 1998), the outcomes were uncertain and had to be monitored.

Within an adaptive management framework, considerations about actions do not relate only to what you plan to do to achieve objectives. They also must determine which activities can be modified in the future. Changeable elements of practice are direct links to management and must be a focus of monitoring activities. For example, we expected that amounts allocated to the Old-Growth stewardship zone were unlikely to change, but locations could be changed. It was thus important to assess whether modifying locations of the Old-Growth zone would provide more complete coverage of poorly known species and functions. Similarly, there are several features of retention that can be modified – amount, patch or group size, type, spacing, ecological features used to anchor patches. Because these features can be modified, they are important elements of monitoring to assess effectiveness. They also specify comparisons that are part of the monitoring design – across levels of retention, across patch sizes (or edge effects), across retention types, or across different kinds of anchor points. Both intended actions and actions amenable to change were determined to address the question "How do we get there?"

3.4 Evaluating Success

The third of the managers' questions – "Are we going in the right direction?" – is the most difficult to answer. This question could be phrased "How do we know the direction is wrong?" The criterion provides obvious guidance – apparent reductions in native species richness are the wrong direction. However, planning and practising forestry may be humanity's most complex undertaking (Bunnell 1998b). Combining planning to sustain biodiversity and to harvest wood creates a complex wicked problem (§1.1). The task of sustaining biological diversity in managed forests is not tidily constrained but leaks into areas of genetics, climate change, wintering grounds thousands of kilometres away, and events at sea. Moreover, in this case, it involved a novel combination of forest practices, as yet untried, with all their associated uncertainties. It is exactly this uncertainty that compelled the adaptive management program and the need for monitoring. But there is simply too much uncertainty. Ideally, the problem would be well bounded and focused before we begin designing and creating the adaptive management program. Below, we introduce four large ideas related to evaluating success: bounding the problem, monitoring questions, determining kinds of monitoring, and creating structured learning.

3.4.1 Bounding the Problem

The real world is not ideal. We so rarely bound the problem before undertaking to solve it that Holling (1978) considered bounding the problem as part of adaptive management. When corporations or governments decide to make a major change, they expect that decision to be followed quickly by actions. No time is available for gradual change. In the case study, six teams were formed to address different portions of the question of whether or not the company could stop clear-cutting and still make a profit. The CEO allowed them ninety days to prepare their answers. Uncertainties and risks were identified but only in general terms that provided little guidance about which uncertainties posed the greatest economic or ecological risk. There were both benefits and costs to proceeding simultaneously on the tasks of bounding the problem and developing an adaptive management program. A major benefit was that the early effectiveness of monitoring helped to expose unanticipated uncertainties. A significant cost was that pilot projects began without being focused on areas of greatest uncertainty.

A monitoring problem can be bounded or constrained in two broad ways. The first is the scope and definition of the problem; the second is the questions that will be asked. Scope and definition were addressed at the outset. Two large issues were the spatial extent of the program and the conceptual extent of biodiversity. The entire tenure of 1.1 million hectares was selected as the necessary scale at which the maintenance of biodiversity would be addressed. The operational definition relied on both the criterion and indicators (§3.2.2). The criterion of native species richness and its associated values was augmented by measures of ecosystems and habitat to bound conceptual extent.

Bounding the problem eliminates some potential features or organisms as indicators. Our interest in the effects of forest planning and practice on species richness excluded some taxa as effective indicator species. Some species (e.g., salmon) spend most of their life at sea, where forestry practices have little impact. Among species restricted to land, some are more influenced by features other than forest practices, or their response to forest practices is so poorly known that it provides no reliable guidance. Bounding the spatial and functional connections of the problem encourages a focus on those elements most likely to provide guidance and control.

The process of bounding the problem and the monitoring program is facilitated when the following steps are taken.

- *Determine major issues.* Our issues were to reserve more old growth and implement a harvest system that sustained native species richness.
- *Clearly define objectives.* Along with associated indicators of success, these help to focus the monitoring on appropriate variables. We used a single criterion and three broad indicators of success (§3.2.2). Specific measurements are discussed subsequently for each indicator.
- *Identify the management plan and practices.* The company decided that stewardship zones and variable retention were appropriate for addressing the major issues. The approach to planning and practice became hypotheses to be evaluated by the monitoring process (Davis et al. 2001).
- *Bound the problem.* Establish the physical, functional, and conceptual boundaries. The goal is to delineate the problem such that extraneous influences over which the manager has little control, and which could mislead monitoring results, are reduced. Large issues were noted above. Issues specific to individual indicators are discussed in the following chapters.

- *Identify the major questions.* A monitoring design must ask the right questions to reduce statistical uncertainty, properly estimate parameters from noisy data, and assign probabilities to alternative hypotheses (Mulder et al. 1999; Noss 1990, 1999; Noss and Cooperrider 1994). Questions selected for monitoring are listed below (§3.4.2).
- *Identify data needs.* Although an assessment of current conditions indicates broad knowledge gaps, specific needs must be identified in terms of data best suited to answer the specific monitoring questions.
- *Rank the objectives or questions and data needs.* Available resources for monitoring will be limited. Ranking or setting priorities is critical because it focuses on questions that present greater uncertainty and risk. Questions selected for monitoring should be ranked and assigned to specific objectives. Which practices and objectives require immediate attention and are more likely to have an impact on biological diversity? Data needs may then be ranked accordingly.

There is nothing tidily linear about these steps, and the process is generally iterative.

3.4.2 The Major Questions

Once the physical and conceptual nature of the problem is bounded, the most compelling questions for the monitoring program can be determined. Monitoring questions depend on the objectives, the practices, and what we know of the current conditions. They follow from the managers' questions (§3.1) on objectives, actions, and effectiveness. The questions provide a set of hypotheses and focus monitoring on areas where management requires information to adjust activities and avoid unplanned and undesirable outcomes. In this sense, the link between monitoring and decision making begins with agreement on the questions.

Inevitably, there are more questions and uncertainties than the monitoring budget can accommodate. Uncertainties around management decisions and actions were phrased as questions and gradually winnowed down to six major questions. The second-year review by the International Scientific Advisory Panel emphasized the importance of evaluating some questions over large areas. These major questions are stated below in an order that reflects increasingly large areas and longer time frames. Priorities assigned to the questions are provided in italics.[7]

1. What is variable retention providing as habitat? *(very high)*
2. Are there major edge effects within aggregated variable retention? *(very high)*
3. What is the best way of implementing variable retention (e.g., types and amounts of retention)? *(high)*

4 Is stand restoration effective at creating desired structures and ultimately restoring species distributions or numbers where old growth is rare? *(moderate)*
5 Are stewardship zones established in the most appropriate locations? *(very high)*
6 Is biological richness maintained over the tenure, given the mix of zoning, variable retention, and operational constraints? *(high)*

These questions were considered to present the areas of greatest management concern, greatest uncertainty, and greatest ecological risk in implementing the new approach to forest planning and practice. Our ability to answer the questions determines how well the monitoring and modelling program addresses the major issues of managers. These questions do not fit the traditional view that monitoring is a management or regulatory activity unrelated to scientific research. They reflect growing awareness that monitoring and research are inseparable and joined by four features – design, quality control, interpretation, and assessment of effectiveness of the monitoring (Franklin, Harmon, and Swanson 1999; Noss and Cooperrider 1994; Stem at al. 2005). It is clear that all questions cannot be answered at once. Some effects of forestry require decades to become obvious. Questions are broadly ordered as answers should become apparent. More generally, the longer-term questions are addressed by simulation.

All three major indicators contribute to answering the questions. For most questions, Indicator 1 (ecosystem representation) helps to stratify and focus the monitoring effectively. Indicators 2 and 3 contain the response variables of interest (habitat structure and organisms). The indicators provide insights at different rates. Indicator 1 can reveal potentially troubling issues most quickly. Indicators 2 (habitat) and 3 (organisms) are frequently linked, but data are acquired more quickly for Indicator 2. There are efficiencies in this approach because results from Indicator 1 can focus efforts of Indicators 2 and 3, and results of Indicator 2 can focus potentially costly efforts of Indicator 3.

3.4.3 Kinds of Monitoring and Adaptive Management

The most widely accepted typology of monitoring is that of Noss and Cooperrider (1994). Since they proposed their typology, more experience has been accrued in monitoring to assist adaptive management, and the typology can be refined. It is helpful to consider the forms of monitoring in terms of the questions that they attempt to answer.

Have we done what we were told to do? Compliance monitoring. Compliance monitoring assesses actions relative to external targets. It records whether a particular standard, regulation, or proposed "best practice" is met. It is not directly useful to adaptive management because it does not evaluate the effectiveness of the standard and thus provides no guidance for improvement. In fact, when initially proposed, variable retention could not be in compliance because it was novel – new definitions and regulations had to be developed (existing regulations addressed clear-cutting). Compliance, in this instance, would have led to further clear-cutting and eliminated a growing edge of knowledge from forestry.

Have we done what we said we would? Implementation monitoring. Implementation monitoring is undertaken to assess whether internally derived targets or plans are being met. The monitoring may record rates of adoption of new practices and whether they were implemented as planned. Examples could include monitoring operations to see whether 80 percent of cutblocks use variable retention within the planned time period or whether 10 percent retention blocks actually retain 10 percent of the trees. Implementation monitoring is not relevant to adaptive management unless it records changes specific to implementing an adaptive management program (i.e., can guide adaptive management). Because new practices were being introduced, we used implementation monitoring to assess whether practices were being introduced on schedule and as planned (see Chapter 4). Many adaptive management programs do not require implementation monitoring.

Did our actions achieve our objectives? Effectiveness monitoring. Effectiveness monitoring is used to determine whether the practices implemented operationally are actually meeting their objectives: that is, are the practices effective? In this case, the objectives are to maintain the indicator components of biological diversity. Effectiveness monitoring, by definition, must sample operational blocks, with the goal of obtaining the most precise estimate of operational effectiveness. Ideally, monitored blocks are a representative sample of typical operations. Because resources for monitoring are always limited, rare atypical practices are largely ignored (e.g., variable retention blocks with high-percentage retention in a dispersed pattern that currently are rarely used operationally). Effectiveness monitoring is a necessary part of operational feedback in adaptive management, but it provides limited opportunities to refine future operations. We consider the baseline monitoring of Noss and Cooperrider (1994) to be a specific form of comparison within effectiveness monitoring.

Can we achieve our objectives better, faster, or more cheaply? Refinement monitoring. Refinement monitoring samples beyond common practice, usually with very specific questions in mind. It may sample the widest range of available practice, including rare but informative extremes or combinations. Creating learning opportunities through experimental treatments beyond the normal operational range is also part of refinement monitoring. Issues of design, quality control, and interpretation are intrinsic to all monitoring, but potential boundaries between monitoring and research are least evident in refinement monitoring. Both effectiveness monitoring and refinement monitoring are part of adaptive management. Whereas effectiveness monitoring evaluates current practice, refinement monitoring asks questions such as "do I have sufficient confidence in the underlying relations to apply them in novel ways?" or "can I attain the same ends more cheaply?" Refinement monitoring is most helpful when the learning is focused on causal mechanisms of response or ways of increasing cost effectiveness. When properly incorporated and extended, the validation monitoring of Noss and Cooperrider (1994) is part of refinement monitoring. Because effectiveness monitoring and refinement monitoring require different monitoring designs and bear different costs, a conscious decision needs to be made on the effort allocated to each in designing the monitoring program.

Monitoring, or assessing proximity to objectives, is only one part of adaptive management. To make adaptive management work, we used three kinds of monitoring: implementation, effectiveness, and refinement monitoring. In the following chapters, we tend to group effectiveness and refinement monitoring because we are interested more in the questions than in the specific approach to monitoring. The need for implementation monitoring follows from the fact that novel forest practices were introduced, and the way they would actually be implemented was unknown, as was what they would achieve (see Chapter 4). Actual implementation had to be assessed before any predictions of consequences or evaluations of effectiveness could be made. The two major forms of monitoring, implementation and effectiveness, have helped to inform each other.

3.4.4 Creating Structured Learning

The complexity of the problem impedes learning. The objective of sustaining biological diversity cannot apply to specific cutblocks but must be evaluated for the entire tenure. Because assessments everywhere are too costly, they must be designed so they can be "scaled up" to reflect larger areas. That scaling up relies on computer projections of structural attributes and patterns of retention. Large numbers of species are involved, for most of which we know little or nothing. We learn fastest, or reduce uncertainty fastest, when we organize current information in a way that can be directly challenged by new information: that is, by making predictions and evaluating their accuracy. The rationale for the use of indicators and subindicators is intended to invoke mechanisms that create predicted outcomes. Unfortunately, most predictions that we feel confident in making address limited groups of organisms and are nearly trivial in terms of the overall objective. Nonetheless, to the degree possible, ways of assessing proximity to an objective or its component features should be assessed by predictions, even though they may be limited to specific groups (e.g., cavity-nesting birds) or processes (e.g., windthrow). While ecological relations are too complex for predictions to be universally applicable, challenging even qualitative predictions can be revealing. Predictions should specify some mechanism(s) through which their outcomes are enacted. If predictions fail, then modifying practices will either use knowledge about the mechanisms specified (e.g., provide more large snags) or invoke new mechanisms (e.g., interspecific competition). In either case, learning occurs. By helping to reveal mechanisms of change, monitoring can provide a more quantitative basis for a risk assessment that helps to inform managers of the likelihood of success, given changes in planning or practice.

Creating a system for structured learning is critical. When designing the monitoring program, we accepted that prediction is not always possible. For large groups of organisms (e.g., invertebrates and fungi), we know too little to make useful predictions for more than a very small portion of the group. However, we can make comparisons among different treatments. Vertebrates currently offer greater utility in guiding forest practices because of readily detectable habitat selection, the wealth of available studies, and their often direct connections to stand elements changed by forest practices (e.g., cavity sites) (Bunnell, Kremsater, and Boyland 1998; Bunnell, Kremsater, and Wind 1999). We recognize vertebrates as a relatively well-known group among forest-dwelling organisms but also monitor other organisms and indicators (e.g., ecosystem representation and habitat structure) that are intended to address all forest-dwelling organisms. Expanding the range of organisms reduces the opportunity for comfortable prediction and limits learning to comparisons among treatments but is much more directly connected to the goal of sustaining biodiversity.

Lack of knowledge also prohibits credible predictions about the degree to which different forms of ecological representation serve to sustain poorly known species or poorly understood ecological functions. Often, undesirable conditions can be identified, but no unequivocal target or threshold can be specified (e.g., the total commitment of an ecosystem type to harvest is undesirable, but the amount that should be reserved from harvest is equivocal). Any monitoring system intended to assess proximity to a criterion of success provides guidance rather than an absolute target. That is one of the features of a wicked problem (§1.1) but is not greatly limiting provided that we can distinguish better from worse or more from less effective.

The spatial and temporal scales through which forestry operates mean that not all predictions can be cast in an experimental setting. Some require too long a time to measure; others extend over areas too large to sample well. Some would require replication of landscapes or populations that is impossible. Some predictions are necessarily word models addressing large areas (e.g., create a range of habitat conditions or don't do the same thing everywhere) or computer simulation models covering large areas over long periods. However, it is sometimes possible to break assumptions, models, and predictions into components and to test at least some of those components. The challenge is to clearly identify those components and to select those that account for the greatest variation in outcome and that are testable within a reasonable time frame and budget.

We gain knowledge in several ways, and the structure for learning that we created is not limited to a single form of questioning. Scientists are most familiar with "active adaptive management," or knowledge gained through controlled experiments with randomized allocations of treatments and controls, and statistically valid sample sizes (Marcot 1998). Active adaptive management can be specifically designed to better inform forest practices, address very specific questions, and expand the available range of specific practices. Where possible, it should be designed to increase the accuracy of computer projections over large areas or long time periods. "Passive adaptive management" indicates only the absence of a designed experiment and includes retrospective comparisons and comparative computer simulations. The sampling still requires careful design but is restricted to sampling what is or has been produced operationally. Passive adaptive management has the great value of sampling actual practices, but the lack of randomization in the assignment of controls and treatments limits statistical inference. Other sources of information, such as expert judgment, anecdotes, and literature reviews, also provide valuable information but have less inferential value than well-designed experiments. We used all of these approaches to gain knowledge.

3.5 Linking Findings to Actions

Answering the final question – "how do we change if the direction is wrong?" – closes the loop in adaptive management. We were aware that the most frequent cause of failure in adaptive management is the lack of subsequent action based on findings of monitoring (Lee 1993, 1999; Ludwig, Hilborn, and Walters 1993). Given that the will to enact change is present, we believe that the best way to implement this step is to provide a set of known ways to modify practices if objectives are not attained. That is, a fundamental question when developing a monitoring program is "what would we do with the data if we had them?" We noted above (§3.3) how determining which management activities can be modified lends focus to the monitoring. New information is of little use to practitioners if it does not link to management practice. Linkage is provided by both the evaluation system developed for each indicator and descriptions of potential management actions that are expected to help correct any failures. The evaluation system should assess the success of current management activities and track improvements in management through time. The ability to change may require a formal mechanism for accepting results and associated management or policy changes. It was for this reason that new structures, such as the Variable Retention and Adaptive Management Working Groups, were formed within the overall strategy (§2.4).

For some elements of biological diversity on the tenure, we are attempting to develop a credible procedure for assessing risk (see Chapter 12). Relative risk will reveal those deficiencies that are the most serious (e.g., high risk resulting from availability of large snags compared with lower risk from availability of shrubs). Provided that the risk assessment procedure is sufficiently transparent, managers can evaluate changes in practices to correct any deficiencies. Specific feedback to management is presented for each of the major indicators discussed in the following chapters. The nature of this feedback and its relationship to potential management actions is summarized in the final chapter. Acceptance by management of the need for change is fundamental to success of the program.

3.6 Summary

Our answers to the four broad questions stimulated by the introduction of new practices incorporated adaptive management.

- "Where do we want to go?" was answered by defining biodiversity as "native species richness and its associated values" and by establishing three broad indicators that reflect success in sustaining biodiversity. These indicators are intended to encompass the various components contributing to biodiversity and to include ecosystem representation, habitat, and organisms.
- Answering "how do we get there?" included not only the planning and practices anticipated to produce desired outcomes but also determining which elements of them were subject to future modification. The "road map" produced not only yields directions for getting there but also identifies alternative routes if the initial one fails. Adaptive management provides course corrections to the road map.
- The question "are we going in the right direction?" reflects the effectiveness monitoring component of adaptive management. The design of a monitoring program begins with bounding the problem and phrasing the critical questions and ends by linking back to management. Assessing whether the direction is correct includes various ways of learning, including active (research) and passive adaptive management. To be useful to managers, the information gained by monitoring must connect directly to specific management decisions and practices.
- Answering the question "how do we change if the direction is wrong?" begins with identifying practices amenable to change and monitoring responses within them. It can be completed only if there is the will to change.

Most of the book is devoted to the rationale, implementation, and interpretation of the indicators used to assess success: that is, effectiveness monitoring to evaluate success in sustaining biological diversity. Implementation monitoring is discussed in Chapter 4. Chapters 6 through 11 present the three indicators used to assess success in sustaining biological diversity. Chapter 12 summarizes the design of effectiveness monitoring to sustain biodiversity. Lessons learned are reviewed in Chapter 13.

4
Implementing the Approach

Fred L. Bunnell, William J. Beese, and Glen B. Dunsworth

4.1 Change in Midstream

The requirements for planning on forest tenures in British Columbia vary with the kind of tenure. Most of the study tenure discussed here was within two tree farm licences (TFLs). Planning requirements for TFLs required that the company create five-year and twenty-year spatial harvest plans that addressed harvesting and protection of all forest resources as part of a Sustainable Forest Management Plan. These plans are a condition of the licence, are reviewed annually, and are renewed every five years. When MacMillan Bloedel announced its intention to phase out clear-cutting in June 1998, it had already planned clearcut blocks at least four years into the future. Change had to be introduced to a system going in a different direction without stopping the system. The planning for different areas was at different stages, depending on how far into the future the harvest had been planned. Because of the complexity of forestry, considerable planning must be undertaken before a tree is felled. For some areas, planning had begun with determining where gravel pits could be located to aid road construction and how best to locate access roads in rugged terrain to meet both environmental goals and reduce costs. For other areas, roads were already constructed, cutblock boundaries were established, and the required pre-harvest silvicultural prescription was complete. These prescriptions are detailed and include site-level planning for wildlife, regeneration, road construction or deactivation, stream management, residue management, forest health, windthrow, and other environmental features (see Zielke and Beese 1999). The plans had been prepared with clear-cutting in mind.

Changes required to adopt the new approach to forestry were massive and ranged from strategic to operational levels of planning and practice. At the strategic level, three new zones with different intensities of harvest had to be defined, located, and implemented. Practices in the Timber zone were different from practices in the Old-Growth zone (Table 2.2). Some planning had to be overturned. At the operational level, clear-cutting was to be phased out in five years (by 2003). That was not as simple as cutting fewer trees within a cutblock whose boundaries were already laid out. An immediate barrier was that there were no regulations to describe or guide variable retention. Existing silvicultural systems terminology was inadequate to describe the objectives and the spatial and temporal requirements of variable retention. Government personnel reviewed the ecological rationale for the new approach (Bunnell, Kremsater, and Boyland 1998) and met with the authors to help develop regulatory standards. As a result, a new silvicultural system was defined in BC regulations for the first time in over fifty years; it allowed use of the new terminology in legal prescriptions.[1] Previously, most regulations in the BC Forest Practices Code were based on clear-cutting, such as rules governing the timing and adjacency of cutblocks. Widespread application of the retention system challenged the relevance of these rules. To attain its goals, the company had to begin phasing in variable retention before all standards were formalized. Phasing in itself was a massive change because forest planners, engineers, contractors, fallers – in fact, everyone with a role in extracting wood from the forest – had to understand what this new system was and how to implement it. Simple but effective operational guidelines had to be created and communicated through training sessions, videos, and manuals. Over 250 people took a three- or four-day training course during the first two years of implementation.

The required changes provided the first stimulus to creating the new structures necessary to enact the adaptive management program (§2.4). Researchers who had crafted the initial ecological rationale for the new approach formed the nucleus of what became the Adaptive Management Working Group. Their initial task was to work with government agencies to create a way in which novel practices, not encompassed by current regulations, could proceed. Company personnel who had worked with these researchers in the initial design had to create operational guidelines so that company employees and contractors could enact the new approach. Those activities formed the nucleus of what became the Variable Retention Working Group.

4.2 Progress in Adopting the Approach

To implement the new approach to forestry, changes had to occur in both planning and practice.

4.2.1 Implementing Planning

Knowing where we are going is essential to getting there. Planning provided the road map to where the new changes would lead. That was particularly true because practices differed among the stewardship zones. Given the differences in practices among zones, the locations of zones had to be established quickly. Although the company was applauded for its commitment to the new approach, there was considerable suspicion about where the zones, particularly the Old-Growth zone, would be located. The issue of ecological representation within the land base that would not be harvested was raised immediately, before a formal indicator to address the issue was adopted.

The company engaged in consultation with the government and stakeholders, then made its decisions following three simple principles. First, the zones had to be relatively large areas if they were to achieve the potential benefits of separating objectives that were least compatible. For example, the Old-Growth zone was intended to sustain species that required larger areas of late-seral conditions. Similarly, if potentially disruptive activities associated with timber harvest (e.g., road construction and maintenance) were to be concentrated, then those concentrations had to be in areas large enough to permit economic harvest. Land units of about 5,000 to 50,000 hectares were sought.

Second, the zones were to be as compatible as possible with development history in the area and uses specified by higher-level plans created by government processes. Watersheds with more than 70 percent old-growth cover were candidates for the Old-Growth zone, and watersheds with a long history of logging and a high proportion of younger age classes were candidates for the Timber zone, including most of the company's private land (Beese et al. 2003). To a considerable extent, the uses specified by higher-level plans[2] had an ecological rationale. Areas designated for "low-intensity" harvesting or "medium- to high-biodiversity emphasis" in existing plans provided initial candidates for the Habitat zone and to a lesser degree the Old-Growth zone. The presence of abundant old-growth forest cover and the associated lack of roads were the primary determinants for locating Old-Growth zones.

Third, in locating the Old-Growth zone, the principles of conservation biology were adopted. They reduced to designating some larger areas of low impact by adjoining areas to existing parks, encompassing a range of forest types and geographic distribution, and including a range of sizes within the zones (Figure 4.1).

By employing simple guiding principles, the company was able to complete its initial location of stewardship zones quickly enough that implementation of practices could proceed in a timely fashion, but it had not attained its initial target of 10 percent for the Old-Growth zone. In total, thirty-seven areas were designated as Old-Growth stewardship zones. Most of these areas were on Vancouver Island – twenty-two areas ranging in size from 500 to 40,000 hectares. Ten were designated on the mainland coast (500 to 30,000 hectares), and five were designated in the Queen Charlotte Islands Haida Gwaii (100 to 25,000 hectares; see Figure 4.1).

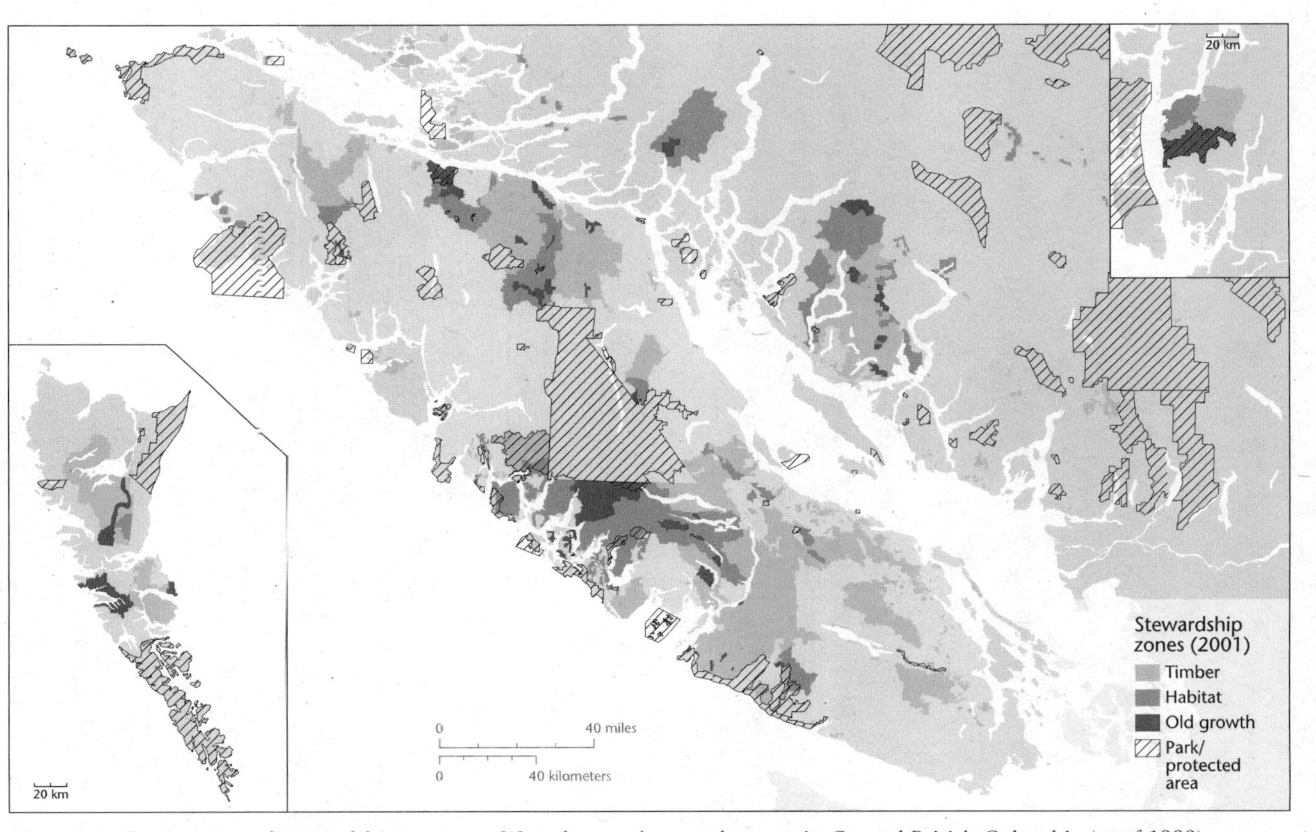

Figure 4.1 Designation of stewardship zones on Weyerhaeuser's coastal tenure in Coastal British Columbia (as of 1999).

Note: One old-growth area abuts a major park (vertical stripes).

Source: MacMillan Bloedel Limited.

Initial targets for amounts in zones had to be set, but there was no practical or theoretical guidance on the proportions of the tenure that should be allocated to each zone. Zones were selected based on an arbitrary, preliminary target distribution of 65:25:10 (Timber:Habitat:Old-Growth). The intention was to make adjustments to initial targets based on feedback from effectiveness monitoring. Among the social, ecological, and economic factors considered in the distribution of zones were capturing key contentious areas in Old-Growth zones, reducing timber-harvesting impacts on a significant portion of the landscape (e.g., over one-third for the Habitat and Old-Growth zones combined), and keeping the economic impact of the strategy at a level acceptable to the company's Board of Directors as a cost of maintaining market share. Of all zones, the Old-Growth zone received the greatest external scrutiny and created the largest economic impact on the company. It was apparent that indicators of success in sustaining biodiversity would have to evaluate contributions of all zones but particularly the Old-Growth zone.

4.2.2 Implementing Variable Retention

Implementation monitoring answers the question "did you do what you said or thought you would?" It is an important step in any assessment of forest practice but is particularly important when the practices are novel and there is little related material to draw upon. The company had two broad goals for the implementation of variable retention. The first was a schedule – phasing out clear-cutting within five years (by 2003). The second was the application of principles designed to meet the ecological rationale of variable retention.

The Schedule

Monitoring performance relative to a schedule was straightforward. The target was known – the phase in of variable retention at 20 percent increments per year until 2003. Each timberlands unit simply had to report the relative proportions of clear-cutting and variable retention that were being implemented. Summing them by area indicated progress toward the specified goal.

Figure 4.2 illustrates overall progress and reveals that goals for each individual year were exceeded and that full implementation was nearly completed on schedule. Although all normal harvest planning in 2003 was done using variable retention, the shortfall in actual harvested area was due to deferrals in contract logging operations and the carry-over effects of salvage from a major storm event in December 2001. The company monitored progress toward the goal by division or geographic location within the tenure. That permitted early exposure of potential trouble spots so that efforts could be allocated effectively toward solutions. The total amount of variable retention completed over the five-year phase-in actually exceeded the goal.

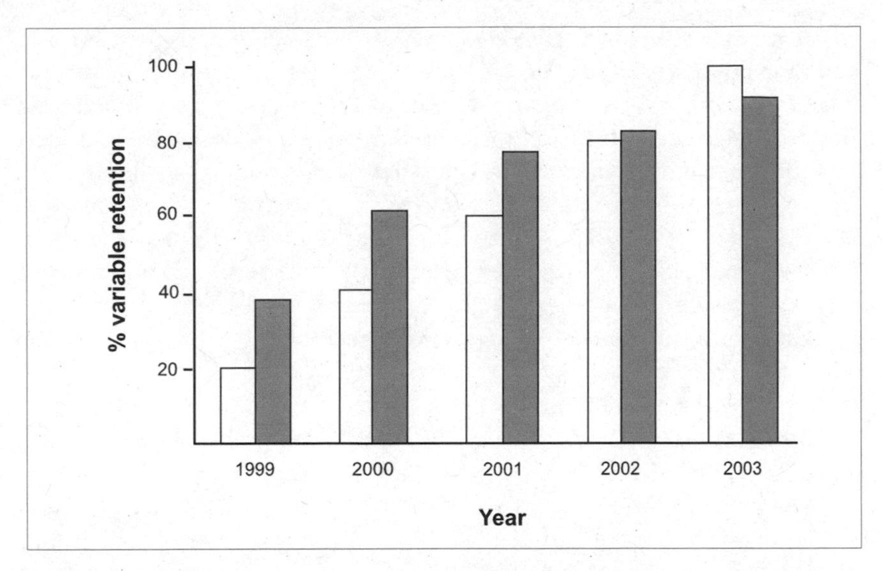

Figure 4.2 Progress in implementing variable retention on Weyerhaeuser's coastal tenure in British Columbia.

Note: Clear bars are the goal or target; dark bars are what was attained.

Principles

Documenting the rate of application of variable retention does not reveal whether the principles designed to guide application of VR are being met. Before the decision to invoke the new approach to forestry was taken, relevant literature was reviewed and synthesized (Beese and Zielke 1998; Bunnell, Kremsater, and Boyland 1998). From that review, an approach to forest planning and practice was derived that was believed likely to sustain all desired values. A necessary first step was to translate underlying ecological principles into operational statements that field foresters could implement. Although the guidelines themselves are relatively detailled (fifty-five pages, see Zielke and Beese 1999), they were intended to attain just three major goals:

1 to leave a biological legacy of old-forest attributes, well distributed in each stand, and to maintain and promote biological diversity over the landscape of the tenure;
2 to ensure that each cutblock was defensible, using a sound ecological rationale (e.g., maintaining forest influence on the majority of the block) to distinguish it ecologically from a clearcut; and
3 to ensure that each cutblock would be viewed by the public as an example of good forest stewardship using a non-clearcut approach.

The first two goals reflect the ecological rationale of variable retention – to maintain biological legacies and forest influence. Anchor points of retained patches and group size influence the kinds of biological legacies retained. Desirable anchor points include features that serve to sustain native species richness, such as riparian areas, wetlands, deciduous patches, or wildlife trees and snags. Appropriate minimum group or patch size was estimated as 0.25 hectares, to be evaluated by monitoring. That size was intended to ensure that the smallest groups retained undisturbed attributes of understorey vegetation, organic matter, and stand structure and allowed sufficient unharvested buffers around snags to ensure worker safety. Forest influence for many properties was determined from the literature to extend at least one tree length (Keenan and Kimmins 1993; Kimmins 1997; Kremsater and Bunnell 1999). This translated into practical guidelines recommending that retained groups be no more than four tree lengths apart and that small groups or dispersed trees be no more than two tree lengths apart. These guidelines ensure that greater than half of the original forest area is within the influence (i.e., one tree length) of surrounding trees or trees retained in the harvested area.

The third goal, good forest stewardship, includes safety concerns and forest health issues, avoiding damage to remaining trees, and visual aesthetics. Although minimum levels of retention were specified for each stewardship zone, it was expected that the kind and level of retention would reflect characteristics of individual sites in a fashion that exemplified good forest stewardship. A secondary consideration to the three major goals was to avoid a "cookie cutter" approach and not do the same thing everywhere. This guideline served not only to emphasize the flexibility and "variable" portion of variable retention but also to acknowledge that under any form of retention, different groups of organisms would be advantaged or disadvantaged. Variability is important to sustaining as many organisms as possible. Because variability was sought, it was recognized that even with clear guidelines (see Zielke and Beese 1999) it was unclear how VR would be applied, particularly which structures would be retained. Both broad and specific outcomes of applying principles had to be monitored. Implementation monitoring assessed broad outcomes. More specific outcomes were assessed by effectiveness monitoring (see Chapters 9 and 11).

4.3 Assessing the Outcomes of Guidelines

A "scoring" system was developed to assess the degree to which operational guidelines met the three broad goals of variable retention. The system asked three questions related to the three broad goals.

Goal 1: Does the retention on this block leave a legacy of old-growth attributes, well distributed within the block, using structures and features present at the time of harvest?

Goal 2: Does this block meet or exceed the required 50 percent forest influence?

Goal 3: Does this block meet the intent as an example of "good forest stewardship" using a non-clearcut approach?

Each goal was evaluated independently and annually using ratings of 1 (poor) to 10 (excellent) but could receive 0 if below standards specified by the company. Because implementation monitoring had to assess a large number of blocks across the entire tenure, measures had to be simple. Random samples of the operational blocks were distributed across the tenure each year, averaging 17 percent of the harvested area over the five-year phase-in (248 blocks and over 7,000 hectares under prescription).

4.3.1 Biological Legacies

Goal 1 (biological legacies) was evaluated by two broad measures: anchor points of retention groups and group size. Anchor points are the ecological attribute(s) that characterize a particular retained patch. These attributes were considered the ecological anchor or, in engineering terminology, the ecological "control point." Both anchor points and group size are factors amenable to change, so they offer opportunities to link to management. Implementation monitoring and effectiveness monitoring were initiated simultaneously, so specific effects of different anchor points were unknown. Literature and field experience suggested that different anchor points would yield different mixtures of structures and contribute differently to sustaining biodiversity, but it was unclear which anchor points would be used. We predicted that some anchor points (e.g., riparian areas, snags or wildlife trees, deciduous patches, wetlands) would make greater contributions to sustaining diverse organisms than would others (e.g., gullies, rock outcrops). These expectations were subsequently affirmed by effectiveness monitoring (§9.3.6). The simple assessment of success in retaining biological legacies evaluated whether the choice of anchor points exploited available site features to retain desirable habitat and whether adequate variability was achieved.

Based on the assessment of available anchor points used, the choice of retention was judged on the basis of its ability to provide a range of habitat. Initially, only 40 percent of retained groups contained large snags (>8 m tall), but by 2002 and 2003, about 80 percent of groups contained large snags (Figure 4.3). The proportion of groups anchored around riparian features increased from 21-25 percent in 1999 and 2000 to 35-43 percent during 2001 to 2003. Retention along small streams that did not require treed buffers under provincial regulations was assessed from 2001 to 2003 along 193 kilometres of stream within and along the boundaries of VR blocks. Over this three-year period, some retention occurred along over 50 percent of the total stream length that otherwise would have received no

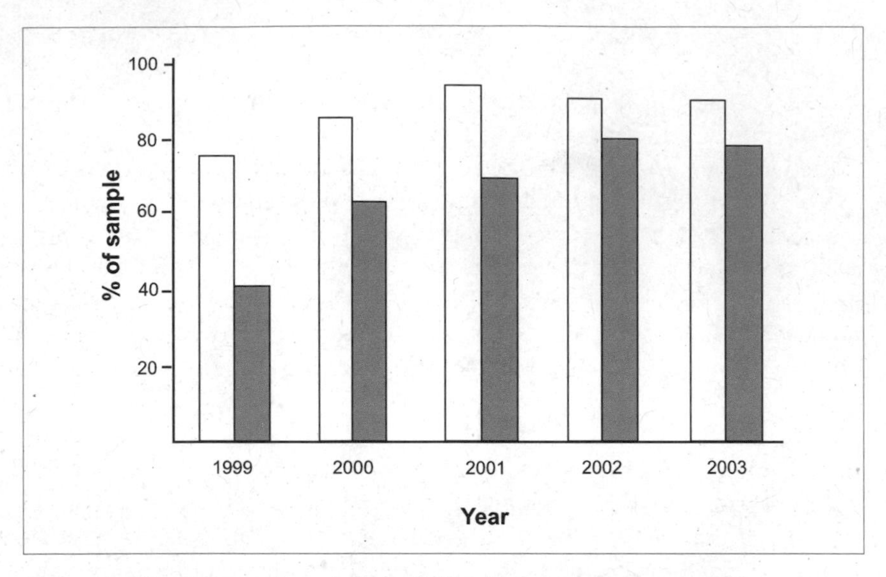

Figure 4.3 Proportion of sampled blocks in which retention was judged to meet desired criteria. Clear bars are blocks assessed to be of good to optimal choice to provide a range of habitats for forest-dwelling organisms; dark bars are blocks in which retained groups had larger wildlife trees.

regulatory protection. The majority of these buffers exceeded 20 metres. Conversely, proportions of anchor points that contributed less to biodiversity declined during the phase-in period. For example, the proportion of groups anchored to scrubby timber declined gradually from 16 percent in 1999 to 6 percent in 2003. The proportion of retention assessed as representing good to optimal choices from options available (see Zielke and Beese 1999) increased from 75 percent in 1999 to over 90 percent in 2001, after which it stabilized (Figure 4.3).

We also expected that the size of groups retained would influence their ability to sustain different organisms. Given the amount of flexibility granted to operational foresters during implementation of retention, it was not known what distribution of group sizes would actually be retained.

Group size gradually increased over the first three years of the phase-in period as the proportion of groups <0.25 ha declined and the portion >0.50 ha increased (Figure 4.4). The arbitrary lower limit for group retention provided in operational guidelines was met or exceeded by 70 to 80 percent (or more) of groups in all years. Smaller groups still contribute to retention but are typically classed as dispersed retention. Mean size of group increased from 0.8 hectare in 1999 to 0.9 hectare in 2000, then stabilized at about 1.0 hectare.

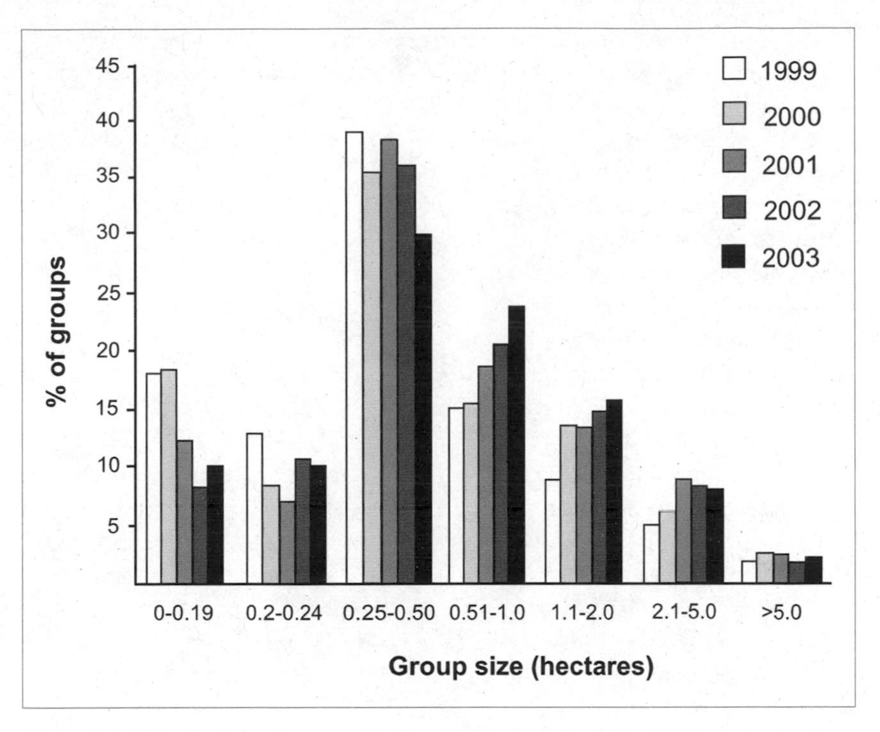

Figure 4.4 Distribution of group size for 1,591 groups in 248 variable retention blocks over the period 1999 to 2003.

4.3. Forest Influence

Forest influence and distribution of long-term retention throughout the harvested area are the main features distinguishing variable retention from traditional harvesting approaches. The simplified "one tree length from an edge" criterion for measuring forest influence required planners to use the actual height of dominant trees along cutblock edges, retained groups, or individual trees for measuring influence. A GIS tool facilitated this calculation. In most cases, the operational guidelines governing spatial distribution of retention (e.g., groups spaced less than four tree lengths apart and dispersed trees less than two tree lengths apart) are sufficient to ensure greater than 50 percent forest influence over the entire block, thus differing ecologically from a clearcut (Keenan and Kimmins 1993; Kimmins 1997; Kremsater and Bunnell 1999). The extent of forest influence actually achieved was assessed directly by estimating the proportion of blocks meeting the >50 percent criterion. We also assessed the actual level of retention in individual blocks to determine the range of retention being attained by operational practice.

Across the five years of assessment, the average amount of forest influence within a cutblock was well above the 50 percent minimum required for retention blocks, ranging from 68 to 79 percent over the assessment period with no apparent trend. Nonetheless, in each year, a number of blocks failed to meet VR standards. In most cases, this failure resulted from poor distribution of retention rather than amount of retention. The proportion of blocks failing to meet retention standards ranged from 0 to 11 percent across the five years, again with no apparent temporal trend.

4.3.3 Amount of Retention

A related but separate requirement from maintaining biological legacies and forest influence was the amount of retention. Each cutblock had to meet the minimum long-term retention level specified for the zone in which it occurred. Group retention was measured as a percentage of the area under prescription. Dispersed retention was measured as basal area[3] of the retained trees as a percentage of the original stand basal area. For mixed approaches, dispersed basal area was converted to equivalent hectares and added to the area under group retention.

Despite the fact that some blocks failed to provide 50 percent forest influence and meet variable retention standards, average amounts of retention were consistently higher than the minimums specified. Long-term retention includes group and dispersed retention plus other reserves (e.g., riparian and wildlife tree patches) within or on the edges of cutblock boundaries. Retention level averaged over 20 percent for the five-year phase-in – well above the 10 percent and 15 percent minimums specified for the Timber and Habitat zones, respectively (Table 2.2). These values represent total retention. The incremental retention or that portion of the total that comes from the harvestable land base was about half of that and very close to the original business case estimates.

Almost 75 percent of the blocks had retention in the 11 to 30 percent range; only 2 percent of the blocks were over 50 percent (Figure 4.5). Retention levels typically exceeded the zone minimums because of the combination of leaving riparian reserves and forested rock outcrops and the spatial distribution requirements of the retention system. The vast majority of retention was left as groups or patches rather than as dispersed single trees, although a mixed approach was common. Independent monitoring of cutblocks confirmed the reported retention levels.

Concentration of retention toward the lower end of the range was intended to maintain structural attributes while minimizing reductions in growth and yield. The large number of blocks between 5 and 50 percent retention provide an adequate sample for effectiveness monitoring. Retention was implemented at similar levels in both second-growth and old-growth forests.

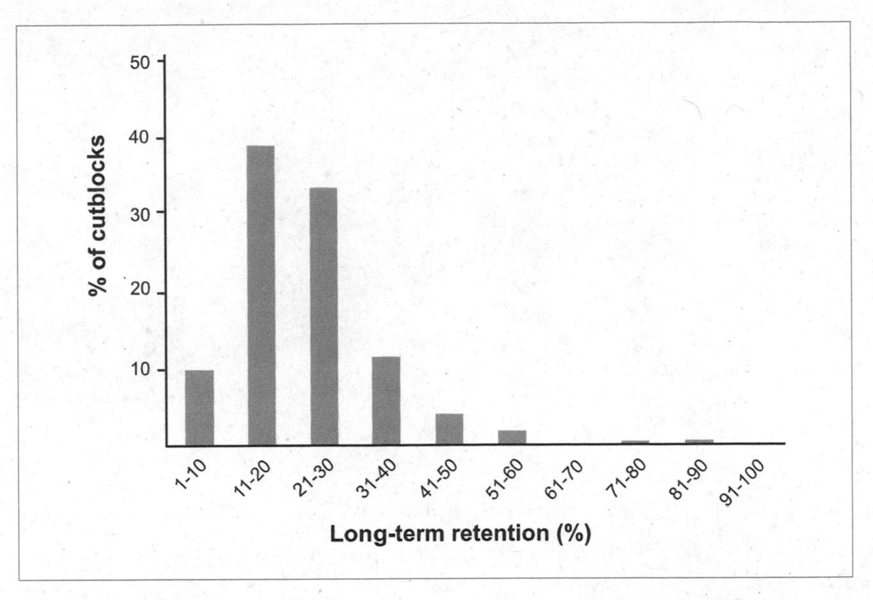

Figure 4.5 Distribution of long-term retention levels for all variable retention blocks (2000-02).

Note: The distribution was similar each year.
Source: Based on annual reports from timberlands units.

4.3.4 Forest Stewardship

The degree of good forest stewardship was scored from 1 to 10 for each sampled retention block by addressing eight features. They included safety concerns, potential "high grading,"[4] leave-tree damage, forest health and windthrow risk, harvesting efficiency, and visual appearance. The last one was simple, assessed by subjective visual inspection, and was referred to as the "BC-TV test" – would the public, seeing the block on television, consider it a clearcut or not? Over the course of the phase-in period, the aggregate stewardship score for retention blocks varied relatively little. The proportion of blocks scoring good to excellent gradually increased from 1999 until 2001 but has since decreased somewhat (Figure 4.6).

By including different features in the stewardship score, it was possible to determine specific areas most in need of improvement. For example, in 1999, the proportion of retention blocks with excessive leave-tree damage was 26 percent. By 2001, it had been reduced to 2 percent and has stayed low since. Similarly, the major reason for the recent decline in the proportion of blocks showing good stewardship (Figure 4.6) appears to be failure to meet the visual objectives. When variable retention was first introduced in 1999, 71 percent of blocks had significant portions deemed too open to meet the

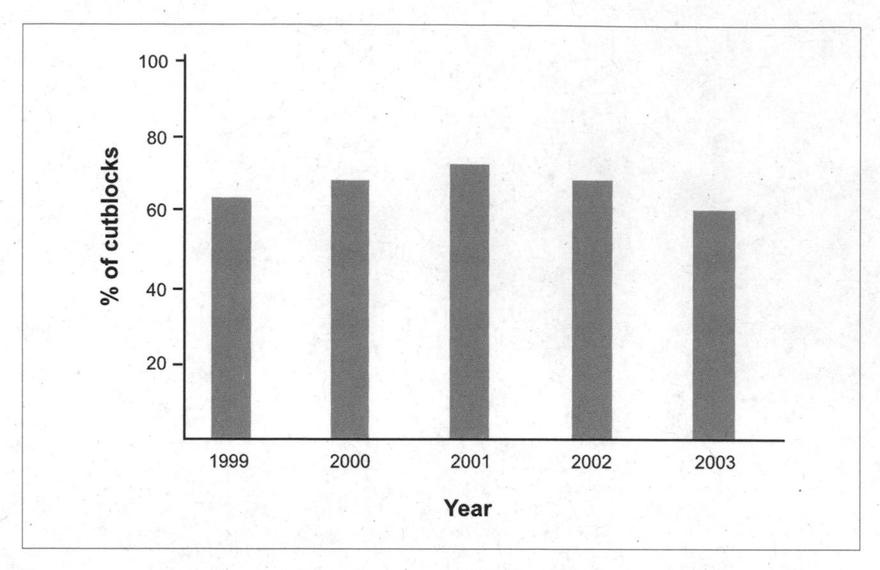

Figure 4.6 Aggregate score for good stewardship (blocks rated as good or excellent) on retention blocks during the 5-year phase-in period (1999-2003).

visual goal. The percentage declined to about 32 percent in 2001 and 2002 but had increased again to 45 percent in 2003.

4.3.5 Requisite Variety

Although ensuring a range of practices on the ground was deemed important, it was considered less necessary to assess because terrain and other ecological conditions would impose variety. Moreover, shelterwood and selection systems were also employed in some zones (Table 2.2).

In practice, shelterwood, selection, and other systems have comprised less than 10 percent of the area harvested (Figure 4.7). Dispersed retention declined from 21 percent of the area harvested in 1999 to 4 percent of the area harvested by 2002. Group retention showed a concomitant increase, from 29 percent in 1999 to over 60 percent in 2002 and 2003. Multi-pass systems were employed on about 25 percent of the cutblocks. The most common use of two or more entries was where, under the Forest Practices Code, a cutblock had adjacency restrictions that required retention of at least 40 percent of the basal area of a stand until the surrounding area was of sufficient height to reach green-up.[5] This policy was directed at preventing progressive clear-cutting. In these circumstances, an initial harvest is possible with a second entry after "green-up" height targets are reached on the harvested portion. In some instances, windthrow or visual concerns encouraged a two-pass approach.

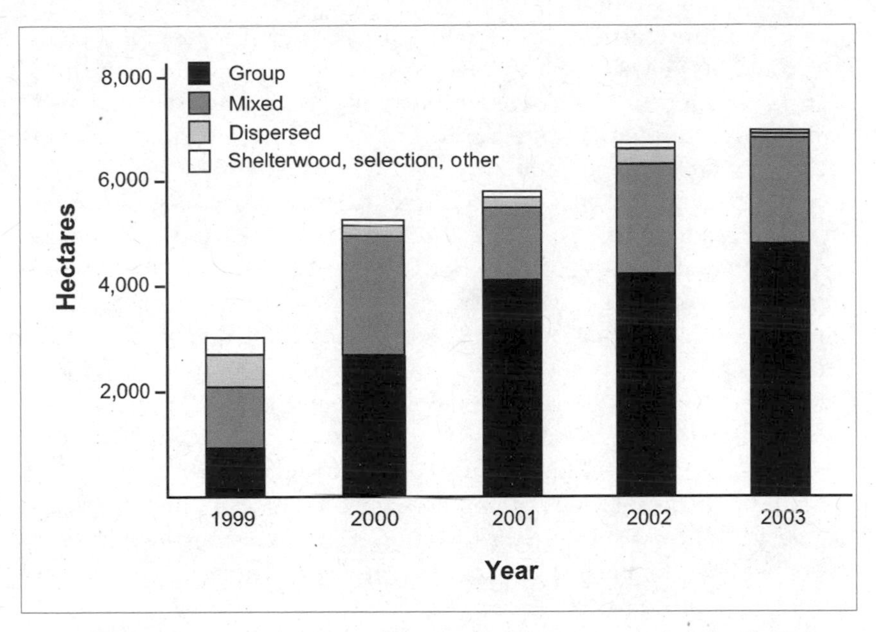

Figure 4.7 Area harvested under different retention systems during the 5-year phase-in period (1999-2003).

4.4 Lessons from Implementation Monitoring

Implementation monitoring served to evaluate whether major goals of the corporate philosophy were being implemented as described. That is, it asked the question "did you do what you said or thought you would?" By itself, it contributes little to evaluating the effectiveness of variable retention in sustaining specific values. Yet, without knowing what was actually done, there would be no way to properly assess effectiveness. Effectiveness monitoring addresses the question "has what you have done achieved the ecological objectives?" In fact, the two forms of monitoring are not tidily discrete; each informs the other. The operational guidelines, for example, were proposed to meet the ecological rationale for VR.

We noted that the fastest way to learn is by making predictions based on what you already know and evaluating the outcomes of those predictions (§1.4). In a general way, this was done at the very outset of the project by the synthesis and reasoned estimates of the structure and composition of the resulting forest when the new planning and practices were applied (Beese and Zielke 1998; Bunnell, Kremsater, and Boyland 1998). Failure to attain the outcomes anticipated could occur either because of inaccurate relations being derived from synthesis or because the anticipated forest or landscape structures were not actually attained. The first step in assessing the likely consequences of new forest practices is to assess the kinds of forest being produced by the practice. We did not know whether the operational guidelines could be applied effectively. Nor did we know what amounts and kinds of structure different forms of retention would retain when implemented.

Implementation monitoring assessed the kind of forest structure being left in very broad terms – amounts and types of retention. That step was necessary because it was not possible to envision accurately what widespread VR would create as forest structure at the stand level. We knew the desired and predicted direction but were not confident about the way in which practices would change. As implementation monitoring continued, three broad patterns became apparent. First, the proposed schedule for implementation was attainable (Figure 4.2), although operational units showed different rates of progress. Second, field foresters were making reasonably good choices in locating retention (Figure 4.3) and were attaining desired amounts of retention (Figure 4.5). Third, changes could be, and were being, introduced into practice in desirable ways – group size (Figure 4.4), amounts of retention (Figure 4.5), and overall stewardship rating (Figure 4.3) were increasing. At the broad level, implementation monitoring was informative and helpful in guiding improvements.

Because implementation and effectiveness monitoring began simultaneously, initially there was little guidance from effectiveness monitoring to affirm changes in direction. Some were obvious. For example, dispersed retention was encouraged in the Timber zone by lower minimum retention levels, hence less reduction in timber supply. In practice, dispersed retention led to higher yarding costs, while timber growth models predicted that the shade created from well-distributed trees would have higher growth impacts than similar levels of group retention. Windthrow losses were also found to increase with dispersed retention (Mitchell, Hailemariam, and Kulis 2001). The result was a decline in the amount of dispersed retention (Figure 4.7). At the time that this outcome was becoming apparent, effectiveness monitoring was suggesting that dispersed retention was not serving some groups (e.g., songbirds) as well as group or mixed retention.

The "operational" Variable Retention Working Group directed implementation monitoring, while the more "scientific" Adaptive Management Working Group directed effectiveness monitoring. Communication between the groups was facilitated by some shared membership. At this broad level, effectiveness monitoring provided little more than gentle nudges to practice that appeared in the implementation monitoring during the five-year phase-in period. Similarly, when practices changed for operational reasons and there was no apparent ecological reason not to change (e.g., amount of dispersed retention), there were no countersuggestions that change be stopped.

While effectiveness monitoring was nudging or affirming the directions apparent in implementation monitoring, the latter was helping to refine the former. For example, Figures 4.3 through 4.6 indicate that changes can be made to practice that would likely affect the ability to maintain biological diversity. They argue that influential and changeable features, such as anchor points for retention groups, are a necessary form of stratification in effectiveness monitoring. As the phase-in was completed, the primary agent for change became effectiveness monitoring, and implementation monitoring played a lesser role.

4.5 Summary

Novel planning and practices had to be introduced into a system with as little disruption as possible to ongoing business. Implementation monitoring evaluated internal, corporate goals for the schedule of change and the application of new principles to implement variable retention.

- Stewardship zones had to be located quickly because they determined the location of practices. The location of zones required effectiveness evaluation (Indicator 1, Chapter 6).
- The phase-in of variable retention was completed on schedule.
- Because the ecological principles underlying VR were specified, it was possible to construct broad indices of the quality of retention (e.g., Figures 4.3 and 4.6).
- Guidelines for implementation were being followed, and practices showed improvement in meeting underlying principles.
- Changes in implementation were guided primarily by a sense of what was the better direction because effectiveness monitoring had just begun.

Indices used in implementation monitoring were qualitative and do not reveal whether specific benefits anticipated from variable retention were being realized. Indicators to evaluate effectiveness are discussed in Part 2. Concerns other than those noted here arise when new practices are implemented (e.g., windthrow, growth and yield, insects, disease, and other pathogens). Their monitoring is reported elsewhere (Beese et al. 2003; Mitchell, Hailemariam, and Kulis 2001).

Part 2: The Indicators

Part 2 discusses the three major indicators of success in sustaining biological diversity and information derived from each of them. Chapter 5 provides an overview of effectiveness monitoring. Specifically, it introduces four questions that must be asked of any monitoring program. What would we do with the data if we had them? How do we ask the question? How do we discern what is better? Where does the answer apply? These questions provide context for determining what to monitor as well as how to monitor. After the overview in Chapter 5, each indicator is treated in two chapters – one providing the rationale for the indicator and associated measurements, another reviewing major findings.

Chapter 6 discusses the broadest indicator used to assess success in sustaining biological diversity. Indicator 1 (ecosystem representation) states that ecologically distinct ecosystem types are represented in the non-harvestable land base. Non-harvestable areas play critical roles in maintaining biological diversity, including helping to sustain poorly known species and functions, providing safeguards against the inevitable uncertainty in maintaining species on the managed land base, and acting as benchmark sites for comparisons. Chapter 6 discusses indicator measurements used to assess the ability to meet these roles and the rationale for these measurements. The approach used to monitor Indicator 1 is presented as a series of steps.

Learning from Indicator 1 is summarized in Chapter 7. Although the overall percentage of forest reserved within the tenure was high (25.7 percent), some ecosystems were poorly represented (drier, productive ecosystems). Areas designated as Old-Growth stewardship zones made only minor incremental contributions to ecosystem representation. Potential edge effects are also a concern, with 62 percent of non-harvestable area occurring within 50 metres of harvestable stands.

Indicator 2 (Chapter 8) addresses sustaining forest habitat by considering habitat elements (e.g., large snags, shrubs), habitat structure (e.g., vertical distribution of vegetation), and landscape features, such as the effects of adjacent stands and roads. The main ecological reason for using variable retention systems instead of clearcuts is the greater ability to retain habitat attributes that are difficult or impossible to retain in conventional management cycles without retention but that are required by a wide range of organisms. Because Indicator 2 monitors these habitat attributes, it has the most direct link to forest management of all major indicators.

Findings from monitoring forested habitat are reviewed in Chapter 9. Analyses compared retention types, retention patches to uncut benchmarks, amounts of habitat elements to amounts of retention, edge effects on habitat elements into and out of retention patches, and habitat types used to locate or anchor retention patches. Early findings suggested that initial retention practices could be modifed to sustain better habitat, so the resulting habitat retention was compared over six years. Overall, monitoring habitat elements led to a number of specific recommendations for practices, some of which were attained at least for a period. Over the first six years, markets and corporate direction have had more impact on practices than have recommendations from monitoring.

The persistence of organisms or species (Indicator 3) is the ultimate test of efforts to sustain biodiversity. There are, however, too many species to monitor more than a minute portion. Focusing on a few well-known species greatly reduces the scope of organisms considered to represent biodiversity. Chapter 10 reviews vascular plants, bryophytes, lichens, fungi, invertebrates, and vertebrates to identify species that could expose deficiences in habitat elements or forest planning. Informative species are forest dwelling, sensitive to forest practices, and practical to monitor, and they provide information that can guide management. Species identified as potential candidates for monitoring encompass a range of life history characteristics – slow to fast dispersers, small to large home ranges, and reliance on a variety of different habitat elements or forest ages.

Of the three broad indicators, we acquire helpful information most slowly from organisms. In Chapter 11, we outline how the monitoring of organisms was intended to develop, including the design of pilot studies, and what actually happened. We also present initial results from comparisons that have been addressed by two or more organism groups.

5
Effectiveness Monitoring: An Introduction

Fred L. Bunnell, David J. Huggard, and Glen B. Dunsworth

5.1 Context

Effectiveness monitoring asks the question "did our actions achieve our objectives?" The growing literature addresses spatial scales from microhabitats (Gibb et al. 2006) through an entire province (Stadt et al. 2006) to national status (Gillis et al. 2005). A few workers have attempted to address the uncommonly long temporal scale of events in forests (Kimmins et al. 2007). Although there are exceptions (Rempel et al. 2004; Thompson 2006), much of the literature addressing biodiversity focuses on a specific group of organisms with little or no explanation of how or why that group was selected for monitoring. We outlined a process for evaluating outcomes that begins with the initial, problematic issues and noted seven steps necessary to enact successful effectiveness monitoring (§3.4.1). In Part 1, we discussed how we accomplished the first five steps.

- *Determine major issues.* Our issues were to reserve more old growth and implement a harvest system that sustained native species richness.
- *Clearly define objectives.* Along with associated indicators of success, our objectives focused monitoring on appropriate variables. We selected a single criterion and three broad indicators of success.
- *Identify the management plan and practices.* Stewardship zones and variable retention were believed appropriate for addressing the major issues. The approach to planning and practice became hypotheses to be evaluated by the monitoring process.
- *Bound the problem.* Establish the physical, functional, and conceptual boundaries. We determined that the scope included all native species over the entire tenure.
- *Identify the major questions.* The major questions relevant to the new planning and practices were winnowed down to six.

Additional work is necessary to bound the problem (indicator measurements must be limited to those believed to be most informative), and there are two remaining steps.

- *Identify data needs.* Data needs to answer specific monitoring questions, which must be identified through literature review or pilot studies.
- *Rank the objectives or questions and data needs.* Because resources are limited, ranking to focus on greatest uncertainty and risk is critical. When it is clear which practices and objectives are more likely to have an impact on biological diversity, data needs can be ranked accordingly.

Before treating specific indicators, we provide context by examining these basic questions.

- How do we ask our questions?
- What would we do with the data if we had them?
- How do we discern what is better?
- Where does the answer apply?

5.2 How Do We Ask Our Questions?

There are two quite different aspects to this question: what is the mechanism for learning, and from which potential examples and comparisons do we draw our answers? We learn from the monitoring program by making comparisons and by understanding mechanisms. The two approaches extract their answers from a different pool of examples – typical operational practices or a wider range of treatments, either carefully selected or experimentally created. To some extent, this distinction in ways of learning reflects the difference between effectiveness and refinement monitoring.[1]

We consider the typology of monitoring (§3.4.3) secondary to the kinds of questions asked during monitoring. Simple comparisons among treatments are a powerful way of learning. In many instances, we wish to know only whether operational plans or practices are creating the desired outcomes or which practice attains desired outcomes best. Those questions can be answered by comparisons among actual practices, with the goal of obtaining the most precise estimate of operational effectiveness, given the available budget. Monitored sites are then a representative sample of typical management practices, and atypical sites and practices are avoided. This approach is basic effectiveness monitoring. In other instances, the actual mechanism may be important to facilitate "scaling up" or guiding future practice. The two alternatives reflect the major distinction between passive and active adaptive management (§3.4.4). Assessing such mechanisms may sample the widest range of available practice, including rare but informative extremes or combinations. Creating learning opportunities through experimental treatments beyond the normal operational range can also contribute. The approach is then more similar to validation or refinement monitoring and may be considered research.

Approaches to learning are complementary and have distinct advantages (§12.2.1). Most simply, learning by comparisons is usually simpler and more precise but restricted to current practices. If these practices are determined to be ineffective, then little guidance is provided on new directions. Affirming mechanisms requires more complex studies and is less precise but has greater generality that permits "scaling up" to larger areas than those studied and can clarify new directions. Both approaches are useful. Because each draws on a different pool of samples, requires different monitoring designs, and bears different costs, a conscious decision needs to be made on the effort allocated to each in designing the overall monitoring program.

In the following chapters, potential comparisons and causal mechanisms are treated together. Some candidate measures are suitable for simple comparisons, others need to be scaled up, and some may be easily confounded. Small amounts of retention are an example of the latter. Among the consequences are high amounts of woody debris, low canopy cover, and likely an increase in shrub cover. If it were desirable to separate the particular mechanism of response – effects of shrub cover and amounts of debris – then appropriate conditions might have to be created experimentally because operational conditions would not provide informative differences (e.g., low debris and high shrub cover). Similarly, efforts to define a general, repeatable relation that reliably predicted some desired outcome may also require an experimental approach that carefully selects a range of conditions for measurement. Conversely, it is useful to recognize which kinds of measurement could be informative but are so poorly understood that experiments cannot be designed to define causal relations, which are better evaluated by broad comparisons. Examples include the responses of many non-vertebrate species to broad habitat types.

In short, some questions or variables are better treated in an experimental setting (e.g., refinement monitoring or active adaptive management), while others are revealed in operational settings that are less costly to establish (e.g., passive adaptive management). More specifically, experimental settings have the advantages of a greater range of treatments, control of single variables, controls and pre-treatment measurements, and randomized blocks. If these advantages appear particularly useful for a specific measurement, then the more costly approach may be warranted. For example, evidence of strongly confounded features (e.g., harvest type and stand type) would encourage randomized blocks and thus experimentation. Pre-treatment measurements can be very useful prior to assigning experimental treatments to replicate units. Conversely, if the variables of interest change rapidly with time but vary less dramatically through space, pre-treatment measurements are of little use, and simultaneous spatial controls are more helpful. Asking the question in an operational context has the obvious advantage of direct relevance to current practice and permits sampling more units. Although operational blocks are not truly randomized replicates, there are many of them from which to derive comparisons.

Comparisons and mechanisms are considered jointly for specific candidate mechanisms in the following chapters. Pilot projects (§5.5) were intended to reveal how best to ask the question for selected groups of candidate measurements.

5.3 What Would We Do with the Data if We Had Them?

This is the most fundamental question of any monitoring program. It must be asked of the three indicators selected to assess whether biological diversity is being sustained.

Indicator 1: Ecologically distinct ecosystem types are represented in the non-harvestable land base of the management area to maintain lesser-known species and ecological functions.

Indicator 2: The amount, distribution, and heterogeneity of stand and forest structures important to sustain native species richness are maintained over time.

Indicator 3: The abundance, distribution, and reproductive success of native species are not substantially reduced by forest practices.

Many kinds of data are relevant to these indicators – far more than can be collected in a cost-effective fashion. Bounding the problem by focusing on a few questions helps to reduce the kinds of data (§3.4.1), but we still need to know what we would do with the data if we had them. We want to ensure that we have invested monitoring funds in a way that helps to guide management decisions. In the following chapters, for each broad indicator, we address what we would do with acquired data in three related ways. First, we present the rationale or need for the indicator and its associated measurements. Second, we evaluate whether candidate measurements connect to forest planning and practice. Third, we document the potential kinds of feedback or guidance that could be attained from relevant data. That is, we specify how the adaptive management loop can be closed. Kinds of feedback should be envisioned before the monitoring program is designed because feedback is critical in determining the value of monitoring. In that way, we answer what we would do with the data if we had them. For each indicator, we also provide findings to illustrate how such data can contribute to adaptive management and influence practice.

When considering what to monitor, it is prudent to think broadly rather than to focus on the familiar. We evaluated many more measurements than were finally selected. To be considered for monitoring, candidate measures had to meet four criteria:

- be part of the forest or forest dwelling;
- be sensitive to forest practices used;
- be practical to monitor in terms of sampling, identification, and cost; and
- be able to provide information useful in guiding forest practices.

We review candidate measures in terms of these criteria, noting reasons for omitting some candidate measures from the final monitoring design. Each indicator is treated separately, so overall ranking of the utility of specific candidate measures cannot be determined until the overall design is created (see Chapter 12). Selecting specific measures begins by asking "what would we do with the data if we had them?"

5.4 How Do We Discern What Is Better?

Determining which is the better management approach involves comparisons of management alternatives and comparisons with targets or benchmarks. Both are required. Simply measuring how much of something is present does not help us. For example, knowing that we have five large trees per hectare in one retention type and only three large trees per hectare in another tells us little (see Figure 5.1). Both levels may be adequate to attain our objectives, or neither level may be adequate.

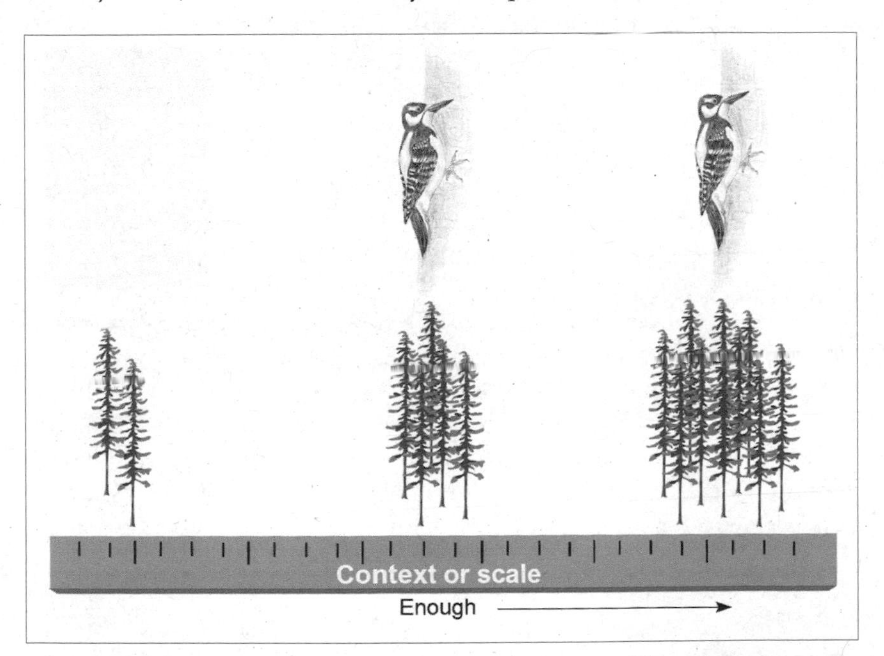

Figure 5.1 Knowing how much of a habitat component is present is insufficient; there must be some scale or context to which that amount can be compared.
Source: Prepared by I. Houde.

Some target or standard is implied in most comparisons. However, even when targets are agreed upon, comparisons to targets alone may not help to improve practices (§12.3).

Comparisons of management alternatives allow direct learning about the relative contributions of different practices and suggest possible solutions to problems. If we know the desired direction of response for each variable, then direct comparison may be adequate but cannot evaluate effectiveness by itself. Comparisons to targets are needed to provide a measurement scale for the comparisons (see Figure 5.1) – for example, is an observed difference substantial, or are both options adequate or highly inadequate?

Criteria and indicators described by a sustainable forest management plan identify values that will be sustained. To determine the effectiveness of the plan, some idea of targets or decision thresholds is needed for the indicators or attributes of the resources being monitored. Targets allow us to judge how good a particular management option is, relative to the target, and to evaluate the importance of an observed difference in a direct comparison. Targets can be derived externally, as with regulatory requirements, or they may require field measurements under some concept of benchmarks. Thresholds specify amounts or levels of different resources that would trigger a management action and are most clearly derived from known discontinuities in ecological responses.

Where targets have been specified or thresholds derived, it is easy to discern the better approach among comparisons. Most relations in nature, however, are curvilinear, without evident thresholds other than near zero (Bunnell and Dunsworth 2004; Huggard 2007). Assigning a decision threshold value is thus somewhat arbitrary. One goal in the chapters reviewing candidate measurements of major indicators is to assess the likelihood of deriving useful thresholds or natural benchmarks. Potential sources for targets and thresholds include regulations, older growth benchmarks, and species' requirements. The last source emphasizes the difficulty in discerning what is better even between two quite disparate forest treatments. The variety of niches inhabited by forest-dwelling organisms ensures that any disturbance to the forest, human made or otherwise, benefits some species while disadvantaging others. This implies that a wide range of organisms of various natural histories must be employed in monitoring, which in turn implies a wide range of habitat features. Consequently, the reviews of candidate measurements for the broad indicators must also be wide ranging.

5.5 Where Does the Answer Apply?

Results from any specific sampling site represent conditions at that site. To generalize to larger areas, we use strata or blocks of similar conditions that minimize the variation between measurements not attributable to the factors being studied. That reduces confounding between variables and site or treatment. The blocks or strata help to reveal the conditions to which the answer applies. The factors that should be blocked are those that appear to have a strong influence on outcomes. Blocking and stratification by these factors help to indicate the conditions under which similar results are likely to be attained (§12.6.2).

At least four related variables could be used to define blocks or strata of similar conditions to which particular results apply: ecosystem type, harvest type, age of stand harvested, and operating division (for implementation monitoring). A review of candidate indicator variables needs to focus on appropriate blocks and strata. For example, strong stand age effects on a candidate variable (e.g., some lichens) might reveal a useful indicator of old-growth conditions. Similarly, effects of elevation might be captured well by blocking using the ecological classification system or might have to be treated separately.

5.6 The Role of Pilot Studies

The preceding questions provide context for the review of variables that might serve the three major indicators in the following chapters. In the pilot phase that began this adaptive management program, we employed field studies as well as a literature review. These pilot studies were intended to assess the adequacy of groups of organisms or other features for the overall monitoring program and to provide information needed to design monitoring. To aid the overall design of the monitoring program, scientists responsible for each organism or feature were asked to use the pilot study phase to provide answers to the following four sets of questions.

1 *Suitability:* For which comparisons,[2] including monitoring overall operational effectiveness, is the organism or feature a suitable indicator? Suitability means that the organism or feature

- responds to the factor(s) being compared;
- occurs at a scale suitable for the comparison;
- uses habitats amenable to comparison (e.g., the organism does not require particular habitat features that confound the comparison and is not too rare to allow comparisons); and
- can be sampled by practical field methods to produce biologically meaningful results.

There are other questions relevant to comparisons. Should the comparisons be done in operational or experimental settings? Would pre-treatment information be worthwhile? Can several of the comparisons be done with little additional expense?

2 *Blocking and sampling:* What are the appropriate blocking units (e.g., what ecological classification units should be combined versus sampled separately) and sampling strata (e.g., do samples need to represent site series,[3] edge distances, different patch sizes, different tree species, etc.)?
3 *Precision and design:* Based on results from the pilot study, how precise will estimates be for the various possible comparisons, given different sampling efforts (budgets), and how should the sampling best be designed?
4 *Modelling:* How can the monitoring results be generalized to management options that are not measured directly and to project future conditions? Generalization requires a predictive model – how will it be generated and tested?

It was anticipated that scientists responsible for various pilot studies would answer these four sets of questions. The answers would then help to determine the overall design of the monitoring program. In fact, we underestimated the commitment of time and effort necessary to achieve effective oversight and answer the design questions. Not all pilot projects provided answers. Nonetheless, each contributed material that aided the selection of candidate variables and approaches to monitoring. Some findings of pilot studies that contributed to monitoring design are included for relevant indicators in the following chapters.

5.7 Summary

Part 1 introduced the generic problem and outlined major details of the overall approach to learning how to sustain biological diversity in managed forests. This chapter is a bridge between the introductory material and the actual learning process. It notes four major questions that must be asked of each indicator.

How do we ask the question? exposes distinctions between experimental approaches and monitoring operational practices and documenting causal mechanisms versus making comparisons. Each has advantages.

What would we do with the data if we had them? is the must fundamental question and defines the link to management.

How do we discern what is better? reveals the need for some contextual scale such as natural benchmarks.

Where does the answer apply? reveals the need to stratify monitoring so that findings can be effectively and accurately scaled up to larger areas.

In Part 2, we consider these four questions for each of the three major indicators and provide major findings of those indicators. Our consolidated approach, combining all indicators, questions, and comparisons, is summarized in Chapter 12.

6
Ecosystem Representation: Sustaining Poorly Known Species and Functions

David J. Huggard and Laurie L. Kremsater

6.1 Rationale

Indicator 1 is the coarsest filter used to monitor the effectiveness of planning and practices in sustaining biological diversity. It complements Indicators 2 and 3 and provides context for them. Monitoring responses of particular species to habitat structure and pattern (Indicator 3) is useful for species whose requirements we know well or whose trends we can track readily. Habitat structures and patterns that are monitored within Indicator 2 are designed as a "medium filter" to capture habitat requirements of many species. To a large extent, Indicator 1 addresses issues about which we know least. Providing representative non-harvestable[1] stands can help to

1 sustain poorly known species;
2 sustain poorly known functions;
3 provide a buffer against management errors; and
4 provide an ecological baseline for comparisons.

There are many species about which we know little but that are likely to be restricted to particular ecosystem types or geographic localities. Examples of little-known taxa include many lichens, most bryophytes, microfungi, soil organisms, worms, and most arthropods. These comprise the majority of species. Ensuring that some portion of each distinct ecosystem type is represented in a relatively unmanaged state is an important contribution to sustaining species, particularly the majority of species for which knowledge is absent or sparse (Higgins et al. 2004; Margules and Usher 1981; Pressey 2004; Scott et al. 2001a).

Areas without active management also help to sustain at least some poorly understood ecological functions (Molnar, Marvier, and Kareiva 2004). For example, some natural disturbances can occur that would otherwise be suppressed or reduced, although some, such as large fires, undoubtedly will also be suppressed in the non-harvestable land base. While some aspects of natural disturbance can be mimicked in managed stands, other aspects cannot be (e.g., large patches of burned snags or large untreated areas with mistletoe). In any case, we cannot be sure that any managed system can mimic all aspects of natural disturbances that might be relevant to organisms. Species that benefit from such features of natural disturbance may not be productive in managed landscapes.

A third role of non-harvestable stands is as a precautionary buffer against errors in sustaining species in the managed forest (Lindenmayer, Margules, and Botkin 2000). While we can develop management practices intended to keep many forest-dwelling species in managed forests, we also recognize that we have insufficient knowledge to ensure that proposed practices will meet all species' requirements. Unmanaged stands are an ecological safeguard against the inevitable errors that occur during management. They can also provide refuges for species that are directly harmed by human disturbance or presence.

A final function of non-harvestable areas in the landscape is to provide an ecological baseline against which the effects of human activities can be compared (Arcese and Sinclair 1997; Noss and Cooperrider 1994; Figure 5.1). This role as a benchmark is especially critical in the long-term monitoring required to assess the effectiveness of forest practices. Non-harvestable benchmarks provide context for the comparison of alternative practices and are the only way of assessing the degree to which managed areas differ from natural areas.

Setting aside a large percentage of the land base as non-harvestable forest to ensure that species richness is sustained is not compatible with current economic and social objectives of managed forests. Fortunately, forest tenures in British Columbia and other jurisdictions typically have 20 percent or more of the forest in areas constrained from harvesting (DeVelice and Martin 2001; Noss et al. 1999; Strittholt and DellaSalla 2001). On the case study tenure, wholly constrained areas represented 25.7 percent of the forest, not including the Old-Growth stewardship zones. Partially constrained areas, having 50 percent to 90 percent of the timber volume constrained, represented 7.3 percent of the forest area. This non-harvestable area was of two types: (1) areas that are not harvested or are harvested only lightly because of concerns other than conserving biological diversity (e.g., operability, visual quality, watershed protection, favoured-species management),[2] and (2) areas intentionally set aside to protect biological diversity (e.g., riparian buffers). This non-harvestable portion of the land base exceeds most jurisdictions' administrative objectives for protected areas (typically 12 percent, a value institutionalized from the loose recommendations of the Brundtland Commission; see Brundtland 1987) and is comparable to many recommendations derived by conservation biologists (e.g., 33 percent to 50 percent) (Noss 1993; Sætersdal and Birks 1993; Soulé and Sanjayan 1998; Stokland 1997).

The critical question for monitoring is thus not the total amount, which approximates values suggested by conservation literature, but whether most or all ecosystem types are represented in the non-harvestable land base. Secondarily, the size, shape, age, and spatial distributions of non-harvestable areas are important features. Monitoring habitat structures and organisms can be used to test assumptions of this indicator (e.g., definition of ecosystem types or edge effects), while monitoring at the overall tenure level tests that harvest practices and the non-harvestable land base, when combined, are meeting the objective of sustaining species on the tenure (see Chapter 12).

6.2 What to Monitor

To assess the current level of representation, we need to evaluate areas that are currently non-harvestable or partially constrained to determine:

- the proportion of the forest currently in non-harvestable or lightly managed areas;
- the ecosystem types comprising those areas;
- the patch size and geographic distribution of those areas;
- the amount of "forest interior";
- special ecosystem types or features of ecosystems (e.g., site productivity) not included in the ecosystem classification system used;
- other indices of spatial pattern; and
- current age (though this is less important in the long run).

This ordered list follows the typical progression of questions asked by a conservation biologist evaluating set aside areas.

- How much do you have?
- Is it just rock and ice, or does it represent all forest types?
- Are they just little bits and pieces, or are there larger areas? And are the larger areas all in one place, or are they well distributed?
- Are the large areas long and thin, or is there some interior, away from edges and disturbance?
- Are the areas less productive or aberrant in some other way not captured by your system?
- Are they "connected"?
- Are you able to reserve older stands?

Monitoring for Indicator 1 is intended to answer these questions and adjust management where needed and when possible. The seven bullets and questions listed above reflect a gradient from primary to more subsidiary issues. Each is discussed below, along with two additional features that could assist interpretation. An additional question is how well ecosystem types act as surrogates for all biological diversity. The answer requires direct monitoring of a range of organisms and is discussed in Chapter 10 and § 12.2.1.

6.2.1 Amount of Non-Harvestable or Lightly Managed Area

Conservation biologists use a simple rule to determine the desired amount of unmanaged or non-harvestable forest – more is better. A rule for the minimum amount required is less simple, but values of 10 percent to 12 percent of the land base are often used for official protected area strategies. These numbers represent social targets. Ecologically based conservation assessments often suggest higher levels: 30 percent to 40 percent for Norwegian fauna (Stokland 1997), 49 percent for representation and connectivity in Oregon (Noss 1993), 71 percent for regionally rare plants in Norway (Sætersdal, Line, and Birks 1993), 99.7 percent for endemic island species in the Gulf of California (Ryti 1992), but only 4 percent to 6 percent in other areas (Margules, Nicholls, and Pressey 1988; Pressey and Nicholls 1989; Rebelo and Siegfried 1992). Noss (1996) reviewed factors influencing how much reserve area is enough and concluded that roughly 50 percent of the land base should be reserved or managed for non-human species, though he recognized the inherent value dependence of any target for "enough." The standard 10 percent to 12 percent minimum amount is not relevant in most of British Columbia because many forest tenures have more non-harvestable area due to the many constraints on harvesting. A desirable amount (other than as much as possible) is extremely difficult to estimate, and upper limits of non-harvestable forest are often set by economics, policy, or other concerns external to efforts to sustain biological diversity. In the near term, there will likely be little change in the total amount of non-harvestable forest on the case study tenure. Moreover, the total amount of non-harvestable forest appears within the range of estimates derived for other regions.

In general, we believe that fixed targets for the overall amount of non-harvestable area are not helpful. Instead, we emphasize improving the weakest points (poorly represented ecosystems, modified by degree of responsibility [§6.2.8], and secondarily edge-dominated or poorly distributed non-harvestable areas). The focus of monitoring is therefore on other features – primarily the ecological distribution of these areas. Improving performance in these features is amenable to near-term adaptive management.

6.2.2 Ecosystem Representation

Given a specific amount of non-harvestable forest, the most basic goal for conserving biological diversity is to have some non-harvestable forest in each ecosystem type (Austin and Margules 1986; UNESCO 1974). This concept is fundamental to ecological gap analyses (Caicco et al. 1995; Duffy et al. 1999; Fearnside and Ferraz 1995; Scott et al. 1987; Scott et al. 2001a), reserve selection strategies (Belbin 1993; Johnson 1999; McKenzie et al. 1989; Mendel and Kirkpatrick 2002; Pressey et al. 2000; Pressey and Nicholls 1989; Rebelo and Siegfried 1992; Sætersdal and Birks 1993; Stokland 1997), evaluations of existing reserves (Mackey et al. 1989; Powell, Barborak, and Rodrigues 2000), and forest planning to sustain biological diversity (Franklin 1993; Lindenmayer, Margules, and Botkin 2000). Although representation has been overshadowed in the conservation literature by more academically attractive debates about reserve size and shape, an ecologically representative system of non-harvestable areas is more fundamentally important than well-sized or shaped reserves (Margules and Pressey 2000; Stokland 1997).

Our knowledge of most species is far too limited to determine exactly what suite of ecosystems is required to represent all species or even which ecosystems different species "perceive" as different. Some analyses of representation have been limited to a well-surveyed subset of organisms in discrete ecosystem types, such as vascular plants in remnant eucalyptus patches (Pressey and Nicholls 1989), or species of family Proteaceae in Mediterranean habitats (Rebelo and Siegfried 1992). Limiting our assessment of what constitutes distinct ecosystems to the few well-known species for which reliable information is available would defeat the purpose of the coarse-filter approach – ecosystem representation is the primary conservation strategy for *poorly known* species. In the long term, well-designed monitoring of various groups can help to define distinct ecosystems for a range of organisms. In the meantime, we must rely on surrogates to define ecosystem types. When complete species lists are unavailable, Austin and Margules (1986) suggest using vegetation associations as a surrogate for small areas and climatic, topographic, and other environmental variables for broader scales (see also Belbin 1993; Johnson 1999; McKenzie et al. 1989). A minimum set of representative areas based on these surrogates will miss some species, but multiple areas in each ecosystem type will help to overcome this limitation (Margules and Stein 1989; Pressey and Nichols 1989). A greater problem can occur if the non-harvestable areas are a biased subsample of the ecosystem types that they are intended to represent, such as the least productive sites in those ecosystem types. Monitoring this possibility is discussed in §6.2.5.

The challenge is to find units of classification that will reflect most components of biological diversity and that are widely used and mapped. In British Columbia, the biogeoclimatic ecosystem classification system (Pojar, Klinka, and Meidinger 1987) provides a good basis for monitoring representation. The BEC system is hierarchical, with "zones" of broad forest types divided into "subzones" that reflect major climatic variables. Subzones are divided into "variants," which have the same general climatic conditions

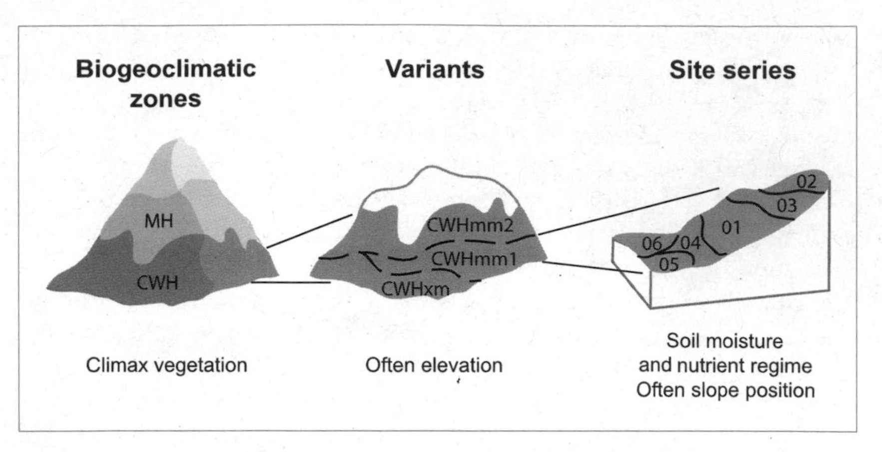

Figure 6.1 Schematic depiction of the hierarchical nature of the biogeoclimatic ecosystem classification system.
Source: Prepared by I. Houde.

but different plant communities because they are in different geographic areas or at different elevations (Figure 6.1). The variant level corresponds closely to the broad-scale environmental variables of Austin and Margules (1986). However, the variant level is not ideal for assessing ecological representation because there are considerable ranges of different ecosystem types among sites within a variant. The variation due to site moisture and soil productivity is reflected in the fine-scale classification of "site series." Different assemblages of indicator plant species, corresponding to the vegetation assemblages of Austin and Margules (1986), identify site series (§2.1.2).

In the BEC system, distinctions among site series are intended to reflect relatively stable differences of soil moisture and productivity (e.g., due to slope position) and are distinguished by differences in the cover of vascular plants, bryophytes, and lichens. We generally expect these relatively immobile indicator species to have more restricted ecosystem associations than more mobile species and therefore to be conservative surrogates for defining ecosystem types for mobile species. Typically, there are ten or more site series per variant. The few studies available have found that, even though many organisms have distinct habitat requirements, no lichen, bryophyte, or vascular plant was limited to a single site series (Kremsater 2000).[3] We therefore chose groupings of similar site series as the finest level of analysis for ecosystem representation, where these units were mapped. Published data allowed us to group site series using cluster analysis of the relative abundance of the indicator species used to identify them (see §6.3 and Chapter 7 for more detail on ecosystem types).

6.2.3 Size of Non-Harvestable Patches and Geographic Distribution

After representation of ecosystems, concern arises for the size of the patches of non-harvestable forest and for their geographic distribution. Lessons from the topic of reserve design can inform our evaluation of existing non-harvestable areas (Diamond 1975).[4] Larger patches are considered desirable because they can sustain larger populations of organisms, which tend to reduce the risk of local extinction (Wilson and Willis 1975). All else being equal, larger patches also tend to be less influenced by surrounding habitats (Williamson 1975). At the same time, geographically distributed patches are also desirable because many organisms have restricted geographic ranges (Lomolino 1994). Genetic or subspecific diversity also increases with distance, particularly for less mobile organisms, favouring representation across a geographic range (e.g., Ritland et al. 2000; Scott et al. 2001b). Additionally, patches of an ecosystem type limited to a single area are at risk from local catastrophic disturbances.

For a given area of non-harvestable forest, the goals of larger patches and wider geographic distribution oppose each other. There is no single optimal solution to this trade-off, in part because the best answer depends on which organism is being evaluated (Simberloff 1986). Large, highly mobile animals typically have large geographic ranges with less subspecific differentiation over larger areas, so they require larger areas to sustain a given population. A few large patches are best for these organisms. In contrast, small, poorly dispersing organisms can have very limited geographic ranges and show substantial genetic differentiation over small distances. Provided that they are large enough to sustain viable populations, many small, well-dispersed patches better represent these organisms. Small, well-dispersed patches may also make the managed land base more hospitable to larger organisms (Lindenmayer and Franklin 2002). The predominant natural disturbance in the ecosystem type may also influence the trade-off between size and distribution of patches. Infrequent catastrophic events favour several, thus smaller, widely distributed patches, so that a single disturbance event does not eliminate all older forest in that ecosystem type. Frequent small-scale events might favour fewer larger patches, so that the areas can have a natural dynamic equilibrium (e.g., the gap-phase dynamics present over much of the case study tenure; §2.1.2).

The lack of specific optimal targets for patch size and geographic distribution does not mean that these are not useful measurements for effectiveness monitoring. As with any management practice for which there is no optimum, a range of conditions is preferable to homogeneity. In particular, extreme cases – any single or a few patches of an ecosystem type with poor geographic distribution or many very small patches – suggest changes in management. Such changes would seek a mix of larger and smaller reserved patches.

6.2.4 Edge and Interior

Representative areas of non-harvestable forest are intended to serve three major functions: low-risk reserves for little-known species and processes, safeguards for errors in managing for known species in the managed forest, and ecological benchmarks. These functions are compromised if adjacent managed stands have strong effects on the non-harvestable areas. Measured edge effects in western forests generally penetrate little distance into the adjacent forest (Kremsater and Bunnell 1999), with 50 metres being a reasonable maximum estimate of the distance of most biological effects. This distance is also a reasonable *average* for physical effects – some measured edge effects are greater on certain aspects of some stands, such as south-facing edges of Douglas-fir, but shorter or absent on other edges (Chen and Franklin 1992). However, a few edge effects can show larger average distances, and in some cases these effects may increase with time (e.g., progressive windthrow). In our analyses, we use a distance of 50 metres to represent the majority of edge effects and 200 metres as an approximate maximum distance.[5]

Using 50 metres as the distance of edge effects, some patches of non-harvestable forest could have most of their area near edges of managed stands. Most obviously, this would happen if the non-harvestable patches were small. Large patches can also be primarily edge habitat if they are long and thin or convoluted shapes (Laurance and Yensen 1991). For example, riparian reserve areas can form a patch with a large total area but all of it potentially affected by the immediately adjacent managed stands. Many organisms favouring riparian areas respond positively to edges, but if riparian areas are the major source of representation in an ecosystem type, all functions sought from non-harvestable forests are unlikely to be met. Monitoring the amount of interior in non-harvestable forests is therefore useful to identify cases where large patches or large total area do not provide much interior condition. Such a condition may encourage larger patches of retention in the managed land base. Monitoring will also reveal where opportunities exist to improve the shape of non-harvestable patches.

6.2.5 Special Ecosystems and Productivity

Defining distinct ecosystems as site series groupings could miss some special ecosystem types that the ecological classification system was not designed to identify. They should be identified and their representation in non-harvestable areas monitored if possible. Candidate special ecosystems include any identified by regional conservation data centres or equivalent organizations, rare soil types (if soil mapping exists for the management area),[6] rare topography or bedrock types (e.g., forests near limestone caves), and geographically isolated examples of site series or other ecosystem groupings (because geographical isolation tends to produce unique organisms).

Including additional identified ecosystems of concern increases the chance of including a full suite of organisms, as does maintaining multiple examples of each ecosystem. However, the non-harvestable areas representing particular ecosystem types can be a biased sample. For example, non-harvestable areas may be on the less productive sites (Mendel and Kirkpatrick 2002; Stokland 1997) or on particularly steep slopes. Some of these possible biases can be monitored, using existing mapped information. Forest productivity classes, for example, can be compared between managed and non-harvestable stands of each ecosystem type. Other features may require field sampling. Habitat structures, for example, can differ between harvestable and non-harvestable areas within an ecosystem type (Huggard 2001). Direct sampling of organisms in the non-harvestable land base ultimately provides the best test of whether non-harvestable stands are truly representative.

6.2.6 Other Indices of Spatial Pattern

Many landscape indices describe spatial patterns of patches that are not completely assessed by amount, size, geographic distribution, and edge/interior (e.g., contagion, interspersion, fractal dimension) (McGarigal and Marks 1995; O'Neill et al. 1988). It is not clear from the literature how organisms respond to variation in these indices of spatial pattern or even what values of the indices indicate better or worse management. When species' responses to landscape patterns have been documented, it typically is in simplified landscapes, not complex dynamic forests, and the responses are species specific (Lindenmayer et al. 2002). Generic indices of connectivity are also elusive (Richards, Wallin, and Schumaker 2002). Often, apparent relationships with landscape metrics are confounded with differences in adjacent habitat (Lindenmayer and Franklin 2002).[7] For these reasons, other indices of spatial pattern have little immediate value for effectiveness monitoring. However, the indices are easy to calculate and could conceivably be useful in the future for indicating how these aspects of spatial patterning have changed or how they differ from natural patterns. Large changes in some index of spatial pattern, or large differences from natural conditions, may stimulate research into the ecological significance of that index. Future detailed modelling of species may also provide a way of evaluating the ecological importance of landscape patterns. At present, quantitative indices of spatial pattern for the non-harvestable land base are a low priority for monitoring. Simply viewing maps of non-harvestable areas does, however, provide insight into their spatial distribution and an intuitive notion of "connectivity."

6.2.7 Natural Disturbances and Stand Age Distribution

The preceding features of monitoring all assess the distribution of the non-harvestable land base without considering what age of forest occupies the areas. Ignoring forest age contrasts with planning exercises that focus on old-growth forest for social as well as biological reasons. We should not be overly concerned with the current age of the forest in non-harvestable areas as a coarse-filter, long-term approach to sustaining biological diversity. In particular, we should not sacrifice ecological representation to capture currently old stands. However, it is worth examining age distributions for two reasons. First, stands older than rotation age are one of the important potential contributions of non-harvestable areas. They help to sustain a wider range of age classes across a landscape tending toward younger age classes under economic rotations. Second, habitats created by natural disturbance events can be unique ecosystems in their own right. In general, it is a reasonable assumption that some organisms will be adapted to any form of disturbance that is persistent and common, such as fire, windthrow, mistletoe, or root rot patches (Bunnell 1995). At least some of these organisms will be adapted to different successional stages. As a consequence, good representation requires that non-harvestable stands have an age-class distribution broadly similar to the natural distribution. Dominance of very young stands or single age-class cohorts in an ecosystem type are a cause for concern and should prompt changes in management. Analyses of natural disturbances can also be useful in identifying disturbance refugia, which could be treated as special ecosystem types. They would include glacial refugia and shorter-term, smaller-scale areas that have evaded catastrophic fires or windthrow.

6.2.8 Responsibility and Regional Protected Areas

Representation analyses would ideally be done for entire ecological regions rather than study areas with arbitrary administrative boundaries. Within limited study areas, two additional analyses may help interpretation. First, the proportion of the total extent of an ecosystem type that is in the study area gives an index of "responsibility" in the sense used by Dunn, Hussell, and Welsh (1999) for bird species. Ideally, the global extent of the ecosystem type would be used, but practically, this usually means the extent in a large political jurisdiction where the same ecological classification system is used (Wells et al. 2003). Immediate management actions to improve poor representation would focus on ecosystem types having high responsibility. Second, the proportion of the total extent of the ecosystem type that is in officially designated protected areas can also help to set priorities. Representation in non-harvestable areas outside parks is more important when official protected areas are rare.

6.3 How to Monitor

Implementing the kinds of monitoring suggested requires the following steps.

1 *Determine which parts of the land base can be considered "non-harvestable."* When forestry is the only human activity in an area, non-harvestable areas are fairly clearly defined as areas unavailable for forest harvesting. It is, however, worth distinguishing areas that are non-harvestable due to fixed regulations, areas with constraints that are less certain, and areas that are considered non-harvestable because they are less productive or anomalous. Examples of less certain harvesting constraints are areas considered non-harvestable for economic reasons. These areas can become harvestable under different market conditions or with technological changes. In British Columbia, frequently changing government policies also create various types of provisionally constrained areas. Areas constrained from harvesting because they are less productive or anomalous could include low-volume sites, sites that are difficult to regenerate, or in some cases, deciduous stands (if they are not recognized as distinct ecosystem types). These low-productivity or anomalous areas undoubtedly contribute to biological diversity, but we cannot be sure that they represent the communities found in more productive or typical stands. We suggest conducting analyses with and without economically and provisionally constrained areas, and with and without low-productivity or anomalous sites, to examine how important these areas are to representation. A high proportion of less certain constraints, low-productivity sites, or anomalous forest types in the non-harvestable land base of an ecosystem type is cause for concern.

The concept of "non-harvestable" is less clear where there are resource activities in addition to forestry, such as extensive grazing in forests, mining, or seismic lines used for oil and gas exploration. Non-harvestable areas subject to these activities, which all have effects on components of biological diversity, should not be considered "non-harvestable," because we cannot assume that these areas represent the typical biota of non-harvestable areas of those ecosystems. Some formerly non-harvestable areas that are subject to novel harvesting practices, such as the newly developed system of standing-stem removal with helicopters, also should not be considered non-harvestable, because these activities negate the assumption that the biotic communities are typical of unmanaged stands. Development of such innovative harvesting techniques serves to emphasize that analyses of non-harvestable areas are not static and should be revisited as regulations, policies, technology, and markets change. Roads passing through non-harvestable areas should be considered part of the managed land base and included as such in calculating edge buffers.

In British Columbia, there are two kinds of partially constrained areas. Some receive area constraints such that only a portion will be reserved after site assessment (e.g., some slopes with unstable soils); others receive constraints on harvest rates, such that the area can be fully harvested, but slowly and lightly (e.g., some areas with visual quality objectives). We need to assess whether or not such areas still serve the functions of non-harvestable forest. If these areas are not included as "non-harvestable," they still need to be recognized in the monitoring process because they are likely to make important contributions to sustaining biological diversity across the tenure. We must also recognize that a partially constrained area may have reduced harvest levels overall but that harvest rates may be high in specific ecosystem types. Ultimately, we need to evaluate operational planning within partially constrained areas to know what proportions of the different ecosystem types remain unaffected by harvesting. Initial analyses of representation should concentrate on completely non-harvestable areas but track the distribution of ecosystem types within partially constrained forests.

2 *Determine appropriate classification for ecosystem types.* Within British Columbia's BEC system, site series is the finest level of ecosystem classification (Figure 6.1). Data are sparse, but we suspect that organisms are unlikely to be strictly confined to a single site series (in part because they can be so spatially disjunct). Instead, similar site series that are adjacent on the moisture and productivity gradients are likely to exhibit considerable overlap in species represented. At the extremes of site series gradients (e.g., xeric versus subhydric sites), organisms are more discriminating. We therefore provisionally use groups of similar site series for fine-scale analyses of ecosystem representation. With available published information, site series can be grouped based on a cluster analysis of the relative abundance of the indicator species used to identify them. The indicator species are only a surrogate for the full richness of organisms but provide a practical, objective way of grouping the site series into distinct "ecosystem types."

In British Columbia, a practical difficulty with using representation at a level finer than BEC variants is that complete mapping generally is not yet available for site series. However, models have been developed and tested for predicting site series based on mapped characteristics such as topography, slope, aspect, and proximity to water. Although predictions are not highly accurate at the site series level, they should map the broader groups of site series accurately. That is, erroneously classified sites tend to be in similar site series and therefore in the correct *group* of site series. We emphasize that site series groupings and their predictive surrogates are a practical choice and should be continually reviewed as biological information accumulates and improved mapping supports more refined definition of ecosystem units.

Initial classification of ecosystem types will be a first approximation that should be tested with existing or future field data on species' distributions. A thorough review of available information on ecological tolerances of a range of organisms is helpful, potentially suggesting other important distinctions, such as elevation or soil type (see Chapter 10). Any monitoring and research projects, particularly for non-vertebrates, should be encouraged to use designs and data recording that allow improvements in classification (e.g., recording which site series were sampled and where different organisms occurred). Again, this part of the analysis is not static but should be revised as new information allows more ecologically appropriate classifications of ecosystem types.

3 *Assess whether special ecosystem types not represented by the basic classification of ecosystem types should be included.* Ecosystem classification systems that have the necessary wide-scale mapping typically are developed for general purposes (e.g., forestry planning). As such, they usually do not include rare forest ecosystem types that may have unique components of biological diversity. The BC BEC system, for example, does not distinguish different types of forested wetlands well. Conservation data centres or equivalent organizations that track rare organisms and ecosystems should be consulted for their ecosystem types of special interest (e.g., rare soil types). Experts in various taxa, or their publications, should also be consulted for special ecosystem types. We then need to determine what mapping of these types is available for the tenure. In British Columbia, there has been little extensive mapping of rare ecosystem types, which currently excludes them from representation analyses.

4 *Determine classes to use for summaries.* The analysis requires decisions about values for edge distances, patch sizes, and age classes that are ecologically informative but simple enough to interpret. We use edge distances of 50 metres and 200 metres, based on the biological review summarized above. We currently use patch sizes of 0 ha-5 ha; 5 ha-40 ha; 40 ha-100 ha; 100 ha-500 ha; and >500 ha. The vast majority of species are tiny and operate at small scales, but it is not practical to subdivide the smallest class further for analyses over large areas. For age classes, we use 0-40, 40-80, 80-250, >250 years. Dividing the continuous variables of patch size and age class into discrete classes is clearly arbitrary and largely reflects a social decision about how people want to see the results presented. Increasing the number of classes can cause vagaries during analysis.

5 *Analyze the land base to provide information on at least the first four monitoring levels.* Helpful monitoring levels are amount, ecosystem representation, patch sizes, and edge/interior (§6.2.1 through §6.2.4). Separate analyses with and without less certain constraint types, low-productivity

sites, and anomalous forest types will help to determine the utility of the contribution of these non-harvestable areas. Based on our experiences, some unexpected artifacts during the GIS analysis can be avoided.

- Do not double-count areas where harvest constraints overlap. The areas under a particular constraint type should be summarized and then removed from subsequent analysis.
- Proceed from permanently constrained areas (e.g., parks, regulatory reserves) to less certain constraints such as low-productivity or anomalous forest types; this approach allows summary of the additional contribution of these more questionable sources of representation.
- Exclude age from patch size analyses. The intent at this level of monitoring is to assess the size of the non-harvestable areas of particular ecosystem types, regardless of the current age of their constituent stands. If patch analyses do include age, then the interpretation should recognize that the patch size results are largely determined by the arbitrary choice of age-class boundaries.
- Different results are obtained if the edge buffering of the non-harvestable area is conducted inward from the boundary of the non-harvestable land base versus outward from the boundary of the managed and partially constrained land bases. The difference arises from inclusion or omission of edges created by natural non-forested areas. If the intent is to look at only potential anthropogenic edge effects from harvestable stands, then edge buffering should be done outward from the managed and partially constrained land bases.

6 *Assess other indices of spatial pattern.* Currently, assessment is done by visual inspection of maps by ecosystem type. Review of existing landscape-level studies or future research may indicate specific additional indices worth summarizing (see, e.g., Perera, Buse, and Weber 2004).

7 *Review other candidate features for determining representation (e.g., topography, productivity).* For example, the company has examined representation on its more productive lands (site index >25 m at fifty years). Other mapped features should be assessed to determine whether they would be meaningful and feasible to include in the representation analyses.

8 *Integrate Indicators 1 through 3.* Representation monitoring should ultimately be integrated with monitoring of habitat structure and organisms. Points of connection include the following.

- Poorly represented ecosystem types should be a priority for variable retention monitoring.
- Field assessments of the composition of retention groups by ecosystem type can record the amount of stand-level representation.

- Structural monitoring in non-harvestable areas can be compared to the same ecosystem types in the managed land base to assess possible biases in the stand types included in the non-harvestable land base (e.g., different productivity classes).
- Monitoring of organisms can be used to test the assumption that representative non-harvestable areas truly are retaining their associated organisms.

6.4 Anticipated Feedback to Management

Effectiveness monitoring is intended to inform managers of their progress toward specific goals, with a view to altering practices if necessary. It is thus necessary to anticipate the kinds of feedback that the monitoring program should be capable of providing. Forms of feedback presented below are specific to the planning and practices of the case study but require only modest modification to apply to other forests. We anticipated that information from Indicator 1 could help management decisions in several ways.

1 *Assessing amount and location of Old-Growth stewardship zones.* Provided there is flexibility in the location of some Old-Growth zones, candidate areas that contain higher amounts of poorly represented ecosystem types should be favoured. Although the name "Old-Growth zone" implies an emphasis on old forest, representation of ecosystem types is more important than current age in designating such areas, when the primary objective is maintaining poorly known species and functions. If there is little flexibility for designating Old-Growth zones, the representation analysis can inform "mini-zoning" of non-harvestable forest to locate reserves required by regulation (e.g., Old Growth Management Areas in British Columbia).[8] Areas with little old growth represented in non-harvestable areas can be targeted for restoration. More generally, ecosystem representation should help to guide the location of any reserves specifically intended to sustain the range of biological diversity.
2 *Locating Habitat zones and management within them.* A primary function of the Habitat zones is to provide more flexibility in meeting the objective of sustaining biological diversity while still maintaining timber flow. Locating Habitat zones in areas with relatively high amounts of poorly represented ecosystem types can assist this function. Representation analysis can also guide planning areas within the zone that have higher or lower amounts of retention.
3 *Guiding the intensity of harvest – retention patches and retention levels within variable retention settings.* Relying on the non-harvestable land base is unlikely to be sufficient for sustaining all organisms. Settings within poorly represented ecosystem types can be favoured for relatively high

levels of retention and larger retention patches. Patches of habitat retained within retention blocks serve as representative unmanaged forest for many small organisms (e.g., §10.3.2, §10.4.2). Ecosystem types that are poorly represented overall can serve as anchor points for locating within-block reserves. Knowing overall levels of representation of ecosystem types across the tenure provides context to evaluate the measured distribution of retention patches (i.e., are reserves within harvest blocks catching poorly represented types, or are they mainly types that are already well represented in landscape-level reserves?). In this way, monitoring ecosystem representation helps to guide management within cutblocks.

4 *Focusing finer-scale monitoring.* The potential ecological benefits from retention during harvest are greatest within ecosystem types that are poorly represented in non-harvestable forest. In such types, retention will be the major means of sustaining some organisms. These ecosystem types should therefore be a focus for monitoring habitat (medium filter) and specific organisms (fine filter). Ecosystem types represented relatively well in non-harvestable stands are of less immediate concern for monitoring, though they may be useful for comparisons to natural conditions. Assessing Indicator 1 helps to focus the large task of monitoring throughout the management area. It also helps to identify representative non-harvestable areas that are large enough to serve as benchmark controls for long-term monitoring.

5 *Evaluating other reserves.* Many areas are left non-harvestable for reasons other than sustaining biological diversity (e.g., soil or slope stability, visual quality) or single species (e.g., black-tailed deer winter ranges, northern goshawk nesting sites). Knowing which ecosystem types are, or are not, well represented in such reserves helps to evaluate their broader contributions to biological diversity and can guide location and extent. For example, it may be possible to locate deer winter ranges such that they contribute to both their original objective and to ecosystem representation or shift anchor points for patches to better capture desired representation.

6 *Simplifying operational decisions.* There may be a group of species, for example, that appears well supported in non-harvestable areas, so it need not enter into operational decisions for the managed land base.

7 *Evaluating changes in timber supply assumptions and mapping.* Definitions of non-harvestable areas are partly a management decision. Changing markets may change areas considered to be economically inoperable, or riparian reserves on private land may be formally recognized on maps. Such decisions have economic and ecological consequences. Monitoring representation allows assessment of one major ecological consequence and can bring those findings into decision making.

6.5 Summary

Non-harvestable areas play several roles in sustaining biological diversity, the most important of which is sustaining poorly known species and functions. They contribute to that role most effectively when each ecosystem type within the management area is represented within the non-harvestable land base. Secondary influences on their contribution include amounts of forest interior, a range in both size and distribution of non-harvestable areas, and an age distribution that is not biased to younger age classes. We found few spatial statistics that were reliably useful beyond ecosystem representation, size, and age class. Useful analyses can be conducted following eight straightforward steps. This chapter has presented the rationale and general approach to analyzing ecosystem representation. Chapter 7 presents the results of monitoring ecological representation on the case study tenure, and their implications.

7
Learning from Ecosystem Representation

David J. Huggard, Laurie L. Kremsater, and Glen B. Dunsworth

7.1 Context

This chapter presents an analysis of ecosystem representation on the study tenure in British Columbia (see Chapter 2), including methodology, results, management implications, and the company's response. The rationale for this major indicator and discussion of features to monitor were presented in Chapter 6.

The representation analysis was conducted in 2001. Since that time, there have been numerous changes that affect the original results. Chief among them has been the "take-back" by government of 20 percent of the public portion of the company's tenure as part of an attempt to resolve a trade dispute with the United States, the sale of the company, and changes in definitions of constrained areas due to continuously changing land use planning. In addition, landscape planning has increased the non-harvestable land base with the addition of Old Growth Management Areas (OGMAs) and Wildlife Habitat Areas (WHAs) across the tenure. These major changes emphasize that monitoring of even a coarse-filter indicator needs to be considered an ongoing process, to be updated as conditions change. Ideally, representation monitoring would show continuous improvement over time.

Recent efforts to update the analysis for the area did not produce reliable results because

- new company ownership no longer allowed access to data from private lands;[1]
- some constraint information had been lost or become uncertain with company cutbacks for resource analysis;
- stewardship zones could not be used because of uncertain effects of government tenure take-backs;
- information on recent constraints from government landscape planning was not readily available or was considered provisional; and

- the analysis was restricted to only the Vancouver Island part of the tenure.

These difficulties illustrate that even relatively simple monitoring requires a stable corporate and government infrastructure.

7.2 Methods

7.2.1 Ecosystem Representation

The 2001 analysis of representation was conducted at two levels of British Columbia's Biogeoclimatic Ecosystem Classification system (Pojar, Klinka, and Meidinger 1987; see also §2.1.2): (1) BEC variants across the entire study tenure, and (2) site series for a 210,699 hectare area (16 percent of the tenure) on north-central Vancouver Island, where site series mapping was complete and ground-truthed. BEC variants are broad ecosystem types, representing different elevational and geographic expressions of climate regimes. Site series are a finer level, reflecting local site moisture and productivity, as identified by indicator vegetation. Variants, but not site series, have been mapped across the entire province, allowing us to evaluate provincial rarity of variants and their representation in official protected areas province wide.

GIS overlays were used to determine the portion of each ecosystem type (variants or site series) in three land bases:

1. *Non-harvestable land base:* Non-harvestable areas included any forested areas within the tenure where regulatory or operational constraints prevent timber harvesting over the long term, including unstable slopes, riparian reserves, physically inoperable areas, reserves for ungulate winter ranges and other individually managed wildlife species, and some recreation reserves where harvesting is prohibited. Provincial and federal protected areas adjacent to or embedded within the tenure were excluded from the analysis area, as were non-forested areas.
2. *Partially constrained land base:* The partially constrained land base comprises stands available for harvest, but with >50 percent reduction in harvest rates due to various constraints, including some recreation areas and management areas for wildlife species that have restrictions on the amount of forest in young age classes. These areas present opportunities for improving representation of ecosystem types with lower additional economic costs. Conversely, when specific forms of harvest are selectively located within partially constrained areas, even limited harvesting in the partially constrained land base can affect most of the area of certain ecosystem types.
3. *Timber-harvesting land base:* Areas with standard operational harvest rates or up to 50 percent reduction in harvest rates were included in the timber-harvesting land base.

A major component of the non-harvestable land base is forest considered commercially non-productive. Non-productive stands were defined by the company as those with <210 m^3 / ha of merchantable timber. Most of these lower-volume stands are old growth and contribute to the ecological representation of their ecosystem types. However, they were tracked separately in the analysis because these stands may contain a different set of organisms than more productive sites in the same ecosystem type (Stokland 1997). Additionally, some non-harvestable areas are excluded from harvest only for economic reasons; they are productive but too expensive to access. Because these stands may become economic to harvest in future markets, a separate representation analysis was conducted in which stands constrained only for economic reasons were included as part of the timber-harvesting land base.

Within productive forest, stands have greater or lesser productivity, indicated operationally by site index (a measure of the height that trees are expected to reach at fifty years old). Highly productive stands may be important to some components of biological diversity and may be disproportionately rare in the operationally constrained land base (Mendel and Kirkpatrick 2002; Stokland 1997). A separate analysis of ecosystem representation was therefore conducted with only stands that have a site index >25 m at fifty years.

The additional voluntary contribution of Old-Growth stewardship zones to providing incremental non-harvestable forest in each variant was determined as the percentage of the variant that was in the Old-Growth zone and was not already non-harvestable for other reasons.

7.2.2 Edge/Interior and Patch Size

The area of interior non-harvestable stands was indexed by masking out 50 metre and 200 metre buffers into the non-harvestable land base. These distances correspond to a typical maximum edge effect for most biotic variables studied in western forests and a more extreme distance reported for a few biotic variables in some situations (review in Kremsater and Bunnell 1999). Because we were interested in the long-term characteristics of non-harvestable areas, the buffer distances were applied to the entire edge of the non-harvestable land base, regardless of the current age of the adjacent stands in the managed or partially constrained land bases. Interior areas are therefore more than 50 metres or 200 metres from a *potentially* managed stand, whether or not that adjacent stand has actually been harvested yet. These interior non-harvestable areas represent a "worst-case scenario" that would occur only if all the adjacent harvestable stands were in an early-seral stage at the same time.

7.2.3 Other Land Use Designations Emphasizing Conservation

In addition to the company's selection of Old-Growth stewardship zones, two other land use planning processes undertaken on Vancouver Island had identified areas of the case study tenure where conservation is a priority. The BC government's Vancouver Island Land Use Plan (Province of British Columbia 2000) identified "special resource management zones." As well, "areas of interest" were delineated by environmental non-governmental organizations (ENGOs). We assessed the representation of BEC variants in these two conservation designations for the Vancouver Island portion of the company's tenure to assess whether accommodating these other conservation designations would complement ecosystem representation.

7.2.4 Responsibility and Protected Areas

Two additional analyses at a larger scale were used to provide regional context for the representation analysis. First, the company's responsibility in the provincial context for each BEC variant on the tenure was indexed as the percentage of the total provincial extent of the variant that was within the tenure. The greater the proportion of a variant that falls within the company's tenure, the greater responsibility the company has to maintain a representative portion of that ecosystem in an unmanaged state. This concept of "responsibility" follows the approach used nationally for bird species (Dunn, Hussell, and Welsh 1999). Second, the percentage of each variant that was in officially designated provincial and federal protected areas in British Columbia was used to index the extent of official protection outside the tenure.

7.3 Results

7.3.1 Responsibility and Protected Areas

Fifteen biogeoclimatic variants make up most of the case study tenure (Figure 7.1a), including a single variant of the dry Coastal Douglas-Fir zone (CDF), eleven variants in the Coastal Western Hemlock zone (CWH), and three in the high-elevation Mountain Hemlock zone (MH). Five additional variants were omitted because they occupy less than 25 km^2 of the tenure. The company's provincial responsibility is relatively low for three variants that are common on the tenure (CWHvm1, CWHvm2, and MHmm1)[2] because these variants are abundant in the province outside the tenure (Figure 7.1b). Company responsibility is high for two variants that are a relatively small percentage of the tenure (MHwh2 and CWHwh2), because a large percentage of the total extent of these ecosystem types occurs in the company's tenure, and both are restricted to Haida Gwaii. The company also has a relatively high responsibility for four other CWH variants: CWHmm1, mm2, wh1, and xm2. In a broad sense, responsibility is also high for the two driest variants (CDFmm and CWHxm1). Most of the area of these latter variants outside the tenure is urbanized, subject to intensive agriculture or short-rotation managed forest.

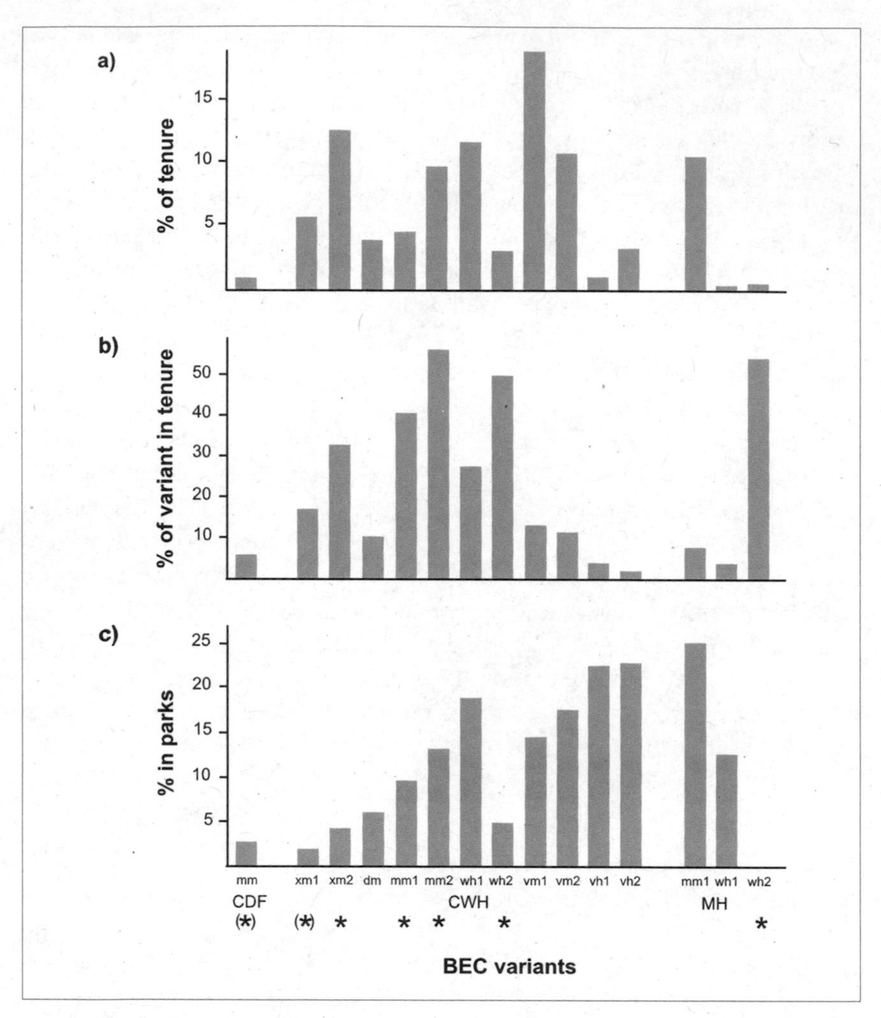

Figure 7.1 Ecosystem representation analysis of the tenure.

Notes: (a) BEC variant composition of the company's coastal BC tenure. (b) Percentage of the total provincial extent of each variant in the tenure, a measure of the company's responsibility for maintaining that variant. (c) Percentage of each variant in official protected areas across British Columbia.

Variants are arranged from dry, southerly, and lower-elevation variants on the left to wet, northerly, and higher-elevation variants on the right. CDF = Coastal Douglas-Fir zone; CWH = Coastal Western Hemlock zone; MH = Mountain Hemlock zone. * = high-priority variants for the company. See §6.2.2 for an explanation of BC's BEC system.

Provincial representation in protected areas shows a strong gradient from low percentages in southerly, dry, and lower-elevation ecosystem types to relatively high levels of protection for northerly, wetter, higher-elevation forest types, usually of lower productivity (Figure 7.1c). The exceptions were the two wet, northerly ecosystem types on Haida Gwaii for which the company has a high provincial responsibility. These types have little or no representation in protected areas. Combining responsibility and regional representation in protected areas, priority ecosystem types for the company include these two Haida Gwaii types, CWHxm1 and xm2, CWHmm1 and mm2, and the CDF zone. Given the limited protection of low-elevation, drier coastal forests in the adjacent United States, company stewardship of the drier ecosystem types on the tenure is probably of global significance.

7.3.2 Representation of Variants in the Non-Harvestable Land Base

Overall, 25.7 percent of the forested area of the tenure was fully constrained from harvesting, while an additional 7.3 percent was partly constrained. Within the commercially productive forested land base, 14.2 percent was fully constrained from harvesting. Non-harvestable percentages by BEC variant showed a clear increasing trend from the driest, southerly variants, with minimal non-harvestable area, to the wettest, northerly variants, with high percentages (Figure 7.2). Less productive, higher-elevation variants

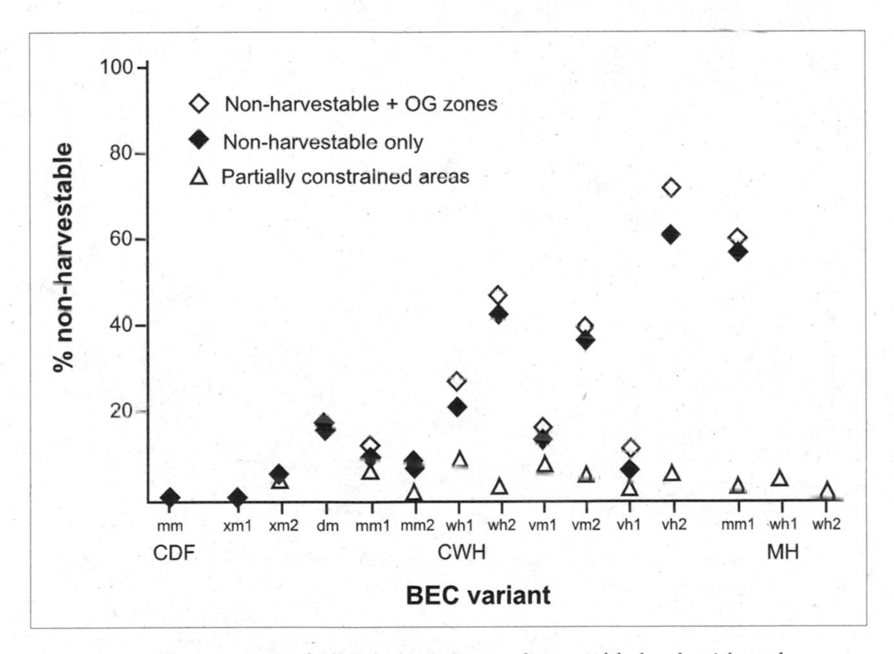

Figure 7.2 Represenation of BEC variants in non-harvestable land, with and without the additional contribution of Old-Growth stewardship zones. Old-Growth zones contribute relatively little additional non-harvestable area. Partially constrained land offers some additional opportunities for improving representation in some variants.

Note: Within a subzone, variants ending in 1 are at lower elevations than variants ending in 2. See §6.2.2 for explanation of BC's BEC system.

consistently had more non-harvestable area than lower-elevation variants within a subzone (Figure 7.2).

Commercially non-productive stands comprised a substantial part of the non-harvestable land base, but the same trends in representation were seen when only commercially productive stands were analyzed (Figure 7.3): colder, wetter sites at higher elevations were better represented in non-harvestable areas. Only 8.8 percent of highly productive stands (site index >25m at 50 years) was non-harvestable, with a less pronounced increase within wetter or more northerly forests (Figure 7.3).

Considering stands that are productive but economically inoperable as part of the harvestable land base had little effect on representation in the non-harvestable land base, with <1 percent drop in non-harvestable area in any variant. Market changes allowing harvest of these stands would have relatively little effect on representation.

With the designation of Old-Growth stewardship zones ongoing, 8.2 percent of total forested area in the tenure had been designated as Old-Growth zones. Within commercially productive forest, 6.8 percent was in

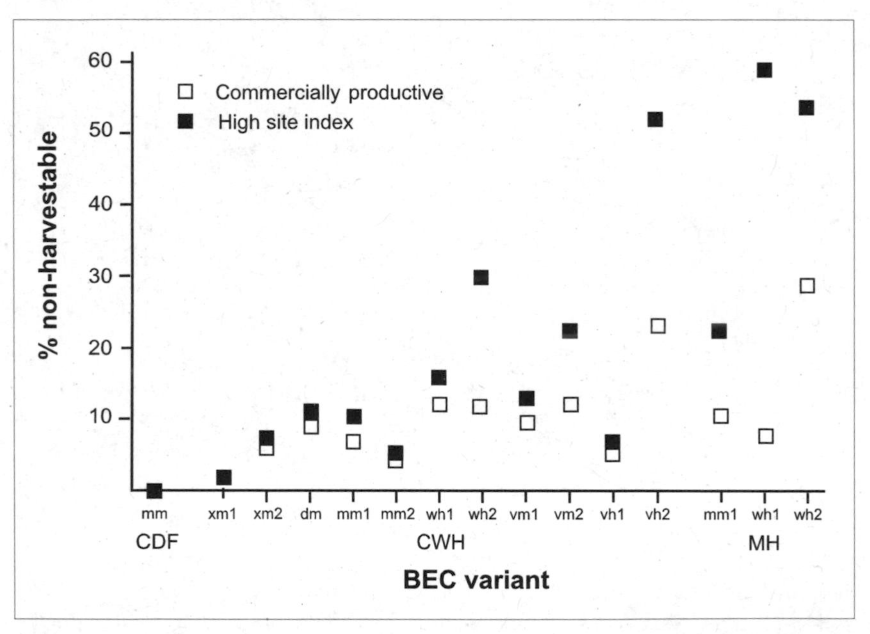

Figure 7.3 Ecosystem representation on commercial sites within the non-harvestable land base.

Note: Representation increases from dry, low-elevation ecosystems (left) to wet, high-elevation ecosystems (right). The trend is similar but less pronounced in stands with high site index.

Old-Growth zones. As with operationally constrained areas, Old-Growth zones were concentrated in the more northerly and wetter ecosystem types. Because the intent of Old-Growth zones was to address social concerns about conserving old forest, the bulk of the Old-Growth zones was located where large areas of old-growth forest existed. Much of this forest had remained as old growth because it was constrained from harvesting for various operational reasons. As a result, 60.5 percent of the area allocated to Old-Growth zones was already non-harvestable. The incremental contribution of Old-Growth zones to improving representation of ecosystem types in the non-harvestable land base was therefore small (Figure 7.2), although it did improve representation of some ecosystem types with moderate percentages of non-harvestable forest. The percentage of forested area that was partially constrained did not show a consistent trend across variants, but higher-elevation variants within a subzone tended to have less partially constrained area (Figure 7.2). The relatively high percentage of partially constrained area for CWHdm was due to grizzly bear management areas in this mainland variant.

7.3.3 Representation of Site Series

Overall, 37.2 percent of the subsample of the tenure that had been mapped to site series was in the non-harvestable land base, with a range from 16.4 to 98.3 percent from individual site series (Figure 7.4b). Three dominant variants were examined: CWHvm1, CWHvm2, and MHmm1. As expected, mesic site series (01)[3] dominated in the three variants, along with richer mesic (05) sites in CWHvm1, submesic (03) sites in both CWHvm variants, and subxeric (02) sites in MHmm1 (Figure 7.4a). Wetter site series (06 and 07) in CWHvm1 were also fairly common. An additional twenty-one less common forested site series in these variants each occupied less than 2 percent of the subsample area.

Site series in the high-elevation MHmm1 were well represented in non-harvestable forest, while the lower-elevation CWHvm1 had a lower overall level of non-harvestable area. Site series 08, 06, and 05 in the CWHvm1 had the lowest proportion non-harvestable. The driest and wettest site series in these CWH variants had greater than average representation, as expected from their associations with either steep topography or difficult operating conditions.

In the CWHvm1, rarer site series (<2 percent of the total area) were generally well represented. The slight under-representation of common ecosystem types and over-representation of rare types could be considered a nearly ideal situation. If this subsampled area is typical of the whole tenure, then biased representation of site series within a variant is less of a concern than the pronounced bias in variant representation across the tenure (Figure 7.2).

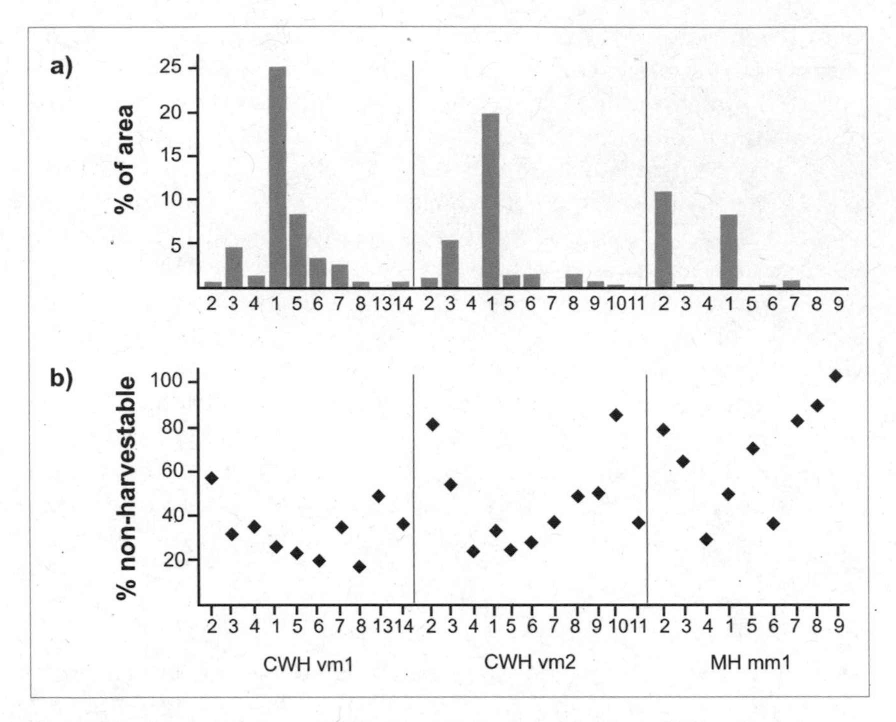

Figure 7.4 Representation of BEC site series analyzed for three BEC variants in a subset area of the tenure.

Notes: (a) The site series composition was dominated by mesic site series ("1"). (b) Representation in non-harvestable areas tended to be greater for dry or wet site series (left or right) than for circum-mesic site series.

Site series within a variant are arranged from driest (left) to wettest (right). Site series 1 = mesic. See §6.2.2 for explanation of BC's BEC system.

7.3.4 Edge/Interior and Other Spatial Aspects

Overall, 62 percent of the non-harvestable area was within 50 metres of areas that were potentially harvestable (Figure 7.5), reflecting the narrow or highly convoluted shapes of many reserves (e.g., riparian, steep slopes, sensitive soils, and inoperable areas).

The percentage of unmanaged forest near edges was higher in the drier ecosystem types, where fully constrained areas were smaller and riparian reserves comprised a greater proportion of the fully constrained land. Most of the non-harvestable area was within 200 metres of potentially harvested stands, while only a small percentage was >200 m from the edge.

The spatial distribution of non-harvestable land is shown for four 240 km^2 areas in Figure 7.6. Narrow riparian reserves were the dominant source of non-harvestable areas on the relatively flat east side of Vancouver Island (Figure 7.6a), whereas other landscapes had a greater diversity of sources,

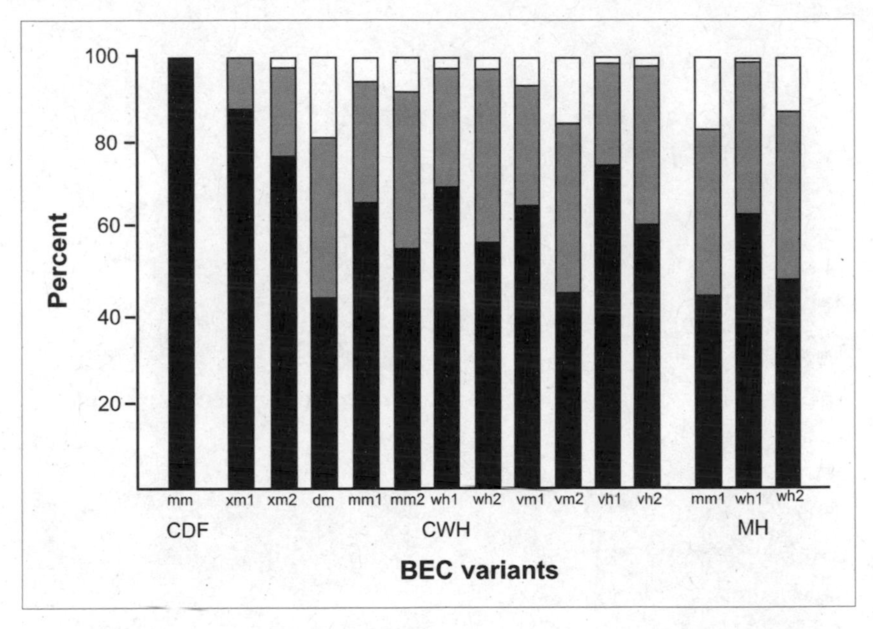

Figure 7.5 Percent of the non-harvestable area in each BEC variant that is <50 m (black), 50-200 m (gray), or >200 m (white) from adjacent harvestable land base.
Note: A majority of the non-harvestable area in most variants was <50 m from a harvestable stand, especially in dry and low-elevation variants.

shapes, and sizes of non-harvestable areas, reflecting the greater diversity of harvesting constraints in complex topography (Figure 7.6b-d).

Non-harvestable areas were generally well spread across local landscapes, even in areas with low levels of non-harvestable forest (Figure 7.6a), suggesting favourable geographic distribution of ecosystem types. The non-harvestable land also appeared to be well connected, though functional connectivity depends on species-specific responses to size and shape of non-harvested areas and to the variable habitat in the surrounding matrix (Beier and Noss 1998). However, small areas of potentially harvestable forest were present within many of the larger non-harvestable patches (Figures 7.6b, c, d), contributing to the low percentages of interior non-harvestable areas (§7.2.2). Partially constrained areas were often adjacent to fully non-harvestable land, including areas around riparian reserves (Figures 7.6a, d) and beside larger non-harvestable patches (Figures 7.6b, c). Partially constrained areas provided potential increased connectivity between non-harvestable patches in some landscapes (Figures 7.6b, c).

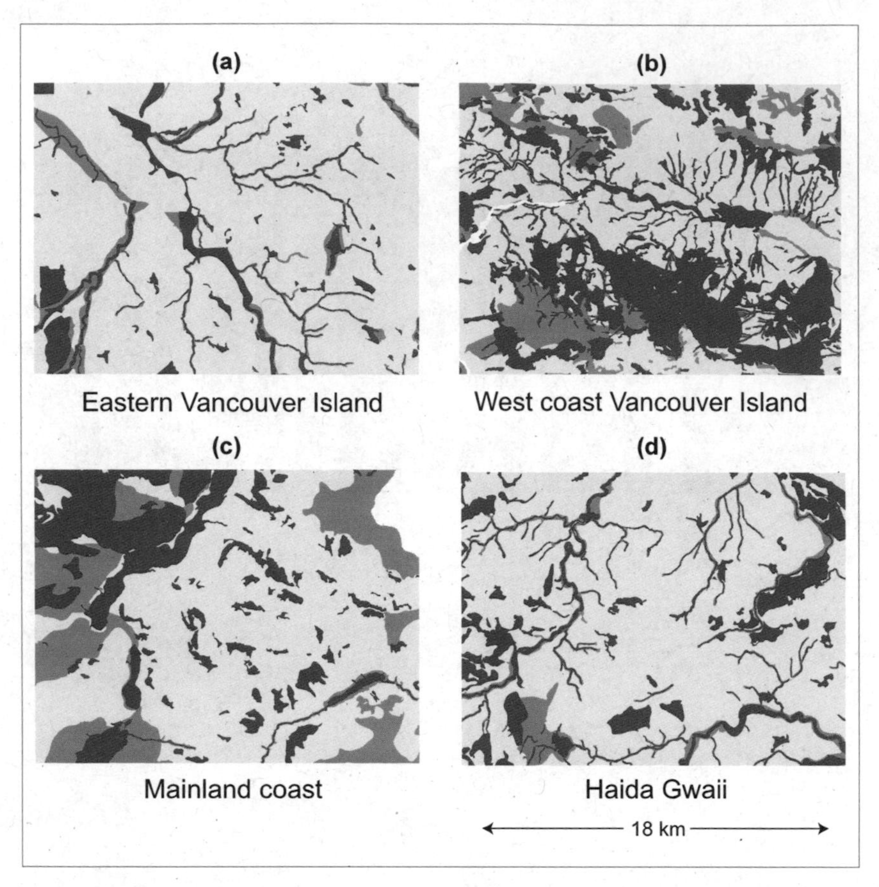

Figure 7.6 Spatial distribution of non-harvestable (dark gray) and partially constrained (pale gray) land in four 240-km^2 areas of the study tenure.

7.3.5 Representation in Other Conservation Designations

Areas with conservation emphasis in the BC government's land use plan for Vancouver Island and areas of interest to ENGOs showed the same pattern as non-harvestable areas and provincial protected areas: increasing percentages in the wetter ecosystem types (Figure 7.7). These trends reflect greater public environmental interest in "rainforest" than in rarer and less protected drier forests. Government planning and past protected area designations were further influenced by minimizing economic impacts on the productive, easily accessed, drier forests. The ecosystem bias was less pronounced for ENGO areas, which did include some areas in CDF and drier CWH ecosystems.

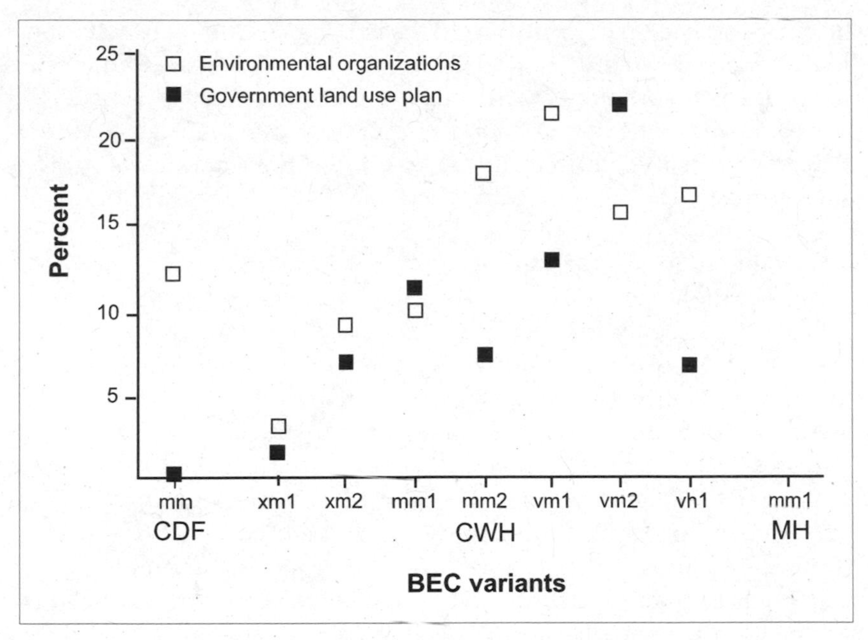

Figure 7.7 Representation of BEC variants on the Vancouver Island part of the tenure in special resource management zones of the government land use plan for Vancouver Island and in areas of interest to environmental non-government organizations.

7.4 Discussion

The 25.7 percent non-harvestable forested area on company tenure considerably exceeded the proportion of coastal British Columbia that was in official protected areas at the time of the analysis (17.2 percent). Even the 14.2 percent of commercially productive forest that is non-harvestable likely exceeds the percentage of productive forest that is officially protected. Non-harvestable areas can therefore have a substantial role in regional conservation. In some priority ecosystem types, such as the two Haida Gwaii variants on the tenure (Figure 7.1), non-harvestable areas contribute almost all the unmanaged area. No provincial or federal parks include these variants. Accounting for the non-harvestable land base is an important coarse-filter component of environmental planning for managed forests, because it helps to indicate where potentially expensive conservation actions would be most beneficial. Limited resources for conservation should not be spent on ecosystems with high *de facto* levels of protection. From a company's perspective, the cost of more conservative management in poorly represented ecosystem types could be offset by fewer restrictions on management in the harvestable part of ecosystems that are well represented in the non-harvestable land base (e.g., block size and adjacency constraints). Examining contributions that non-harvestable land can make toward meeting ecological objectives, and designating them as core areas for an overall conservation strategy (Noss et al. 1999), recognizes the values of non-harvestable land and can help to ensure that the areas remain unharvested as markets, technologies, and social pressures change.

7.4.1 Limitations of Analysis

Because of the assumptions underlying our analysis, ecological representation is intended only as a coarse-filter approach to maintaining biological diversity, to be used in conjunction with finer-filter monitoring of habitat and indicator organisms. The major assumption of the representation analysis is that the ecosystem types we used truly represent the full range of organisms of concern (Belbin 1993; Pressey et al. 2000). BEC variants are unlikely to be a perfect surrogate because at least some organisms will use only parts of the gradient of productivity and moisture within a variant (Figure 6.1). Although the subsample area with mapped site series suggested a lack of bias toward specific site series within the non-harvestable part of a variant, this condition may not apply in all variants, particularly those with low overall representation. Completing site series mapping and regaining the ability to do these analyses for the whole tenure comprise an immediate priority to allow this finer-level analysis of representation.

Special ecosystem types that are not well differentiated in the BEC system are an additional concern for ecological representation. Although most of these types are non-forested, such as wetlands or natural dry openings, a few special forest types are recognized operationally. An example is forest on karst formations, some of which are reserved to protect associated cave systems (Province of British Columbia 1994). Mapping of other special forested ecosystems that have been identified in the region is spotty, often associated with road access and therefore biased toward the harvestable land base. Developing methods for broader-scale, unbiased identification of these rare ecosystem types would allow an important refinement to the analysis.

A further limitation of our analysis is that, while the company's tenure is large, its boundaries usually do not correspond to ecological boundaries. Representation should be evaluated in a regional context because biological diversity does not recognize administrative boundaries (Beier and Noss 1998). In addition to provincial responsibility and official protected areas that have been addressed here, a regional view would analyze the representation of each ecosystem type in unmanaged areas across all tenures in the area, including private lands, public lands, and protected areas. Data availability limits such an approach, particularly as many variants of most concern cross the Canada-US border. A recent attempt to analyze representation across Vancouver Island had limited success, with major stumbling blocks being lack of access to data from private land and the inability of the government to provide coherent data sets for the large portion of public land and parks.

7.4.2 Management Priorities: Under-Represented Dry Variants

Despite these limitations in the analysis, two obvious weak points in representation on the tenure should be the focus of management actions.

1. Under-represented ecosystems are drier, southerly variants on Vancouver Island, several of which should be conservation priorities for the company based on high responsibility in the province and low percentages in official protected areas. This major bias in representation follows a global pattern of poor protection of lowland and drier ecosystem types (e.g., Awimbo, Norton, and Overmars 1996; Powell, Barborak, and Rodrigues 2000; Sierra, Campos, and Chamberlin 2002).
2. Limited amounts of non-harvestable forest are distant from potentially harvestable stands. The lack of interior area is most pronounced in the poorly represented dry, southerly variants, exacerbating concerns about those ecosystem types.

Designations of Old-Growth stewardship zones, while initially intended to accommodate public concern for old forest, provide the surest tool for improving the representation of poorly represented ecosystem types. One limitation to using the zones for this purpose is the expectation that they will contain large areas of forest that are currently old. Such areas do not exist in the drier ecosystem types that are of highest conservation concern because of logging history, fires associated with settlement, agriculture, and current development. Some progress has been made toward establishing smaller Old-Growth zones, allowing a modest increase in non-harvestable area in a few variants that had moderate representation. However, no Old-Growth zones or other reserves have yet been established in the small remnant areas of old forest in the dry southerly variants or in larger areas of mature second-growth forest in those ecosystem types.

Areas of interest to ENGOs in the drier variants focus on the remnant old patches. Designating small Old-Growth zones in some of these areas would improve ecological representation while accommodating a significant social concern. Reserves required for featured species, such as ungulates, marbled murrelets, or northern goshawks, could be located preferentially in the poorly represented variants to provide the additional benefit of improving representation. However, government regulations generally provide little flexibility in locating these species-specific reserves.

A less certain option to mitigate poor representation is to increase the level of retention in managed stands in the drier zones, including more and larger retained patches. This option recognizes that, with low levels of non-harvestable forest, stand-level management is being relied on to maintain biological diversity in these ecosystem types. Greater retention at harvest, preferably in larger groups, should increase the likelihood of maintaining sensitive species in the managed stands, but the outcome is uncertain. More extensive monitoring of a wide range of organisms in managed stands would be required if within-block retention were the sole option used to address the problem of lack of unmanaged forest in these ecosystems. Little progress has been made so far in increasing stand-level retention in the poorly represented variants.[4]

The essential difficulty in improving representation of the drier, southerly variants within the tenure is that the area is private land. Historically, the area was chosen as a private land concession to finance railway building on Vancouver Island in the 1800s. Generally, it is easily accessed, with lower operational constraints than public land, and is profitable for timber harvesting. Consequently, these areas are the most consistently profitable operating areas on the tenure. Although the marginal value of conservation actions is also highest in these poorly represented ecosystem types, there are few market mechanisms to reflect this value. A possible solution involves active restoration of old-growth characteristics in some of the abundant second-growth stands in these ecosystems (Krcmar 2002; Perry and Muller 2002). The recovery of some timber value through commercial thinning as an initial restoration step could offset some costs of setting restoration areas aside as reserves. Conservation covenants purchased by environmental or government groups would further reduce the economic costs of creating permanent reserves in these high-value areas. However, the compensatory costs are high, and no action has yet occurred to implement these ideas other than the studies noted above.

7.4.3 Management Priorities: Edge Effects

The second major concern – low proportions of non-harvestable areas that are distant from potentially harvestable stands – can be mitigated in several ways that have fewer direct conflicts with economic objectives than does harvesting less on the most profitable sites.

- VR patches can be located adjacent to non-harvestable areas as buffers to reduce the high proportion of non-harvestable area near harvestable edges. This action applies especially to riparian reserves, which are a main component of the non-harvestable land base in the driest variants, where almost no interior non-harvestable forest exists.

- Retention stands with higher overall retention levels can also be used to buffer adjacent non-harvestable areas.
- In some cases, forgoing harvesting in a few small harvestable areas embedded in larger non-harvestable areas would produce disproportionately large increases in the amount of non-harvestable interior.
- Areas reserved from harvesting to meet the objectives of partially constrained areas can also be preferentially located adjacent to non-harvestable areas.
- At a larger scale, harvest locations and timing can be planned to maintain older managed stands adjacent to at least part of each non-harvestable area through time rather than isolating non-harvestable areas with surrounding young stands at some point in the harvest rotation.

To date, some progress has been made in using stand-level reserves to buffer small or narrow non-harvestable areas within stands, but larger-scale planning to mitigate edge effects in the larger non-harvestable land base has not been implemented.

7.4.4 Focusing Finer-Filter Monitoring

A third group of recommendations emerging from the results of representation analyses focuses the finer-filter monitoring of habitat structures and indicator organisms on the poorly represented ecosystems, where stand management decisions will have the greatest conservation importance. The finer-filter monitoring may also be modified to test some of the assumptions of the representation analysis, including

- determining whether commercially non-productive stands include habitat structures and organisms similar to those of productive stands, particularly in ecosystem types where these non-commercial stands are a dominant part of the non-harvestable land base;
- adding information on site series to all monitoring plots, including other inventories in the region, to allow the eventual delineation of ecosystem types based more directly on assemblages of organisms;
- assessing whether poorly represented sites with high site index values have distinctive habitat structures and biota; and
- measuring edge effects into non-harvestable areas with adjacent harvested stands of different ages to test buffer distances used to define interior conditions.

Some initial work has been done on the last point, but the others are still pending. Ultimately, continued monitoring of a range of organisms across the tenure is required to ensure that the goal of maintaining all native species is being met by the combination of non-harvestable areas, designated Old-Growth zones, and improved stand-level practices.

7.5 Summary

Analysis of Indicator 1 (ecosystem representation) across the entire case study tenure had to rely on relatively broad ecosystem types. The major findings are listed below.

- Overall, 25.7 percent of the forested area was fully constrained from harvesting, while an additional 7.3 percent was partially constrained. The percent of forest reserved within the tenure considerably exceeded the percent in official protected areas in coastal British Columbia (17.2 percent).
- Representation was poor, with high-priority drier ecosystems on southern Vancouver Island and highly productive sites under-represented in non-harvestable stands (Figure 7.2).
- Areas currently designated as Old-Growth stewardship zones made minor incremental contributions to ecological representation because they occur mainly in areas that are already operationally constrained. Most fully constrained forest was of low productivity (Figure 7.3). Uneconomic areas made little contribution on their own, so market changes that permitted greater access to these areas would have little effect on representation.
- The same trend toward reserving colder, wetter, and less productive sites is apparent in other approaches to conservation developed for the region, but the bias is less pronounced in areas of interest to environmental groups (Figure 7.7).
- Potential edge effects were a concern, with 62 percent of non-harvestable area within 50 metres of harvestable stands (Figure 7.5). The high value arises in part because large portions of the non-harvestable area were often long, narrow riparian reserves. Small harvestable areas embedded within larger non-harvestable areas also contribute to the high amount of edge.
- About 16 percent of the tenure could be analyzed using finer scales (site series groupings) to distinguish ecosystem types. In that subsample, ecosystem representation of site series within variants approximated ideal (Figure 7.4).

Options for planning and practice could mitigate the two major concerns: under-represented drier ecosystem types, and the lack of forest interior away from edge. Relatively little progress has been made in improving representation on the ground. Monitoring Indicator 1 also demonstrated ways of focusing finer-filter monitoring.

8
Sustaining Forested Habitat

David J. Huggard, Fred L. Bunnell, and Laurie L. Kremsater

8.1 Rationale

For biological and practical reasons, we need to identify major structural features of forests that support biological diversity. Biologically, the changes in species composition through time occur primarily in response to changing forest structure rather than to time itself. Trees are particularly long lived and tall, so forests can generate diverse structures. Practically, forest managers are limited to simple approaches (stands are removed in whole or in part; trees are regenerated naturally or by humans) and to simple indices to guide harvest planning (simple enough to cover the large areas necessary in planning or monitoring). To sustain biodiversity more directly, we need a way of selecting a short list of habitat features that connect directly with species and are manipulated by forest practices.

Indicator 2 is the intermediate filter in the three-level monitoring program. Indicator 1 addresses the representation of different ecosystem types in unmanaged and specially managed areas (Chapters 6 and 7). Indicator 2 addresses kinds, amounts, and variability of forest structure believed necessary to sustain organisms in managed stands and landscapes. Like Indicator 1, maintaining diverse habitat and landscape structures is intended to provide for a broad range of organisms, including many that are poorly known.

We discuss three aspects:

1 habitat elements: discrete elements of a stand, such as large snags, shrubs, or downed wood, that provide necessary habitat for specific species;
2 habitat structure: integrative indices of habitat that combine measures of habitat elements; and
3 landscape features: an effect of two or more stands that differs from the effects of the separate stands and influences abundance, dispersal, and distribution of species.

8.1.1 Habitat Elements in Stands[1]

Much of the approach to describing habitat elements at the stand level derives from the comprehensive review of wildlife habitat in the Blue Mountains of Oregon (Thomas 1979). That volume institutionalized a set of habitat elements as "the canon" for describing wildlife habitat. The elements have since been measured in many studies of wildlife habitat and form the basis for most biodiversity monitoring in the Pacific Northwest (e.g., the Northwest Forest Plan; Mulder et al. 1999). This standard set of habitat elements includes large live trees, snags, down wood or coarse woody debris (CWD), canopy cover, shrub cover, and ground cover layers.[2]

The value of the standard elements is supported primarily by studies of vertebrate habitat use (e.g., Table 8.1), but they have been used to describe habitat relations of taxa such as insects, fungi, and bryophytes. For example, most of the 1,457 threatened species of mosses, lichens, macrofungi, vascular plants, invertebrates, and vertebrates in Sweden were related to the occurrence of specific habitat elements, such as old trees, logs, and snags (Berg et al. 1994; see also Bernes 1994; Bunnell et al. 2002; Goward and Arsenault 1997). As more is learned, the set of elements and their attributes desirable for monitoring will likely expand beyond this standard set.

Monitoring the standard set of habitat elements to evaluate effectiveness in sustaining biological diversity has obvious advantages. After hunting and disturbance were eliminated, almost all forest-dwelling vertebrate species designated "at risk" by government agencies in the Pacific Northwest were associated with one or more of the habitat features of Table 8.1 (Bunnell, Kremsater, and Wind 1999). Because so many wildlife studies have used these elements, we can relate a large literature on stand-level habitat relationships to the results of our monitoring (e.g., Bunnell, Kremsater, and Boyland 1998; Bunnell, Kremsater, and Wind 1999; Laudenslayer et al. 2002). We can also compare our results to other similar monitoring in the region or in similar forest types. Additionally, at least some of the parameters needed to project

Table 8.1

Use of habitat elements by native forest-dwelling vertebrates in the three forested biogeoclimatic zones of the study tenure

BEC Zone[1]	Total species	Cavities	Downed Wood	Shrubs[2]	Hardwoods[3]
CWH	169	26.6	16.0	23.1	28.4
MH	109	26.6	18.3	22.0	21.1
CDF	139	30.9	13.0	22.3	30.9

1 Biogeoclimatic zones of the area are described by Green and Klinka (1994). CWH = Coastal Western Hemlock; MH = Mountain Hemlock; CDF = Coastal Douglas-Fir.
2 Birds may nest low in trees (<3 m) surrounded by understorey.
3 For birds, at least two-thirds of reported nests were in hardwoods (hardwoods and associated bird species are uncommon in the MH zone).

these elements through time have been measured by other studies (e.g., for snags, Daniels et al. 1997; Harmon et al. 1986; Morrison and Raphael 1993).

Despite the attractiveness of the standard set, we must recognize some perils in simply adopting a standard set of habitat elements when our objective is to reduce negative effects of forest harvesting on the full spectrum of organisms. One concern is that the standard set of habitat measurements was initially based on requirements of terrestrial vertebrates, so the "short list" of habitat elements will always be incomplete. We may overlook habitat elements that are critical to non-vertebrates that we wish to accommodate. The problem is insidious because our familiarity with the elements defined in Thomas (1979) determines how we design our habitat studies and probably even how we perceive "habitat" in forest stands. That condition reinforces a belief that habitat *is* the canonical set of habitat elements. Reviewing natural history studies of organisms, particularly non-vertebrates, may help to avoid any circularity in using a standard set of habitat elements to define habitat and to challenge the adequacy of those elements. Ideally, we should try to define habitat features through the perspectives of a wide range of organisms. For example, some species of calicioid or stubble lichens are found only on the undersides of leaning snags with no bark (Rikkinen 1995, 2003a). This habitat element has not been identified as important from vertebrate studies and is unlikely to be surveyed in general monitoring of snags, where lean is usually not recorded.

In summary, we should include the standard set of habitat elements derived from Thomas (1979) in our habitat monitoring because it offers the benefits of relating to a large existing literature and conceptual and practical ease of sampling. We should also evaluate other elements (or other attributes of standard elements) that could aid monitoring, based on studies of non-vertebrates.

8.1.2 Habitat Structure in Stands

A second concern with relying on the standard set of habitat elements is that they enforce a reductionist, point-scale view of habitat. That is, the elements themselves are considered as separate, discrete units. Many times, combinations of features are more revealing, such as snags *plus* downed wood for woodpeckers. Natural historians – bird watchers, botanists, mushroom pickers – intuitively use a more integrated sense of habitat when looking for certain organisms. Their "habitat" is larger than the individual habitat element and smaller than a typical stand. The habitat in an area of 0.01 hectare to 1 hectare conforms better to our sense that organisms respond more to habitat *structure* than to separate measurements of individual *elements*.

Because a portion of this variability in stand structure occurs as a consequence of stand development, structural stage is the simplest integrative

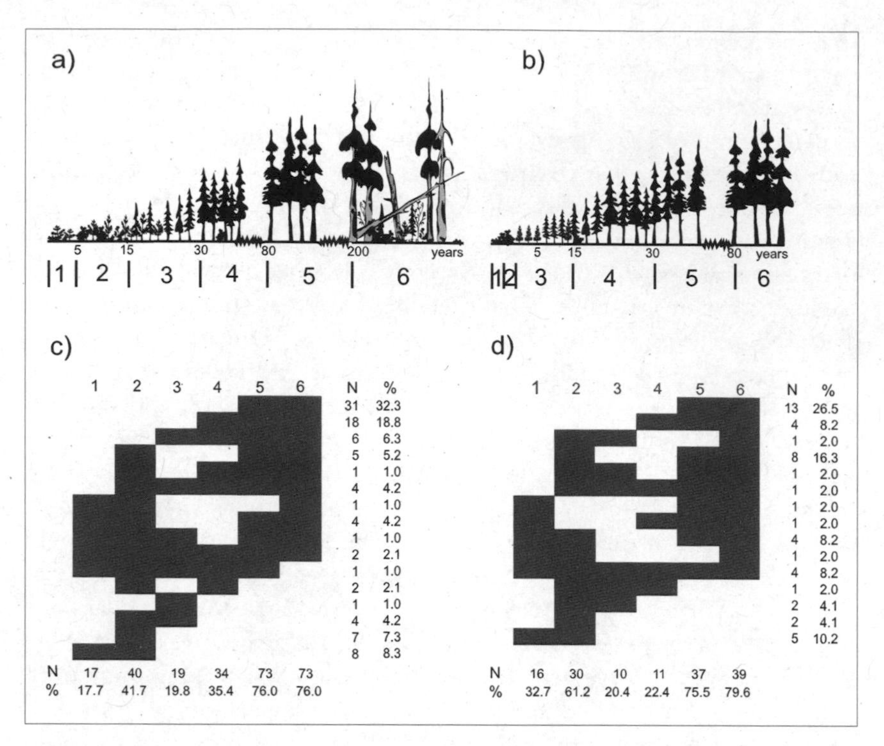

Figure 8.1 Breeding habitats of forest-dwelling vertebrates as a function of seral or structural stage.

Notes: Seral/structural stages in (a) an unmanaged forest after a stand-initiating disturbance; (b) a managed forest subject to even-aged management.

The seral stages are 1 = grass-forb; 2 = shrub-seedling; 3 = pole-sapling; 4 = young; 5 = mature; 6 = old growth. The shaded areas in (c) and (d) represent the seral stages of a Coastal Western Hemlock forest where (c) forest-dwelling birds and (d) mammals are likely to breed at moderate to high abundance. N = number of species.

Sources: Schemata adapted from Thomas (1979) by L. Kremsater; (c) and (d) from Bunnell, Kremsater, and Wind (1999).

measure for even-aged management or stand-replacing natural disturbances. Broad relations between vertebrate species and seral stage have been known for decades (e.g., Bunnell and Eastman 1976). The dynamic nature of vertebrate abundance in forests of different structure is illustrated in Figure 8.1.

Intensive even-aged forest management (e.g., clear-cutting) shortens the duration of early and later seral stages (compare Figures 8.1a and b) and extends the relative duration of stages 3 and 4, which are least rich in vertebrate species (Figures 8.1c and d). Most of the changes in species richness with stand age can be attributed to the diversity of structure in stands of different ages, with the middle age classes contributing the least to structural diversity. Structural retention systems, such as "green-tree retention" (FEMAT 1993; Franklin et al. 1986) or "variable retention" (CSP 1995b; Mitchell and Beese

2002), can maintain some of the structural complexity of old stands in early- or mid-seral stages.

Two points are apparent. First, given the wide range of structural stages to which vertebrates have adapted, it is important not to enact the same practices everywhere. The challenge is to provide the range of structural diversity naturally encountered by species. That means providing heterogeneity within stands, a diversity of stand types in local landscapes, and a diversity of landscape types regionally. Second, stand age is not a consistently useful integrative variable for assessing habitat suitability. Few species respond to age or time itself; rather, they respond to vegetation structure as it changes over time.[3] Neither stand age nor measures of individual habitat elements document all the variability that encourages species richness within stands.

In practice, reliable integrative measures are elusive. The holistic notion of habitat structure invoked by field naturalists is much more difficult to define, let alone quantify, than are separate habitat elements. Nonetheless, we contradict our natural history experience and may mislead ourselves by monitoring and managing *only* for the individual elements. Indices of structural complexity (e.g., foliage structural diversity; MacArthur and MacArthur 1961) and horizontal variability (patchiness) may provide one practical approach to monitoring integrative habitat structure. Particular measurable arrangements of elements may also be deduced from natural history studies. For example, the leaning older snags apparently required by some calicioid lichens must persist for a number of years to allow lichen colonization, growth, and reproduction. That can occur only where the snag is surrounded by larger, stable, supporting trees. Monitoring the number of larger trees and the number of snags in a block will not indicate whether the necessary spatial arrangement of the elements also occurs.

8.1.3 Landscape Features

Landscape is used here in a broad sense, including any features created by two or more relatively homogeneous stands or measurements made over an area larger than a typical stand – often referred to as landscape context. Examples of the former are edge length or percentage of "interior" forest, which only make sense with two or more juxtaposed stands. "Amount of early-seral forest in a 5 kilometre radius" is a landscape variable of the latter type because the scale of measurement is large compared to our typical concept of a stand. The important point in both cases is that we *cannot* predict the importance of the feature from stand information alone. That is obvious in the case of an edge effect, where one stand affects the value of another. In the case of the amount of early-seral forest in a 5 kilometre radius, an assumption is that there is some effect at the broader scale that is more or less than just adding up the effects of the individual early-seral stands. For example, an early-seral species might be more common in any given early-seral stand if there is a lot of that stand type in the larger area.[4] If there were no such additional effects, we could simply add up the values for each stand and ignore landscape variables.

Because stands are taken to be homogeneous units for landscape measurement, stands can be reduced to types, allowing GIS analyses. Computer measurements are easier than field measurements, resulting in hundreds of different landscape metrics (McGarigal and Marks 1995; O'Neill et al. 1988) and infinite possible tabulations over arbitrarily large areas. The problem when selecting landscape variables to monitor as a broad filter for biological diversity is that most available variables are

- organism specific (e.g., percent interior [Murcia 1995] or connectivity [Beier and Noss 1998]), which may not be a problem when managing for individual species but thwarts attempts to use those variables as medium-filter measures for a wide range of species;
- not related to any organism in empirical studies (many examples);
- lacking intuitive natural history basis (e.g., fractal dimension); or
- unsupported by either empirical studies or intuitive natural history (e.g., contagion).

Empirical studies of landscape-level effects on species also commonly fail to account for the fact that organisms are associated with particular habitat types, and these habitat types are not located randomly in the landscape. The confounding of simple habitat selection and non-random spatial distributions of habitat types leads to misleading relationships between organisms and landscape variables. For example, a species that prefers recent cutblocks will tend to be found in landscapes with higher percentages of young stands, more edge, greater contrast, lower fractal dimension, et cetera, simply because each of these features is associated with more intensively managed landscapes. The association will occur even if these other landscape variables are irrelevant to the species. Large sample sizes and designs that avoid known confounding variables allow this design problem to be overcome (McGarigal and Cushman 2002), but it is unlikely that enough suitable landscapes for such a design actually exist in most areas. Studies comparing unconfounded landscapes differing only by pattern are rare and require simplified landscapes. Variance partitioning (Borcard, Legendre, and Drapeau 1992) can assess the confounded effects of stand and landscape variables, but we expect that without carefully designed sampling, independent landscape effects will usually be small compared to stand-level effects (e.g., Cushman and McGarigal 2004).

The biological relevance of landscape variables is further reduced in forested landscapes, where succession occurs after harvest. A successional gradient obscures the concept of stand "type" required for GIS-type analyses.

Some silvicultural systems (e.g., selection or variable retention) deliberately reduce structural differences among stand types and reduce perceived differences between recently cut and older stands for many organisms.

Given the ample opportunities for confusion, it seems unlikely that we can find a set of variables that reliably indicates how well a dynamic forest landscape is sustaining the full range of organisms. Instead, we suggest a combination of two approaches for monitoring landscapes:

1 tracking a few simple variables that have been shown to have relevance to many organisms, such as edge density, age distribution (recognizing the need to track even-aged and multi-aged stands separately), and road density; and
2 using spatial models for a range of indicator taxa to evaluate the dynamic, organism-specific aspects of landscapes.

It is clearly more complicated to conduct spatial modelling for a range of organisms, many poorly known, than to derive simple GIS indices. Both are required to confront the contradiction between two competing concepts: the first, that we need to measure a range of organisms to understand contributions made by stand-level silvicultural treatments (Chapter 10); the second, that we can monitor the biodiversity value of complicated landscapes with simple map indices.

8.2 What to Monitor

Our discussion of what to monitor and appropriate methods treats five broad groups of variables: (1) the standard set of habitat elements and their attributes, (2) more integrative habitat variables describing habitat structure, (3) processes critical to making long-term habitat projections, (4) landscape-scale summary variables, and (5) hypothetical species as landscape indices.

8.2.1 Standard Habitat Elements and Their Attributes

The standard habitat elements are the six broad elements that have become common to many wildlife habitat studies and stand-level habitat-monitoring programs. They include live trees, snags or standing dead trees, coarse woody debris or downed wood, canopy cover and composition, shrubs, and ground cover.[5] These elements and their attributes are important because of their influences on vertebrates and other taxa and because they can be readily related to many other studies. Below, we list habitat elements and their attributes recommended for monitoring. For each element, we suggest specific attributes that should receive initial priority in evaluating forest practices, briefly noting how these attributes provide habitat for various organisms.

Live trees define forest habitat and are the source of deadwood elements. Important attributes of live trees include large trees, hardwood or deciduous trees, advanced regeneration, and a diversity of sizes and conditions. Relations of these attributes to other organisms include the following.

Large (>50 cm dbh[6]) live trees

- have complex crowns providing abundant and diverse foraging sites for foliage gleaners and more sheltered microclimates, particularly in winter;
- have large and old branches providing nesting and perching platforms and long-term substrates for epiphytic plants, lichens, and fungi;
- have rough bark providing foraging for bark gleaners and substrate for epiphytic plants and lichens;
- have large canopies providing greater snow interception for winter ungulate habitat, small mammal and carnivore access beneath the snowpack, and unique early spring conditions for some vascular plants;
- are older, encouraging heart rot decay organisms that produce critical structure in subsequent dead wood;
- are the only source of large-diameter snags and logs, which are needed by some organisms (e.g., black bears, Vaux's swifts) and favoured by many others (e.g., small mammals, epixylic[7] insects, and fungi), and which persist far longer than smaller pieces; and
- potentially produce many more cones required by squirrels, seed-eating birds, and many species of seed beetles.

Deciduous and mixed deciduous-coniferous patches where they occur

- provide favoured nesting sites for several cavity-nesters;
- provide habitat for many folivorous insects and insectivorous birds and small mammals;
- host unique fungal and lichen associates and increase lichen productivity of adjacent conifers;
- are generally shorter lived and faster decaying, thus producing snags and rotting logs earlier in the rotation; and
- produce a rich litter that encourages some shrews, amphibians, gastropods, and other invertebrates.

Advanced regeneration (trees >4 cm dbh but less than merchantable size) retained in harvested areas

- produce variability in the regenerating stand;
- increase the rate of production of dead wood in mid-rotation, when much of the older wood retained at harvest will have decayed;

Figure 8.2 Deciduous trees provide cavity and foraging sites, host unique epiphytes, and produce litter that creates favourable foraging opportunities for shrews (*Sorex*), amphibians (e.g. *Rana aurora*), gastropods (e.g. *Crypromastix germana*), and other invertebrates. | *Photographs by: forest, M. Preston; shrew and frog, F. Bunnell; and snail, K. Ovaska and L. Sopuck.*

- moderate ground level microclimate and provide more vertical structure to help small animals cross harvested areas to recolonize other areas;
- provide substrate for epiphytes earlier in the rotation, improving chances of a "lifeboat effect"; and
- provide shading for small ephemeral streams not otherwise protected.

Although retained advanced regeneration may be highly susceptible to mistletoe or root rot, in some situations, that will not necessarily reduce the ecological benefits of these trees. It may, however, lead to concern for the regenerating crop trees.

A diversity of tree sizes and conditions

- may be important for particular species but also should be encouraged because a variety of size classes will lead to a more structurally complex forest that increases species richness;
- can provide important habitat features associated with disease (e.g., nesting and denning in mistletoe brooms, families of flies and beetles restricted to bracket fungi); and
- can create stand variability and particular deadwood structures.

Snags (standing dead trees) are reduced in abundance and size by intensive forestry. Ecological reasons for maintaining snags in forests are well documented (see the review in Bunnell et al. 2002). Important attributes of snags include some large (>50 cm dbh) and tall snags within a range of species, size classes, and conditions. These attributes should be emphasized during snag retention and monitoring. Relations of these attributes to other organisms include the following.

Large, old, and more recent large snags (plus large live trees to replace them)

- provide dens for large carnivores (black bears, martens, fishers);
- likely contain heart rot, producing hollow snags used by nesting swifts and roosting bats;
- provide woodpecker nest sites when not well decayed and allow higher, less vulnerable nests;
- stand longer, allowing colonization and development of slow-growing lichen and fungal species and contain a buffered microclimate, required or favoured by some insects, including carpenter ants (a common forage species).

Patches with abundant moderate-sized snags

- provide nest sites for smaller cavity-nesters;
- provide foraging sites for woodpeckers (more often limiting than nest sites) in clusters that some species prefer;
- encourage persistence of wood-inhabiting lichens and fungi by providing more substrates and colonization possibilities nearby; and
- provide more widespread CWD later in the rotation than do a few large snags.

A range of tree species produces snags with different structural characteristics, encouraging use by different organisms.

Figure 8.3 Dead trees or snags provide denning, roosting, and nesting sites for many vertebrate species; substrate for insects on which many vertebrates feed; and substrate for various epiphytes, especially mosses and lichens. | *Photographs by: snag with cavities, W. Beese; lichen on snag, I. Houde; and moss on snag, F. Bunnell.*

- Cedar snags persist longer than other species and frequently produce large, hollow snags used by swifts, bats, and denning black bears.
- True fir snags are often hard on the outside and soft inside, ideal for nesting by sapsuckers and other primary cavity-nesters.
- Hardwood snags are ideal for nesting by woodpeckers and smaller cavity-nesters and host unique communities of lichens, fungi, and wood-boring beetles different from those on conifer snags.

Compared to short snags, tall snags (taller than the heights of 3 metres or 5 metres commonly allowed by safety regulations in work areas)

- are preferred nesting sites of snag-nesting birds (short snags or stubs are used for various purposes, but nests near the ground are more likely to be depredated);
- offer more total substrate for epixylic plants and lichens and better opportunities for them to disperse; and
- produce longer pieces of CWD, possibly important for ground-level connectivity later in the rotation.

Snags with special features required by little-known organisms

- are poorly studied but likely exist (e.g., epixylic calicioid lichens noted earlier). Review of the natural history of non-vertebrate snag users may suggest other snag attributes that should be measured during effectiveness monitoring.

Coarse woody debris is used by rodents, small carnivores, and amphibians. The use of coarse woody debris by these organisms is well documented. CWD is also critical for other taxa, both directly as substrate and through its influence on ecosystem processes, such as nutrient cycling, water retention, slope stability, stream morphology, microclimate effects, and contributions to plant-fungus interactions. CWD is the most likely of all habitat elements to appear abundant immediately after harvest but to become a limiting factor later in the rotation. Projection models are therefore particularly important for CWD. Important attributes for evaluating monitoring results include large size, a range of decay stages, and variability in the density or distribution.

Compared to small logs, large-diameter pieces

- provide preferred cover or shelter to small mammals and larger vertebrates;
- provide more breeding substrate for terrestrial-breeding amphibians;
- provide more constant internal microclimate, favouring many species of fungi and invertebrates;
- have slower decay rates, allowing large pieces to persist later into the rotation; and
- have longer persistence, greater stability, and height above other vegetation, thereby favouring many plants and lichens that grow on dead wood (for a given total volume of CWD, smaller pieces of CWD provide more surface area and ground coverage than large pieces, and that can favour other species, e.g., truffles).

A range of decay stages

- is desirable because different organisms use CWD at different decay stages and because succession of communities within individual logs occurs naturally; and
- helps to sustain biological diversity by providing intact, well-decayed wood shortly after harvest and recruitment of new CWD later in the rotation.

Variability in CWD density and distribution is desirable because

- high volumes of well-dispersed CWD are favoured by bryophytes and fungi;

Figure 8.4 Downed wood provides habitat opportunities for invertebrates, mosses, liverworts, lichens, and fungi. Once it becomes well rotted, it hosts specialized plants, such as this Indian pipe (*Monotropa uniflora*). | *Photographs by: fungus, W. Beese; and Monotropa and down wood, F. Bunnell.*

- overlapping logs can provide rest sites for martens and access beneath the snow for some mammals in winter;
- excessively high densities, especially just after harvest, can exclude some early-seral plants from establishing and can physically impede larger mammals (though that may aid some plants sensitive to herbivory);

- patches with low CWD volumes provide foraging areas for some species and exposed soil substrate for others; and
- simply piling and burning destroys nutrient capital and an important habitat element for many species.

Canopy is directly altered by forest practices. Structural variables that involve canopy measurements over an area larger than a small plot are discussed in §8.2.2. In small plots, the influential canopy attributes include dense patches, deep crowns, and hardwood components. Relations of these attributes to other organisms include the following.

Dense patches

- provide favoured cover for several vertebrates, including martens and wintering blue and spruce grouse; and
- intercept more snow to provide winter habitat for ungulates and other herbivores.

Deep crowns

- provide more substrate for foliage- and bark-gleaning birds and foliage-eating birds and insects; and
- provide vertical stratification that permits a complex epiphytic lichen community within individual trees, including sheltered microsites on branches lower in the canopy.

Hardwood components where present

- improve foraging opportunities for many songbirds, some of which are closely associated with hardwood trees;
- provide seasonal variability in canopy cover, which can promote diversity of understorey plants;
- can sometimes enhance lichen growth and diversity on adjacent canopy conifers; and
- produce a nutrient-rich litter favouring many organisms.

Shrubs may be reduced in managed stands by vegetation management in early-seral stages or prevented from re-entering later stages by dense stocking and low mortality of the crop trees. Management may also change shrub species composition by maintaining shrubs only on those sites where they are thought not to compete with crop trees or by targeting particular site types for reserves. Attributes of shrubs that merit monitoring include dense, taller shrub cover and representative species composition.

Areas of dense, taller shrub cover

Figure 8.5 Dense, deep canopy intercepts snow on ungulate wintering ranges, provides a complex environment for rich epiphyte communities, and yields areas of bark and leaves for foraging. | *Photographs by: canopy, M. Preston; stand, A. Inselberg; deer, D. Shackleton; and lichen, F. Bunnell*

- provide nesting and rearing habitat for several songbird and grouse species;
- provide cover for small mammals, substituting for or complementing CWD;
- help to maintain the diversity of ground-dwelling arthropod communities, directly by providing cover and indirectly by discouraging ant colonies (which can reduce diversity of other taxa); and

- provide a moister microclimate with moderated temperature extremes that can promote or maintain some sensitive plant species, particularly some mosses.

Areas dominated by each major shrub species

- provide the range of seasonal forage values sought by ungulates and smaller herbivores;
- support a range of herbivores and sap-sucking insects;
- produce a wide range of berries and other fruits that will benefit a variety of organisms; and
- sustain evergreen, broadleaved shrubs that provide cover in snow-free areas during winter.

Figure 8.6 Shrubs provide cover for small vertebrates, forage for vertebrates such as this chipmunk and many invertebrates, and nesting sites for birds. | *Photographs by: devil's club, W. Beese; chipmunk, F. Bunnell; and Swainson's thrush and yellow warbler nests, R.W. Campbell.*

Note: Swainson's thrush nest above, yellow warbler on nest below.

Ground cover layers are affected directly by forest practices, especially site preparation, but over time probably are more influenced by changes in canopy understorey trees and shrub layers. The abundant debris following logging can also impede development of a diverse ground layer. Ground cover layers undoubtedly have strong effects on many species of small mammals, invertebrates, non-vascular plants, lichens, and fungi on the forest floor and in forest soils, but relatively little is known about what these taxa require. Hence, ground cover layers have received less emphasis in habitat monitoring and evaluation than other standard habitat elements. Chapter 10 addresses potential monitoring of forbs, bryophytes, and ground lichens as individual species (Indicator 3). Here we focus on the structural attributes of these species when grouped as ground cover. Priority attributes include dense patches of forb cover, diversity of forb cover, intact areas of moss or lichen cover, and depth and development of the litter layer.

Patches of dense forb cover

- provide the same benefits of overhead cover as do shrubs, allow local movements of small, ground-dwelling species, and reduce the dominance of species associated with disturbed ground.

Variety of species as dominant forbs

- provides habitat features for different species. For example, ferns can provide forage for ungulates and dense overhead cover for small mammals but appear to inhibit many species of fungi and are unsuitable food sources for the majority of herbivorous insects. No single forb species or group of species should be dominant over large areas.

Figure 8.7 These black-tailed fawns will grow rapidly if their mother has access to highly digestible forage such as fireweed. | *Photographs by: fawns, F. Bunnell; and fireweed, M. Preston.*

Areas with intact moss cover (or terrestrial lichen cover where it occurs)

- may be associated with low gastropod abundance and richness but
- make poorly understood contributions (Brown and Bates 1990; Nadkarni 1984a; van der Wal, Pierce, and Brooker 2005), so they should not be substantially reduced or disrupted everywhere in managed forests.

Deep and developed litter layers

- support high gastropod abundance and richness;
- support insect, millipede, centipede, and arachnid foraging, thus encouraging shrews and some amphibians; and
- provide substrate for complexes of fungal mycorrhizae, truffles, and their associated arthropod communities.

8.2.2 Integrative Habitat Variables

The preceding habitat features are separate elements. Their relations with organisms are most likely expressed through a critical single resource (e.g., suitable nest site) or in a habitat model in combination with other elements (e.g., multiple regression models). A complementary approach to stand-level habitat monitoring is to identify and monitor more integrative habitat structures. These structures may be less directly related to the needs of particular species but help to maintain broader communities or a greater richness of species. Some potential integrative variables synthesizing habitat elements over larger areas can be derived from the habitat-element plots, with minor modifications. Others require additional measurements (§8.3.2).

Vertical structural diversity was one of the first synthetic variables used to explain differences in richness of forest bird assemblages. A diversity index applied to the proportion of different height layers that contained plant foliage was correlated to the number of bird species, independent of plant species composition. The height layers included forbs and shrubs as well as trees.

Horizontal variability in size and density of canopy trees is a feature of older natural forests that is reduced, often dramatically, in intensely managed stands. Canopy gaps, dense patches of trees, and internal soft edges are lost in more homogeneous stands. All three features of natural forests are associated with particular forest-dwelling species and are more generally desired on the basis that variability supports species richness. Monitoring of horizontal variability could use summary measures of spatial variability in tree size and density derived from multiple plots within a stand, provided the same plot size and shape were used in all stand types being compared (measures of variability are sensitive to plot size and shape). A complementary approach is to conduct broader surveys of intact or disturbed gaps and clusters of trees, where these structures are clearly defined so that they can be reliably identified in the field or on air photos.

Horizontal variability of other stand elements besides canopy trees also promotes species richness. Thus, variability in CWD volumes, shrub cover, or other habitat elements within and between stands helps to ensure suitable conditions for the diverse organisms using these elements in different ways. Variability can be measured by indexing the variation of measured habitat elements among subplots within stands and among replicate stands of a particular forest and harvest type.

Special ecosystem types within stands appear to greatly enhance the overall diversity of the surrounding stand. These sites include small wetlands, headwater streams, seepage areas, small landslide areas, patches of noncommercial deciduous species, rocky outcrops, and highly productive mesic sites. The contribution to diversity is not just from the site itself but also from its interaction with the adjacent forest. Monitoring can indicate the extent to which such special sites are retained with intact, or partially intact, surrounding forest cover.

Representation of ecosystem types within stands promotes species richness. We used groups of site series (combinations of site moisture and productivity indicated by unique assemblages of indicator plants) to define ecosystem types (§6.2.2). The site series is a well-recognized synthetic variable, indicating differences in forest growth rates, successional pathways, and vegetation communities (Pojar, Klinka, and Meidinger 1987; see also §6.2.2). Vertebrates often show differences in abundance or use of different groups of site series but no restriction to a particular series. Other taxa (fungi, lichens, bryophytes) may be more restricted to particular site series, though evidence is sparse. Maintaining representative areas of ecosystem types (groups of similar site series) in unmanaged forest is an important part of the coarse-filter approach to maintaining biodiversity (Indicator 1; see Chapter 6 for monitoring methods). However, site series units mapped on the ground are often smaller than cutblock sizes, particularly for the more extreme xeric or hygric types. Hence, any reserved patches within cutblocks may or may not represent site series well, and within-stand representation should be monitored directly as part of the habitat monitoring.

Microclimate is influenced by a combination of vertical and horizontal canopy structure plus understorey trees and shrubs and could act as a useful integrative habitat variable within stands. Wind speed, air temperature, ground temperature, and relative humidity are potentially useful microclimate variables, with direct applicability to many sensitive organisms, although thresholds are unknown for most species. However, monitoring can be informative if it is focused on demonstrated concerns for specific species and designed to avoid the costly equipment and unwieldy data acquisition of many microclimate studies.

8.2.3 Process Variables for Long-Term Habitat Projections

To assess the likely consequences of forest practices, future levels of habitat elements and structures in managed stands have to be predicted (modelling is described in Chapter 12). Process variables are critical to any habitat projection model. Estimates for many parameters can be found in the literature, but high variability suggests that local calibration is required. Growth and yield monitoring can contribute for some variables, while others require additions to the habitat-monitoring plots.

Growth of the regenerating stand has direct or indirect effects on all forest habitat elements and structures of interest. We require reliable information on the development of planted and natural seedlings as well as retained live trees. Information required exceeds summary growth and yield curves and includes effects of variability in growth rates among trees, site conditions and inter-tree competition on producing spatial variability in growth, and the contribution of natural regeneration later in the rotations (even if the regeneration will not produce commercially important trees). The variability that develops within a stand influences future habitat quality.

Mortality of retained and regenerating trees has profound influences on habitat. Growth and yield monitoring and modelling typically ignore retention patches. There is little information to assess or project the fate of retained trees. As the harvested part of a stand ages, the density, types, and distribution of snags become increasingly influenced by the mortality of retained or regenerating trees. Horizontal variability, vertical canopy structure, and shrub distribution later in the rotation are also tied closely to patterns of tree mortality. Estimates of age-specific mortality of trees, related to their site and growing conditions, are required to predict these primary habitat elements and structures. Major agents of mortality should be estimated to predict structural outcomes (e.g., proportions of trees that die standing or are windthrown, whether tops will be broken, whether snags will have a decayed centre, and whether snags will be clustered or dispersed).

Fall and decay rates of snags and decay rates of CWD determine how the input of deadwood elements from live tree mortality meets various habitat requirements. Stage-structured models of dead wood are conceptually simple but require estimates of many decay and fall parameters (Harmon et al. 1986; Mellen and Ager 2002; Morrison and Raphael 1993; Wilhere 2002; Yin 1999). These parameters differ among trees of different species, size, and mortality source.

Shrub, forb, moss, and lichen growth are difficult to predict during the period before regenerating stands are established because they depend largely on which plants survive the logging disturbance and can flourish quickly in the disturbed area. Early-seral ground cover can be measured directly in young stands. Projection models are required for the longer-term changes in mosses, lichens, forbs, and shrubs induced by the growth of the regenerating stand and later changes as mortality occurs in the second-growth stand. Such projections require estimates of parameters relating cover, general composition, and distribution of these ground layers to changes in canopy trees within second-growth stands on different site types.

Decay rates and input to litter cover potentially influence site productivity as well as habitat. Initial decay of retained litter and additional litter from logging slash in the years following harvest can be estimated. Dynamics of litter become more complex in the longer term, with ground-cover layers and the regenerating stand affecting decay rates and contributing input to the litter layer. Neither productivity nor habitat can be projected without some estimate of the parameters that relate litter input and decay to canopy and understorey vegetation conditions.

8.2.4 Landscape Features

Landscape features are broad-scale summary variables. Most available landscape summary indices are unhelpful because they have little apparent ecological relevance (§8.1.3), or they require a simple, "black-and-white" view of habitat. The latter is not appropriate in regenerating managed forests. Moreover, many current forest practices, and especially variable retention, explicitly attempt to make managed stands less different from natural stands. Additionally, different organisms respond to different habitat types in very different ways, obscuring tidy distinctions among habitat classes. Landscape-level monitoring therefore relies primarily on the representation analyses of Indicator 1 and the organism-based approach outlined in §8.2.5. However, three simple landscape summary variables appear to have broad relevance to many organisms. These are the distribution of patch age and size classes, edge-contrast length and forest interior, and road densities and distribution. For all of them, it has proven impossible to derive unequivocal targets from nature that could serve as threshold values for action.

Distribution of patch age and size classes. For stands of similar origin, age is a useful synthetic predictor of organisms that can inhabit the stand. Every organism cannot be retained in every stand at all times, but a range of age classes should be maintained across the landscape to ensure regional persistence (e.g., Figure 8.1). Monitoring changes in age-class distribution can provide a general indication of expected changes in organisms, while spatial modelling of particular groups of organisms can assess the landscape suitability in a more detailed way. The size-class distribution of "stands" (contiguous forest areas of the same age class) is a secondary variable because many organisms appear to be more tolerant of a range of stand sizes if there is an adequate *amount* of suitably aged forest. Patch size is highly sensitive to the arbitrary rules defining a "stand." However, broad changes in size-class distributions of stands across a landscape are of general concern for maintaining regional biodiversity (Lindenmayer, Margules, and Botkin 2000; Noss 1999).

Age-class analyses may make sense for even-aged forests following stand-replacing natural disturbances or clear-cutting. Their application to retention systems is less clear because those systems retain part of the original stand in perpetuity (subject to natural disturbances) and confound the seral-stage relationship with stand age. Monitoring of age distribution therefore needs to separately track stands with no residual structures (clearcuts) and those with different levels of permanent retention.

Edge-contrast length and interior may be more revealing than age or patch size under variable retention. The length of edge between distinct stand

Figure 8.8 The variety in which managed landscapes can be arranged necessitates broad measures of the landscape when measures are to be 'scaled up' across the tenure. | *Photographs by: W. Beese.*

types in a landscape provides an index that combines the amount, size, and shape of stands and has direct ecological relevance for many organisms. That relevance is usually considered not as the total length of edge but as the percentage of stands that are affected by adjacent stands (edge effect). Many edge effects are positive, but some are negative. To assess negative effects, the percentage of suitable habitat that is away from potential negative edge effects (i.e., percent interior) is the critical value. Given that different species respond differently, the only way to interpret potential effects is to employ different distances for edge effects in the analysis.

Assessing edge in dynamic forest landscapes is not straightforward, particularly if regenerating stands have different levels of retention at the time of origin. The concept of "edge contrast" is a useful way to prorate the amount of edge based on how different the stands are on either side of a given edge. For example, the length of edge between a recent clearcut and an old stand may be counted fully, while the edge between stands that are similar in age may be counted as only a small fraction of its actual length. Similarly, an edge between a block with high retention and an uncut stand may also be counted less fully than an edge between a clearcut of the same age and an uncut stand. The distance from the edge at which the "interior" begins may also be reduced when there is less edge contrast across edges. We recognize that distances for edge effects are poorly documented for edges of different contrast (Kremsater and Bunnell 1999), but scattered evidence suggests that edge contrast has a real effect (e.g., Esseen and Renhorn 1998; Marzluff and Restani 1999; Sisk and Margules 1993).

The obvious difficulty with using edge contrast as a general landscape index is that the measure is organism specific. Different species will respond differently to the amount of "contrast" between stands of different age or between uncut and managed stands with different retention levels. Defining "interior" also includes an organism-specific influence on how far the edge effect extends. Details of species' autecology could be assessed for particular focal organisms in the monitoring of Indicator 3 (Chapter 10). For overall indices of landscape edge and interior, we suggest a generic measure of edge contrast based on tree height and percent retention (§8.3).

Road densities and distribution provide another useful synthetic index at the landscape level because they have direct effects on many organisms and ecosystem processes[8] and indicate the extent of past and projected management activities. Road density, expressed as kilometres of road per square kilometre of land, can be calculated readily from maps. The actual area of land in roads and road rights-of-way is also relevant and can be calculated from finer-resolution maps, satellite images, or air photos. Density of stream crossings should be informative. Other potential variables include the degree to which roads bisect areas of land as sampled by random points and contiguous roadless areas. Current literature is inadequate to estimate the magnitude of any effect revealed by most indices. That does not reduce their comparative value (the direction of effect is clear), but it does emphasize the importance of coupling them with information derived from Indicator 3 (organisms).

8.2.5 Hypothetical Species as Landscape Indices

To interpret landscape structure beyond the three features recommended above, we need to know species' responses to stand-level practices, the time to "recovery" of suitable habitat in managed stands, organisms' abilities and propensities to move through non-suitable habitat, and their movement distances. However, there are thousands of species, each responding differently. We suggest interpreting landscape structure by using a set of "hypothetical species" that represents a range of responses.[9] These hypothetical species should range from those that are likely to persist under a wide range of management options to those that will be most challenging to sustain in the landscape. A group of ten to twenty such indices could assess responses to many combinations of landscape variables. As with any index, the approach is intended to assess a *range* of real organisms, not to be a perfectly realistic model of actual organisms. Employing it, we need not measure many parameters for a large number of particular species in the field. Instead, we set values so that the hypothetical "organism" represents a particular set of responses to the landscape features that we believe provide a useful index of landscape condition. Providing we create a range of values for the hypothetical species, we should encompass responses of most species. The danger is in convincing ourselves that the hypothetical species are real simply because they have been modelled on real organisms. We can overcome much of this danger by selecting tractable and comparable real organisms for field-testing the predicted responses (see Chapter 10). These hypothetical species are used as indices of landscape structure, like any other landscape index.

8.3 How to Monitor

We describe briefly how the five types of structural variables can be monitored. A summary "checklist" of the variables and how to monitor them, including any specific points of emphasis, is provided elsewhere (see Appendix 3.1 of Bunnell et al. 2003).

8.3.1 Standard Habitat Elements and Their Attributes

A field protocol and sampling design for measuring habitat elements in variable retention cutblocks was established based on previous work by many wildlife researchers. A nested sampling design was used, with four levels of sampling for group retention: forest type (combining BEC variant and retention type), replicate sites, groups or patches within sites, and transects within patches. Sampling in dispersed retention used three nested levels: forest type, replicate sites, and plots within sites. Sampling layout is illustrated in Figure 8.9.

Two different forms of plots were used.

1 In dispersed retention settings, a 25 m × 25 m quadrat was used to measure live trees, nested within a 50 m × 50 m quadrat in which snags and large trees were recorded (Figure 8.9a). CWD was measured along intercept transects of 25 m (for all sizes) or 50 m (for pieces >30 cm) on two perpendicular sides of the quadrat. Cover variables, dominant plants, and site series were recorded in nine 0.01 ha circular plots across the larger quadrat.
2 In group or mixed retention, transects were used across the boundaries of retention patches, extending up to 50 m into the patch and 50 m into the opening (Figure 8.9b). All trees were recorded within 2.5 m of the transect; snags and large trees were recorded within 5 m. CWD was recorded along the transect and along 10 m-long perpendicular transects every 10 m along the main transect. Trees, snags, and CWD were recorded separately by 10 m transect segments (i.e., 0-10 m from the edge, 10-20 m, 20-30 m, etc.). Every 10 m along the main transect, 0.01 ha circular plots were used for cover layers, dominant plants, and site series. Using 10 m segments allowed examination of edge effects into and out of the retention patches and permitted stratified estimates for the entire cutblock (required because sampling intensity is not equal in all parts of the patch VR blocks – e.g., patch edges versus interior versus the harvested matrix). The same transect layout was used in non-harvestable benchmark sites and clearcut comparisons (instead of the quadrat design) so that measures of variability that are sensitive to plot size and shape could be compared directly to results from transect surveys of patch and mixed retention, the dominant retention types across the tenure.

A pilot study in 1999 refined the field methods. More importantly, the study was designed to measure different components of variance in the sampling design – between ecosystem types, stands, patches within stands, and locations within patches – and results were used to optimize the sampling/subsampling effort. The pilot study results also allowed us to estimate expected precision (different for each habitat element) from different monitoring efforts, which helped to determine which priority questions could be addressed effectively.

Many habitat attributes were sampled in the field plots, and hundreds of different variables can be derived from the data. For simplicity, initial analyses have focused on the "main habitat variables" of Table 8.2. These variables were believed to have the most impact on components of biological diversity.

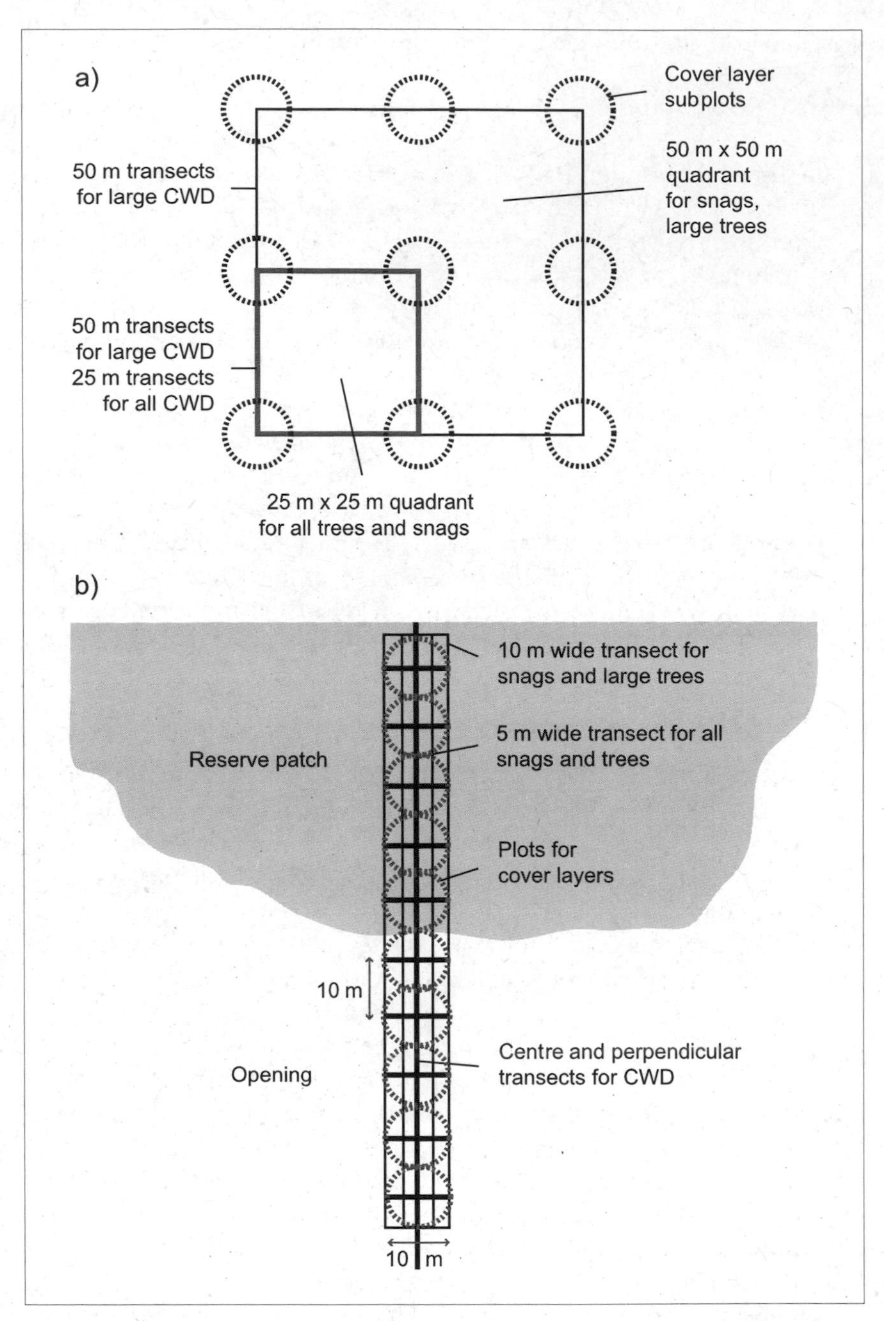

Figure 8.9 Habitat sampling layout: a) in dispersed retention, b) across the edge of group and mixed retention reserve patches, and in unharvested benchmark sites and clearcut comparisons.

Table 8.2

Main habitat variables used in summary analysis

Live trees – density and basal area

- total live trees (trees >12.5 cm dbh)
- by major species groups (Douglas-fir, cedar, hemlock and true firs, deciduous)
- in four size classes
  - small (12.5 cm–30 cm dbh)
  - medium (30 cm–50 cm dbh)
  - large (50 cm–80 cm dbh)
  - very large (>80 cm dbh)
- small trees (trees 4-12.5 cm dbh)
- special tree types (Douglas-fir >50 cm dbh, cedar >50 cm dbh, broken-top live)

Snags – density and basal area

- total snags
- by species groups (Douglas-fir, cedar, hemlock and true firs, deciduous)
- in three groups of decay classes (classes follow Thomas 1979)
  - class 3 (recent, all bark and some branches present)
  - classes 4 and 5 (wood hard, some or all bark and most branches lost)
  - classes 6 and greater (soft, decayed wood)
- in four size classes
  - small (12.5 cm–30 cm dbh)
  - medium (30 cm–50 cm dbh)
  - large (50 cm–80 cm dbh)
  - very large (>80 cm dbh)
- snags >8 m tall
- special snag types (class 6+ >30 cm dbh, cedar >50 cm dbh, Douglas-fir >30cm dbh)

Coarse woody debris (CWD) – volume

- total CWD volume
- by species groups (Douglas-fir, cedar, hemlock and true firs, deciduous)
- in three groups of decay classes (classes follow Thomas 1979)
  - classes 1 and 2 (recent, most bark and some branches intact)
  - class 3 (hard wood, most bark and all branches lost)
  - class 4 (soft, decayed wood, but still maintaining a "log-like" form)
- in four size classes
  - small (12.5 cm–30 cm diameter)
  - medium (30 cm–50 cm diameter)
  - large (50 cm–80 cm diameter)
  - very large (>80 cm diameter)
- special CWD types (class 4 >30cm diameter, cedar >30cm diameter, Douglas-fir >30cm diameter)

►

◄ *Table 8.2*

Cover layers

- canopy closure
- small tree cover (trees <4cm dbh)
- shrub cover
- shrub height
- herb cover
- moss cover
- litter cover
- cover of bare ground

Plant composition

- dominance of a few major species

Results of major comparisons and their implications for management are presented in Chapter 9.

Extensive sampling over the past five years has provided precise results for major comparisons for broad habitat elements (e.g., "live Douglas-fir density," "total CWD volume" – standard errors of 5-25 percent of the mean). Estimates for more specific elements (e.g., "class 3 Douglas-fir snags >30 cm dbh") are more variable (standard errors of 25-75 percent of the mean) but still usefully precise for assessing differences in the adaptive management comparisons. Unfortunately, our interest is often with rarer features (e.g., "hollow class 5 cedar snags with broken tops," as nesting habitat for Vaux's swifts). The nested design of the field plots helps somewhat, but no general-purpose sampling will ever precisely estimate rare features without inordinate effort. The most problematic elements are large snags (which are rare in old, uncut forest and rarer still in VR stands) and very large pieces of CWD (which generate imprecise overall estimates of CWD volume because rare large pieces make a large contribution to total volume). The current protocol is also unlikely to provide good estimates for rare special features identified from the natural history of non-vertebrate taxa (e.g., leaning snags with surrounding trees). Using brute force to increase precision by sampling a huge number of blocks is impractical. We therefore must consider alternative methods of efficiently sampling rare habitat elements over larger areas. Options include plotless methods recording only the targeted type of element or using high-resolution, low-level aerial photos. The reliability of the air photos must be tested against ground measurements before widespread application.

8.3.2 Integrative Habitat Variables (Habitat Structures)

Some of the structural measures suggested below can be derived by combining several measured elements in a habitat plot, others come from combining measurements from several plots within a stand, while some require additional sampling or different survey methods than provided by plot measurements.

Vertical structural variability is based on amounts of cover in different height layers. The two lowest layers can be estimated using the forb and shrub cover recorded in the habitat plots. The upper layers can be approximated from canopy cover, total height, and height-to-crown of trees, with some assumptions about crown shape and relation of crown size to dbh. One unresolved issue is whether this measurement is best calculated at each point (individual plots) or across larger areas (e.g., averaging plots across a patch or stand). There is no evidence about which is most relevant to organisms.

Horizontal variability of tree, snag, and CWD measures, as well as cover layers, can be calculated directly at two scales for dispersed retention (two plots times nine subplots) and at three scales for group retention (three patches times two transects times five subplots in both patch and matrix). Because the size and shape of measurement plots affect horizontal variability, the two retention types cannot be compared using current sampling designs. Current monitoring does not indicate the extent to which intact gaps and tree clumps are being retained within variable retention. The best way to assess this omission would be by comparing pre-harvest and post-harvest aerial photographs.

Retention of ecosystem types within stands is measured by recording anchor types for retention patches (e.g., wetlands, outcrops, seepages) as a comparison that is directly relevant for guiding management practices and as important strata for summarizing results. BEC site series are also recorded in habitat plots. However, determining site series in cut areas immediately after harvest is difficult. Uncertain site series in disturbed areas tend to be called the same as that of the nearest undisturbed patch, biasing results toward representative retention. Site series could be mapped for some blocks prior to harvest to allow more legitimate monitoring of representation within stands after harvest.

Microclimate monitoring has not been undertaken because such detailed monitoring has a reputation for being expensive. Questions requiring numerous precision instruments and recording devices limit intensive studies of microclimate to very few sites. However, large microclimate changes are likely important to many small and sessile organisms, and there are ways to compare basic microclimate effects of stand management more cheaply and simply. An appropriate approach would acquire large numbers of simple measurements at many sites, using simultaneous paired comparisons to factor out weather effects. For instance, simultaneous measurements by three people with simple instruments could compare air and ground temperature, wind speed, and relative humidity in retained patch centres, edges, and uncut forest. Because readings require little time, an extensive design of sampling and subsampling is feasible and relatively inexpensive.

8.3.3 Process Variables for Long-Term Habitat Projections

Long-term projections of live trees, snags, CWD, and shrub cover are being developed (Huggard and Kremsater 2005). They rely on parameters from the literature. However, many parameters show wide variation between sites or studies, making projections for local stands highly uncertain. Local estimates of parameters are needed to update the literature distributions, presumably in a Bayesian framework. Considerations for estimating main projection parameters include the following.

Growth of the regenerating stand is a fundamental consideration from an economic viewpoint and is examined in growth and yield projects by the company. The silvicultural approach in most VR is still the same as in clear-cuts, with a mixture of natural regeneration and open-grown, planted seedlings. The familiar approach allows well-developed stand trajectories for clearcuts to be used for most of the area of current VR stands. Additional research plots examine the effects of shading adjacent to retention patches or in dispersed retention. Because simple growth and yield models often use only stand-average values and a correction factor for non-productive parts of the stand, they are inadequate for projecting habitat features. More complex growth models consider effects of interactions among trees (competition) in producing variability. Realistic models for habitat projection may need to include the spatial variability in growth produced by microsite characteristics. Current models do not handle growth of retained trees well, particularly the response of advanced regeneration (trees smaller than merchantable size) retained in dispersed or mixed retention.

Mortality of regenerating and retained trees is the source of deadwood attributes through the harvest rotation. Silvicultural models typically assume ideal growing conditions and only include self-thinning or suppression mortality. Incorporating mortality of retained trees and the larger regenerating trees is essential for realistic deadwood projections and affects projections of canopy heterogeneity and understorey cover. Old permanent sample plots usually provide poor estimates of mortality rates and patterns for overstorey trees because they were intentionally placed in parts of stands where mortality was low, or the plots were discontinued if there was extensive mortality. Newer permanent sample plots should eventually provide more reliable estimates of mortality rates. However, few if any of these plots will be in stands that are similar to VR systems. In the short term, habitat models rely on best guesses for mortality rates and patterns. In the longer term, revisiting habitat plots with permanently tagged retained and regenerating trees will help to refine these estimates. Windthrow monitoring will also provide information on the initial fate of larger retained trees. However, estimates of live tree mortality are likely to remain the weakest link in long-term projections of many habitat attributes.

Fall rates of snags, decay of snags, and CWD are reported (Harmon et al. 1986; Mellen et al. 2002; Yin 1999), but the majority measure biomass decay rather than structural changes. Studies that do measure structural changes show wide variation due to site types, mortality sources, and, possibly, study methods. In the short term, few local values for these parameters are available because older permanent sample plots generally did not follow stems after they died. As well, the decay classes of snags, which are critical to interpreting their habitat value, were not recorded. Changes in CWD are even less well documented. In the long term, permanently tagged snags can be revisited to estimate fall rates and decay transitions (five years will give crude estimates of initial fall rates, while ten or more years will be needed to estimate decay rates, because decay classes are not tightly defined). Long-term remeasurement of CWD transects on habitat plots may provide information about CWD decay rates, but better information could be obtained by individually tagging logs and following them through time.

Growth and succession of shrubs and forbs have received attention because of their importance as ungulate forage. Retrospective studies have examined early-seral and old forests in several ecosystems, but information is lacking for mid-seral stands. More importantly, retrospective studies cannot account for changes induced by novel silviculture, such as variable retention. Some information is available on relationships between shrub growth and light levels, which could be used with light-based, live tree models to project shrubs. The approach does not account for other poorly known effects, such as root competition. Understorey modelling is initially limited to projecting broad changes in cover layers, until more detailed autecological information becomes available.

Decay rates of litter and how they are affected by different site conditions appear to be fairly well understood. The difficulty in projecting litter layers is the input, which requires good projections of the live trees, shrubs, and forbs.

8.3.4 Landscape Features

Age-class and patch size distributions are easily calculated with GIS once rules for defining a "patch" are decided. What constitutes a discrete patch is species specific and almost a meaningless concept in a complex managed landscape. Therefore, we suggest simply declaring rules for defining a patch and recognizing that any interpretations are limited by those arbitrary rules. At best, several sets of patch-defining rules might be used to represent species more or less restricted to older forest. Even then, arbitrary decisions need to be made about how this definition is affected by retention levels and patterns, different widths of intervening stand types, natural barriers such as rivers, effects of roads or trails, et cetera. To date, we have found no simple approach to patch size that we believe is meaningful as a general landscape indicator.

Edge and interior are also easily obtained from GIS, using buffers of different distances of edge effects. The main challenge is in recognizing the different edge contrasts produced by different adjacent stand types. The degree of contrast between stand types depends on which organism perceives the contrast. These species-specific effects are probably best addressed using organisms with a range of tolerances of stand ages and different degrees of benefit from stand-level retention. For a general index of edge contrast, we suggest a simple formula based on the tree-height difference at the edge and the proportion of old-forest retention in the younger stand:

$$\text{Edge contrast} = (1 - Ht_a/Ht_b) \times (1 - \text{proportion of trees retained in "a"})$$

Where the taller stand is "b" and the shorter (younger) stand is "a." The edge contrast then modifies the edge length or area of interior calculated from GIS. For example, a 1,000 m edge length with a contrast of 0.6 represents 600 m of effective edge. Effective edge depth would be similarly reduced. For instance, if we were using an edge depth (i.e., extent of negative edge effects) of 50 m across a high-contrast edge, then an edge with a contrast of 0.6 would have 30 m of effective edge depth. A 200 m edge effect would similarly become 120 m of effective edge depth, when reduced by edge contrast.

Road density and area are likewise straightforward to measure with GIS, provided road mapping is up to date, at high resolution, and includes widths of roads and rights-of-way. Monitoring the density of stream crossings is also a simple GIS task though sensitive to the resolution of stream mapping. The bisection of areas by roads can be indexed with a GIS model that calculates what proportion of random pairs of points at different separation distances is separated by a road. A fairly simple algorithm can determine the proportion of a large number of pairs of points 100 m, 500 m, or 2 km apart that has one or more roads between them. Buffering roads and recording proportions of the land base at different distances from roads would provide similar information to the bisectional index.

8.3.5 Hypothetical Species for Landscape Evaluation

The idea behind using hypothetical species as landscape indices is that most aspects of evaluating landscapes have species-specific components, but for this medium-filter indicator, we want metrics that are not closely tied to the autecological quirks of particular species. Generally, indicator organisms will be chosen to cover a range of several important variables in combinations that reflect real organisms. These variables could include

- effectiveness of different silvicultural systems in maintaining habitat quality;
- "recovery time" of disturbed habitat;
- dispersal distances and rates;
- effects of treatment or stand age on dispersal distances; and
- utility of different types of unmanaged, non-harvestable stands as habitat or for dispersal.

All possible combinations of different values of these variables are not needed as indices. Some combinations represent species that are of no conservation concern because they can tolerate any possible managed landscape. Other combinations will not correspond to any known organism or be so restrictive that no such organism could evolve. The focus is on representing feasible organisms that are potentially sensitive to possible landscape management options.

We are currently using the hypothetical species concept to index "connectivity" with two approaches. These approaches recognize that there are at least two distinct biological aspects of "connectivity." One approach, based on the biological level of gene flow and subpopulation structure, assesses how many isolated fragments of suitable habitat exist in a landscape for a particular "species." Areas of suitable habitat form one patch if they are closer than the maximum dispersal distance of the species, which varies depending on the stand types between the patches. The second approach is based on the biological level of individual dispersal success, which ties in to spatial demography. It is a modification of the approach used by Richards, Wallin, and Schumaker (2002) and tracks the number of suitable home ranges encountered by members of a large cohort dispersing in all directions from all suitable home ranges. Cohort survivorship decreases with distance dispersed, with the rate depending on the suitability of the underlying habitat. Both approaches produce quantitative indices of connectivity to evaluate different planning scenarios as well as maps of key dispersal sources and corridors or connected fragments to assist detailed landscape planning.

8.4 Anticipated Feedback to Management

Specific feedback to management is a function of the monitoring design or the combination of what is measured with how those measurements are organized. Generally, feedback to management in the adaptive management program attempts to identify the weakest aspects of practices, suggest better alternatives, and monitor long-term improvements. Habitat and landscape features are directly changed by forest practices, so the connections between Indicator 2 and management are the most direct of all indicators. Information gathered by monitoring Indicator 2 can help management decisions in four ways.

1 *Selecting among and refining forest practices:* Although variable retention is the major practice, the company employs a range in the amount and distribution of retention. Indicator 2 documents amounts of habitat elements and structure that result from actual practice. At the simplest level, monitoring Indicator 2 is a form of implementation monitoring refined to reflect requirements of forest-dwelling organisms (not just live trees). Comparing habitat elements among retention options and other silvicultural alternatives can suggest the appropriate mix of retention types. Comparisons to benchmarks provide context for the values measured (e.g., Figure 5.1). By employing both forms of comparison, the monitoring design can help to refine practice. Similarly, sampling across edges helps to reveal how the pattern of retention within a block influences how well habitat elements are maintained and can suggest appropriate changes to layout of retention.
2 *Guiding forest planning:* Forest practices are applied over large areas, producing particular mixtures of landscape features. How these features are distributed through time and space is a product of forest planning and natural succession. Although potential consequences to habitat supply and distribution by inappropriate planning appear to be large, little is known about desired landscape features in managed forests. Monitoring Indicator 2 should at least reveal undesirable trends over larger scales (e.g., age-class distribution, road density). Model projections of the distribution of retention types and their amounts of habitat elements will allow comparisons among options.
3 *Simplifying operational decisions:* Most of the literature suggests that amounts of specific habitat types are more important than distribution (Cushman and McGarigal 2004; Fahrig 1999; Schmiegelow and Mönkkönen 2002). Although some models predict large effects of habitat pattern, five of six empirical studies reviewed by Fahrig (2002) found that effects of habitat amount far outweighed effects of pattern. The other study found near-equal effects (with the fragmentation effects being positive). There may be low levels of habitat amount where pattern does become important (Boutin and Hebert 2002; Fahrig 2002). Because Indicator 2 assesses both amounts of habitat elements within different forms of retention and the consequences of landscape features resulting from the distribution of practice, it should be able to focus operational attention on whichever one appears to be most limiting. Limiting amounts might encourage changes in the amount of retention or the type of retention. Distributions might be modified by changes in the harvest schedule or the introduction of mini-zoning.
4 *Assessing long-term trends:* Simply scaling up the amounts and kinds of habitat documented by Indicator 2 assists planning. However, because

trees grow relatively slowly, the consequences of particular forest planning and practice may not be evident for decades. An additional objective of monitoring Indicator 2 is to expose undesirable trends early by projecting long-term consequences. The ability to project consequences of particular sets of planning and practice should also expose the possibilities of improvement. Both capabilities can help to guide management decisions and actions. Creating models to project changes through time in stand and landscape features is a critical step in applying monitoring to long-term forest management (see Chapter 12).

8.5 Summary

Monitoring Indicator 2 provides the most direct feedback to management because it measures those portions of the ecosystem most directly affected by forest practices. The main ecological reason for employing variable retention rather than clear-cutting is the increased ability to retain habitat structures that are difficult or impossible to retain without retention. Monitoring thus focuses on the manner in which well-known habitat elements are provided under VR and compares them to uncut benchmarks. More integrative measures than single habitat elements are theoretically insightful but difficult to interpret. Three large-scale summary variables have the potential to describe landscape features relevant to many organisms – distribution of patch age and size classes, edge-contrast length and forest interior, and road densities and distribution. These variables are easy to acquire using GIS but are difficult to interpret.

Our monitoring of habitat began with a pilot study that increased the efficiency of the design and allowed estimation of the precision with which different habitat elements could be estimated. The differences between dispersed and group retention meant that two broad forms of sampling were required. There are four broad kinds of feedback possible from the design. In Chapter 9, we discuss results from monitoring habitat elements and the feedback actually derived.

9
Learning from Habitat Elements

David J. Huggard, Jeff Sandford, and Laurie L. Kremsater

9.1 Context

Habitat elements are one component of Indicator 2, which also includes integrative habitat structures, landscape measures, and parameters for projecting habitat elements. The rationale for this indicator, and the particular components included in the monitoring program, are presented in Chapter 8. Here we focus on results for habitat elements within stands, because that has been a large component of the monitoring program to date, addressing most of the high-priority comparisons and associated questions at the core of the monitoring program (summarized in Chapter 12). The major questions are listed below.

1 *How well do the different types of variable retention retain different habitat elements?* Comparison of retention types (group, dispersed, and mixed retention) is the most basic comparison for choosing the best balance of the retention types or, more realistically, for evaluating the mix of retention types prescribed for various reasons.
2 *Do retention patches capture the same amounts of habitat elements as uncut stands? What are the weakest points (lowest levels in retention patches relative to uncut benchmarks)?* Comparison of retention patches to uncut benchmarks provides a baseline for scaling comparisons of retention types, indicating how meaningful differences between retention harvest treatments are. Comparisons to benchmarks are also critical in identifying weakest elements, which helps to simplify operational implications of the monitoring.
3 *What is the relationship of habitat elements with amount retained (percent retention)?* Stated differently, does retention of a portion of a cutblock's area (group retention) or basal area (dispersed retention) retain an equivalent portion of each habitat element? Relationships found help to reveal processes affecting retention of particular elements at different overall retention levels, such as operational choice of rich or poor patches. The

relationships may reveal thresholds that could lead to specific retention targets.

4 *How do levels of different habitat elements change with distance into (or out of) retention patches?* Variable retention creates significant amounts of edge. Comparisons of habitat elements retained with distance from edge have direct implications for designing size and shape of retention patches and for assessing the actual nature of "forest influence" sought by retention systems.

5 *What levels of different habitat elements are captured in retention patches anchored on different anchor types (ecological features such as wetlands, rock outcrops, or typical mesic forest)?* The comparison of anchor types has immediate use in guiding foresters' choice of where to locate retention patches.

6 *How have retention levels of different elements – particularly weakest points identified in initial monitoring – changed over time in operational blocks?* The evaluation of operational progress over the first six years of implementation tests whether the overall monitoring and adaptive management system is working.

We discuss the implications of these results for adaptive management and the results that they have had on practices. Other aspects of Indicator 2, including the critical issue of projecting the long-term habitat consequences of variable retention, are ongoing and not presented here.

9.2 Methods

9.2.1 Field Methods[1]

Field methods to measure habitat elements within stands are presented in Chapter 8. We use nested transects across patch edges in group and mixed retention and nested quadrats in dispersed retention (Figure 8.9). We also use transects in uncut benchmark sites to allow direct comparisons with group and mixed retention, the dominant operational types. The edge transects provide direct measurements of edge effects into and out of retention patches and are used to provide a stratified mean for the group and mixed cutblocks as a whole. Stratification is based on percent retention and the area at different distances from edge into and out of the retention patches.

The range of habitat elements and their attributes that are measured in these plots are also presented in Chapter 8. They include

- live trees (for trees ≥12.5 cm dbh: species, dbh, height, height to base of crown, damage or disease; density of trees 4-12.5 cm dbh; percent cover of smaller trees);
- snags (for snags ≥7.5 cm dbh: species, decay class, dbh, height, top breakage);

- coarse woody debris (for logs ≥12.5 cm diameter at point of transect interception: species, decay class, diameter at interception point, height above ground, length class);
- cover layers (coniferous and deciduous canopy; shrub height, cover, and dominant species; herb height, cover, and dominant species; litter depth, cover, and type; mineral soil cover).

The measurements can be combined to calculate vertical diversity, while the six transects or two sets of quadrats per site allow the calculation of horizontal variability. Plants from a list of rare or introduced species were recorded as part of the vegetation monitoring.

9.2.2 Subsampling Design

A pilot study was conducted in the initial year of habitat monitoring to optimize the allocation of effort to different levels of the sampling design:

- for group and mixed retention – number of transects per retention patch, versus patches per site, versus total number of sites;
- for dispersed retention – number of quadrats per site versus total number of sites.

Variance components for several main habitat elements were estimated from nested plots conducted at fifty sites and used with measurements of time costs in standard subsampling optimization (Krebs 1989). Results suggested two sets of quadrats per site in dispersed retention and two transects in each of three patches per site in group and mixed retention. These designs were used in subsequent years. Six transects were used in benchmark sites to match the total sampling effort of group and mixed retention. Each site required two to four person days to complete.

The pilot study also allowed us to estimate the precision that we could expect for measurements of different habitat elements with different total monitoring effort (§9.3). Subsequent monitoring confirmed that these values for expected precision were close to the actual precision obtained in the monitoring program. That finding has been useful for setting targets for the number of sampling sites for different comparisons and, in some cases, for deciding not to address comparisons where the necessary sampling effort for usefully precise results was unaffordable.

9.2.3 Study Design

One requirement for useful monitoring in an adaptive management program is a designed comparison or set of comparisons to address critical management questions. These comparisons provide a structure for learning that is linked to improving management action. Habitat elements were chosen as the primary indicator for making many of the comparisons in the overall monitoring program (summary in Table 12.2). Our goal was to address the major comparisons with sufficient numbers of sites to permit useful results after five years of monitoring (numbers as indicated by the pilot study).

The monitoring design is made more complex, and the total sampling effort increased, by the diversity of ecosystems on the study tenure, which includes three zones, nine subzones, and twenty variants in British Columbia's hierarchical BEC system (Pojar, Klinka, and Meidinger 1987; see also §6.2.2). Operational and regulatory users of the monitoring desire results at the variant level. Evaluating all variants would distribute sampling effort so thinly that results would be too imprecise for most comparisons in most variants. We addressed this issue by sampling across all variants but using an AIC-based model selection process[2] to assess the most efficient way of combining the results at any time in the monitoring program. We examined a range from the coarsest groupings of one mean for all sites or for each zone, through to separate values for each variant, with various sensible intermediate levels of grouping. Currently, the best-supported option is to report results separately by subzone but with two relatively wet subzones – CWHvm and CWHvh – combined. Increasing sample size over time will support reporting at increasingly fine levels.

One goal of the monitoring program was to sample at least ten sites of each type (group, mixed, and dispersed retention plus benchmarks) in each subzone, except where a specific retention type was rarely used in a subzone.[3] Some comparisons, however, require more effort in a particular combination of stand type and subzone. For example, many group retention sites have been measured in the wetter CWHvm and CWHvh subzones, where much of the harvesting occurs. That allows a more detailed look at relationships with percent retention (question 3 in the list above).

Table 9.1

Number of sample sites for monitoring habitat structure (1999-2005)

BEC subzone(s)	Dispersed VR	Group VR	Mixed VR	Benchmark
CDFmm	10	2	7	14
CWHxm	18	21	21	29
CWHdm	7	11	2	9
CWHmm	3	7	4	8
CWHvm+vh	9	56	6	32
CWHwh		7		5
MHmm		2		1

Note: The number in each cell is the number of sample sites in that combination of BEC subzone and retention type. For example, ten sites were sampled in dispersed VR in the CDFmm subzone.

Within the relatively dry CWHxm subzone, there was interest from environmental and government groups in comparing the dominant second-growth stands to rarer old-growth stands, which led to extra sites being surveyed in that subzone. Assessing operational progress (question 6 above) also required revisiting combinations of stand type and subzone that were previously well sampled. Some areas, such as the MH zone, were de-emphasized because they have large areas of non-harvestable forest, making effective stand-level retention less important. Together, these factors have resulted in an unbalanced overall design (Table 9.1) but one that allows each of the main comparisons to be made in at least several subzones.

9.2.4 Approach to Summaries

Thousands of variables can be summarized from the monitoring (e.g., "basal area of 40-50 cm dbh broken-top decay class 4 cedar snags"). These detailed results are available from a database containing the accumulated monitoring data (Huggard 2004). Such specific elements can be important for individual species contributing to biodiversity, but reporting the details overwhelms attempts to extract general conclusions, particularly for operational users. Instead, summaries of the results focus on a set of main habitat elements (listed in Chapter 8), ranging from basic measures ("live tree density," "canopy closure") to somewhat more specific variables ("basal area of cedar snags," "volume of CWD >50 cm diameter") and a few more detailed variables with particular ecological relevance ("density of Douglas-fir trees >80 cm dbh"). Interpretations emphasize identifying weakest points (elements with the lowest amounts in retention compared to natural benchmarks) or elements with the greatest differences between retention types.

Here we report comparisons within subzones with adequate sample sizes. Comparisons among retention types, retention versus benchmarks, and operational progress rely on simple comparisons of means and standard errors. The site is considered the sample unit. Transect results are weighted by the area of the block in each distance-from-edge category; then transects or quadrats within a site are averaged. Comparisons of anchor types also rely on simple means and standard errors, but the individual patch is the sample unit because patches within a site differ in anchor type.

Relationships of abundance of various habitat elements and percent retention (benchmarks represent 100 percent retention) were analyzed using AIC-based model selection (Huggard 2004). Simple models included a single mean across all retention levels, a proportional relationship between percent retention and levels of a habitat element, and linear regressions (with an intercept other than 0). More complex models included asymptotic or non-linear relationships to capture possible effects due to edges or selection of different reserve types at different levels of overall retention. Some models allowed the possibility that the relationship across the retention levels in the retention sites did not extend to the benchmark sites (i.e., benchmark sites were not simply "100 percent retention" sites but a distinctly different habitat).

Relationships of habitat elements and distance from patch edges were also analyzed with AIC-based model selection. Models ranged from a single mean at any distance from the edge (i.e., no edge effect), distinct values for 0-10 m from the edge compared with further distances, distinct values for 0-10 m and 10-20 m from the edge, and simple curvilinear relationships becoming asymptotic with distance from the edge. Models were applied separately to edge effects into, versus out of, patches. That is, they allowed different edge effects in patches versus openings. Small patches reduced the number of samples available from >30 m into patches, so edge effects within patches could be examined only for 30 m.

9.3 Results and Implications

9.3.1 Expected Precision

Pilot study results showed the expected increase in precision as total sampling effort increased: that is, the standard error (SE) declined as a percentage of the mean. However, values for main habitat elements showed a tenfold range in expected precision at a given level of effort. For example, within patches in ten group retention sites sampled in the current subsampling design, the volume of large-diameter CWD (which is highly variable) had expected SEs 55 percent of the mean value, while canopy cover (which is much more uniform within patches) had a SE of about 4 percent of the mean. More specific elements, which are rarer and even more variable, had still less precision and likely cannot be estimated well by any feasible amount of sampling using the general-purpose methods employed here. As results accumulated from subsequent years of monitoring, the precision attained generally matched the precision predicted from the initial analysis of the pilot studies.

9.3.2 Comparison among Retention Types

We examined sixty-three main habitat elements in five subzones, from driest to wettest: CDFmm, CWHxm, CWHdm, CWHmm, and CWHvm+vh. Results for individual habitat elements are presented elsewhere (Huggard 2004). We present summaries here and consider differences meaningful when they exceed 20 percent among retention types. The summary of results where both group and dispersed retention occurred showed that group retention had higher levels of the habitat element in about two-thirds of the comparisons (Table 9.2). In about half the remaining comparisons, there was no difference between the two retention types (levels <20 percent different). Only 10 percent of comparisons showed higher levels of retention in dispersed retention.

Table 9.2

Number of differences in amounts retained for sixty-three comparisons of habitat elements between group and dispersed retention

	CDFmm	CWHxm	CWHdm	CWHmm	CWHvm+vh
Group > dispersed	43	37	39	49	27
Dispersed > group	8	10	3	2	11
No difference (<20% different)	12	16	21	7	12

Note: Results are stratified by BEC variant. Totals differ between variants because some elements are based on particular species (e.g., "cedar snags") not found commonly enough in some variants to allow comparisons. Results for specific elements are found in Huggard (2004)

Among individual elements, density of live trees was almost always higher in group retention blocks, reflecting higher overall retention levels by group retention. Western redcedar in drier subzones was an exception, with higher levels in dispersed retention, reflecting operational preferences for leaving this species where it is a minor component of the stand. Retained live trees tended to be larger in dispersed retention, however, so that basal areas were less different though still generally higher in group retention.

Snags showed a pattern similar to live trees, almost always being retained at higher densities within group retention. The few exceptions were again the largest snags, which were at similar levels in the two retention types or slightly favoured in dispersed retention. However, because of safety constraints, most of the snags retained in dispersed retention were very short. Retention of intact tall snags was much greater in group retention.

Amounts of CWD were more variable between the two retention types, as high levels of CWD were retained or created in the harvested parts of both types. There were no consistent differences in which types of CWD were favoured by either retention system across subzones.

Total canopy cover was lower in dispersed retention than in group retention in all subzones (e.g., Figure 9.1). Other cover layers showed inconsistent differences between retention types. Dispersed retention tended to retain more understorey vegetation cover in the drier subzones, while group retention retained more in the wetter zones.

Group retention was compared to mixed VR, using the same sixty-three habitat elements in three subzone groups (CWHxm, CWHmm, and CWHvm+vh). In CWHxm, group retention retained greater amounts of more elements than did mixed retention, while the opposite was true in CWHmm (Table 9.3). The two systems were approximately equivalent in the CWHvm+vh. Although mixed retention retains additional habitat elements in the harvested matrix, mixed retention in the CWHxm had a lower percentage area in patches, lowering the overall amounts of elements retained.

Table 9.3

Number of differences in amounts retained for sixty-three comparisons of habitat elements between mixed and group retention

	CWHxm	CWHmm	CWHvm+vh
Mixed > group	14	29	15
Group > mixed	23	19	15
No difference (<20% different)	26	10	23

Note: Results are stratified by BEC variant. Totals differ between variants because some elements are based on particular species (e.g., "cedar snags") not found commonly enough in some variants to allow comparisons. Results for specific elements are found in Huggard (2004)

In the CWHmm, retention in the matrix made a substantial contribution to overall retention levels, while relatively little structure was retained in the matrix of mixed retention in CWHvm+vh. These results may change over time because mixed retention in particular is implemented in highly variable ways from place to place and at different times.

These comparisons of individual elements do not consider the importance of the different spatial distributions of the elements. Both aggregations of elements (in group retention) and distributed structure (in dispersed retention) can benefit different species, again emphasizing the importance of a mix of practices.

9.3.3 Comparison of Retention Patches with Uncut Benchmarks

Comparisons of the same sixty-three habitat elements in the patches of group retention versus uncut benchmark sites showed differences between ecosystems (Table 9.4). For the three drier subzones (CDF, CWHxm, CWHdm), more habitat elements were more abundant in retention patches than in benchmark sites (Table 9.4). CDF in particular had higher levels of

Table 9.4

Number of differences in amounts retained for sixty-three comparisons of habitat elements between group (or mixed) retention patches and uncut benchmark sites

	CDFmm	CWHxm	CWHdm	CWHmm	CWHvm+vh
Patches > benchmarks	37	21	24	8	10
Benchmarks > patches	16	16	20	27	15
No difference (<20% different)	9	26	18	21	22

Note: Results are stratified by BEC variant. Totals differ between variants because some elements are based on particular species (e.g., "cedar snags") not found commonly enough in some variants to allow comparisons. Results for specific elements are found in Huggard (2004).

Group retention

Mixed retention

Dispersed retention

Figure 9.1 Overall, group or mixed retention retained greater amounts of many individual habitat elements than did dispersed retention. However, some important elements, particularly large trees and large (but short) snags, are retained well by dispersed retention, suggesting that it has a useful role in the overall mix of practices. To maintain heterogeneity across the landscape, dispersed retention better complements group retention than does mixed retention.

Source: Adapted from Bunnell (2005: Figure 2).

many elements in retention patches than in uncut benchmarks. In contrast, in the two wetter subzone groups (CWHmm and CWHvm+vh), retention patches did not do as well as benchmark sites in the amounts of habitat elements retained. This pattern suggests that better choices of retention patches were being made in drier subzones. However, the comparison considers just the retention patches, not overall retention levels, which were lower in the drier subzones.

In drier subzones, retention patches had higher densities of live trees than did benchmark sites, but the trees were smaller on average, so that basal areas were sometimes lower in the retention patches.[4] Deciduous trees were strongly favoured for retention in the drier subzones. The trend toward higher densities of smaller trees was also seen in the wetter subzones but was less pronounced. That retention patches were anchored on wetlands or rocky outcrops could explain these results.

Densities of snags were also considerably higher in retention patches than in the benchmarks for CDF and the other two drier subzones. Densities were more typical of benchmark levels in the two wetter subzones. CDF retention patches included high levels of the largest snags, so that snag basal area was also relatively high in this subzone. In the other four subzones, however,

Figure 9.2 Retention patches in the drier subzones have habitat elements that are fairly typical of benchmark sites, except for higher levels of deciduous trees and snags that were emphasized as important features to retain. However, the retention patches tend to be in sites with lower productivity, producing smaller trees and snags. The retention patches in the CWHmm and CWHvm+vh were less successful at capturing representative levels of habitat elements such as shrub, herb, and moss cover. | *Photograph by: W. Beese.*

large snags were retained at low levels compared with benchmarks. The reduction in very large snags reflects safety hazards associated with large snags in relatively small retention patches.

CWD levels were generally the same in retained patches and benchmark sites. CWD in retention patches is affected little by logging operations and is not directly used to select retention patches.

Retention patches had less shrub cover and lower shrub heights than benchmarks in all subzones. In the wetter CWHmm and CWHvm+vh subzones, moss and herb layers were also at lower levels than in the benchmarks.

These vegetation differences reflect a preference for establishing patches on sites with less understorey vegetation or lower productivity rather than post-harvest declines of vegetation in patches. Lower understorey vegetation in retention patches is less of a concern than lower levels of snags or large trees, because herb and shrub layers should recover quickly in the harvested parts of retention blocks.

9.3.4 Relationships of Habitat Elements with Percent Retention

The retention level of cutblocks is determined by the percentage of the cutblock area retained in patches for group retention and the percentage of live tree basal area retained in dispersed retention. The relationship between percent retention and the amounts of seventeen habitat elements could be examined effectively for group retention blocks in four subzones and for dispersed retention in two subzones. The data set included benchmark sites, representing 100 percent retention. Most results fell into one of three types: (1) relationship close to proportional, (2) one mean value across all retention levels, and (3) one mean value for retention blocks, another for the benchmarks.

Relationship close to proportional: A proportional relationship is expected in group retention if the retention patches are typical of benchmark forest and the habitat element is not retained in the matrix.

Such relationships include cases where the AIC analysis chose the proportional model as the best, where it chose another linear relationship with a small intercept value (i.e., close to proportional), or where the best model was curvilinear but only very slightly different from a proportional relationship. Although scatter among sites was high, approximately half the group retention comparisons showed this type of relationship. Most elements showing this pattern were trees, snags, or canopy cover. Examples for canopy cover and snags are shown in Figure 9.3.

Other than for live tree basal area itself, a proportional relationship is not generally expected for habitat elements in dispersed retention. In dispersed retention, operational activities affect the whole cutblock, so elements negatively affected by operations could be absent from any level of retention. Similarly, elements that can be accommodated by dispersed retention may remain at relatively high levels. As expected, a proportional or similar relationship with percent retention was only found for about 15 percent of habitat elements examined in dispersed retention.

One mean value across all retention levels: This relationship is expected whenever operations had no effect on a habitat element. It is also the "null" expectation when results are completely variable from site to site. This pattern was seen for about 20 percent of the elements in both group and dispersed retention, primarily CWD variables and a few cover layers. The example of CWD is shown in Figure 9.4.

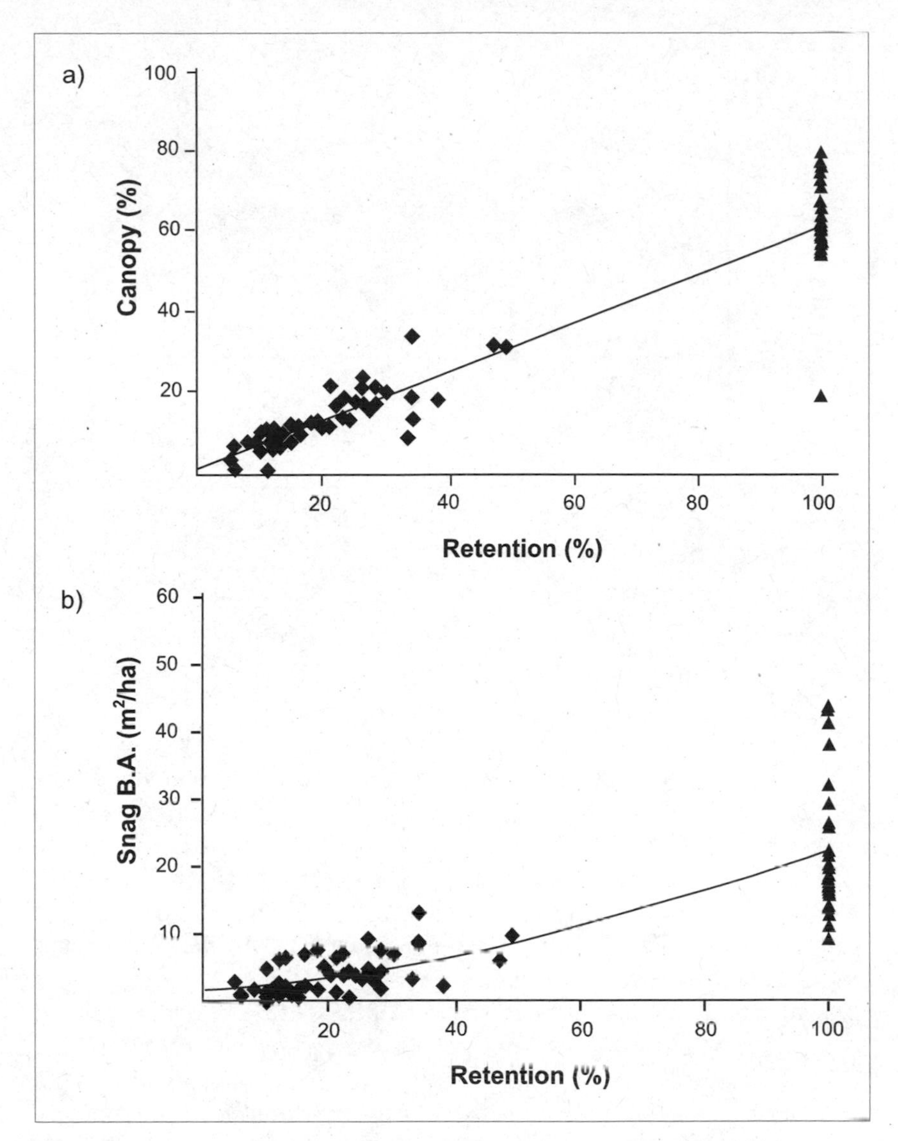

Figure 9.3 Patterns of group retention of habitat elements. Examples of (a) a proportional relationship; (b) a relationship close to proportional for habitat elements in group retention. Diamonds are samples from retention patches; triangles from adjacent, older natural stands.

One mean value for retention blocks, different value for benchmarks: This relationship is expected if harvesting has an effect on a habitat element regardless of the level of retention, which is particularly likely in dispersed retention. This form of relationship occurred in almost half the analyses for habitat

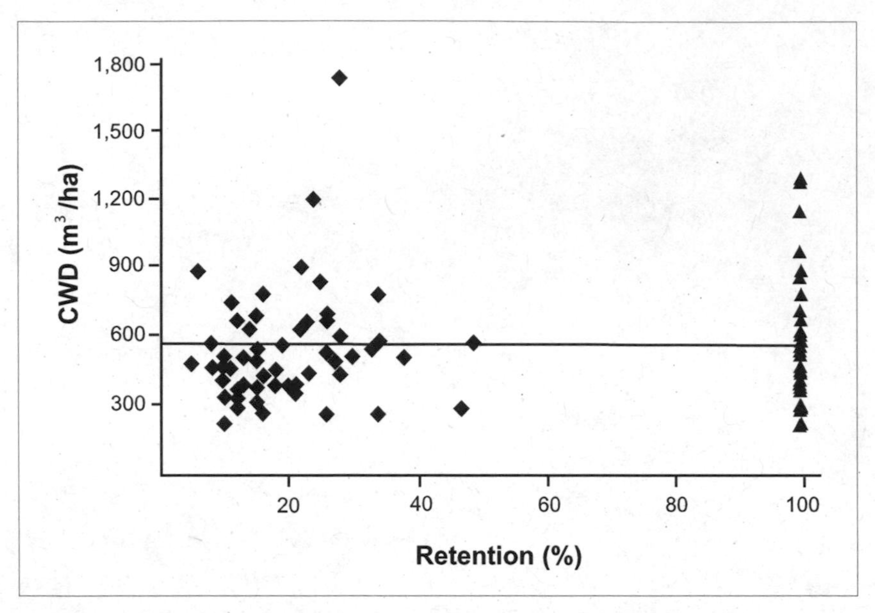

Figure 9.4 Example of one mean value of a habitat element at all retention levels.
Note: Diamonds are samples from retention patches; triangles from adjacent, older natural stands.

elements in dispersed retention, often in cases where an element such as tall snags was nearly absent at all retention levels but present in the benchmarks (Figure 9.5). This form of relationship also occurred 20 percent of the time for group retention, including many cases for CWD as well as some cover layers. In group retention, such a response either represents cases in which the harvested sites differed systematically from the benchmark sites or the retention patches were a non-representative subset of areas.

Relationships that did not fall into one of these three categories tended to be odd curvilinear models that most likely reflect chance effects (mainly for rare, variable elements).

The most common relationships for group retention suggest that, if habitat elements are affected by harvesting at all, they are retained at about the same percentage as the overall percent retention. For group retention, percent retention is a reasonable index of the percent retention of many habitat elements that are affected by harvesting. The most direct way to increase habitat elements is simply to increase overall percent retention. In dispersed retention, proportional relationships are less common, and there is usually little relationship between percent retention and retention of many habitat elements. Improving retention of elements that are negatively affected by harvesting using dispersed retention will require changes in practices, not just greater overall retention.

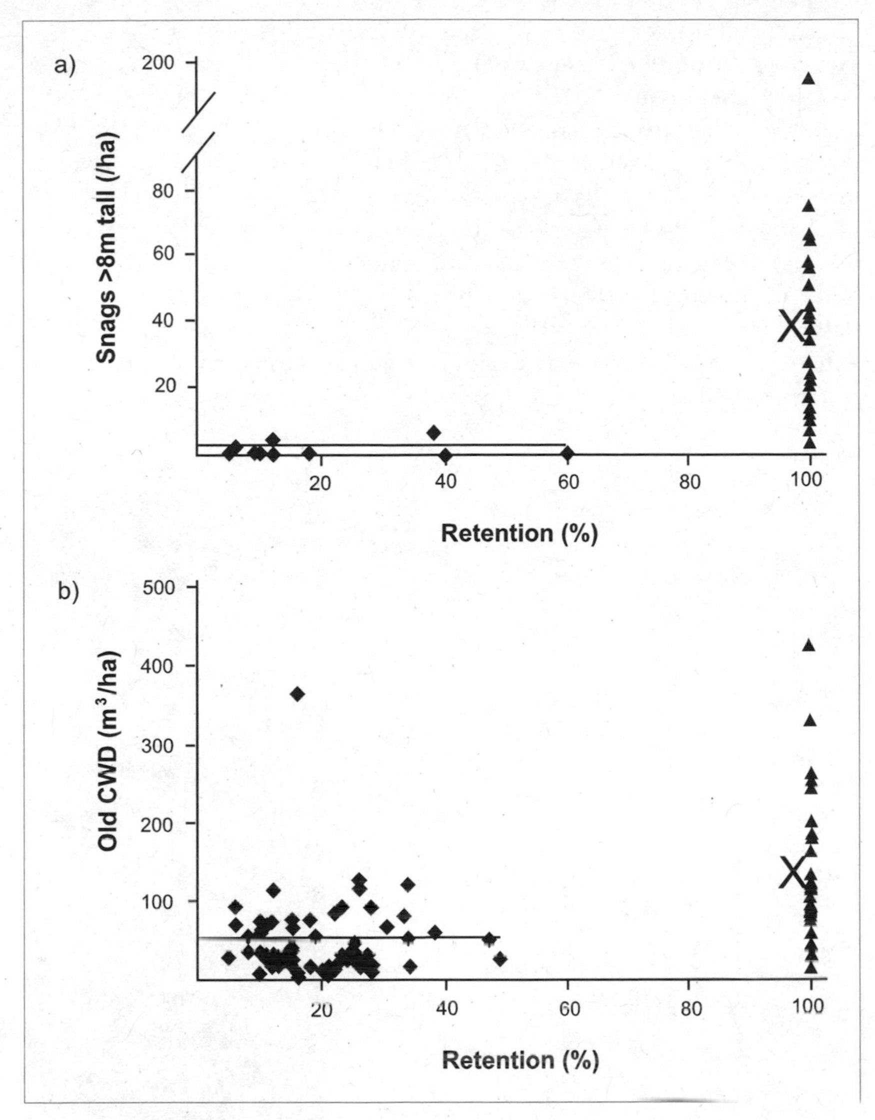

Figure 9.5 Examples of one mean value of a habitat element at all retention levels (dispersed retention only for tall snags), but at a different level in uncut benchmarks. *Note:* X represents the mean value in uncut benchmarks. Diamonds are samples from retention patches; triangles are samples from adjacent, older natural stands.

A main limitation of these analyses, besides inherent site-to-site variability, is that operational cutblocks with >40 percent retention are nearly absent, so a large area of interpolation is required between operational cutblocks and the uncut benchmarks. As well, natural processes will alter patterns over time.

9.3.5 Edge Effects

There are clear differences in habitat structures between retention patches and the harvested matrix, but edge effects were absent for almost all habitat elements – levels of elements 0-10 metres from the edge were the same as levels farther away. When edge effect models other than the "null" were selected in the AIC analysis, the most common result was a slight decline in the habitat element 0-10 metres into the retention patch (Figure 9.6 top, middle). Tree and snag variables most commonly showed this effect, due either to cutting of a few stems at the patch edge during harvest or to post-harvest windthrow. Only tall snags showed a more pronounced edge effect, with about 40 percent fewer tall snags 0-10 metres into the patch than farther in (Figure 9.6 bottom). The reduction in snags near the edge is expected

Figure 9.6 Examples of edge effects on habitat variables at patch edges. | *Photograph by: D. Huggard.*

Note: (top) live tree density; (middle) snag density; (bottom) tall snag density. Most variables showed no edge effect.

from safety rules that require snag falling along patch edges. Only 5 percent of the analyses showed edge effects into the matrix, and they were very small differences in the 0-10 metre range.

The negligible edge effects, except for tall snags, suggest that small retention patches can retain the same levels of habitat elements as larger patches. Long, thin patches would also be as effective as round patches.

Concern about negative edge effects for tall snags was one reason for the company modifying the definition of variable retention to include "large-patch variable retention," which uses larger, but fewer, retention patches than did initial practices.[5]

Documented edge effects for tall snags, plus simple geometry, suggests that a square patch of 0.25 ha, or a larger patch with more complex shape, would retain 37.3 tall snags/ha. A square 1 ha patch (or larger patch with more complex shape) would retain 41.9/ha, while a 2 ha square patch would retain 43.7/ha. All else being equal, this represents a 12.3 percent and 17.1 percent improvement in retention of tall snags by switching to larger retention patches.

These are immediate post-harvest results, and edge effects may develop over time. Edge effects into or out of the patch may be present for organisms, even if habitat elements do not show effects.

9.3.6 Comparison of Patch Anchor Types

Habitat elements in patches anchored on three primary types of ecological features – wetlands or headwater streams, rock outcrops, and wildlife or habitat features (e.g., bear den, eagle nest, cluster of snags) – were compared to patches without obvious anchor points. The latter patches were usually in mesic forest and were often established to meet rules about the spatial distribution of retention patches. Patches anchored on rock outcrops, wetlands, and mesic forest were monitored in four subzones; patches with specific wildlife habitat features occurred in two of these subzones.

Wetland-anchored patches had more, smaller trees in drier subzones, more tall and old snags in some cases, and greater herb cover, but otherwise they showed few differences from "normal" forest. In contrast, patches anchored around rock outcrops had anomalous levels of many habitat elements, including different tree species (more Douglas-fir, less western redcedar), consistent very low densities and basal areas of snags, and low volumes of CWD. Well-decayed snags and CWD were particularly lacking in rocky patches. Vegetation composition was most distinct in patches anchored on outcrops. Patches anchored on special wildlife or structural features were similar to normal forest patches, probably because the special features identified as anchoring features were only a small proportion of the total area of the patch.

Figure 9.7 Patches anchored on rocky outcrops tend to contribute reduced levels of several important habitat elements. | *Photograph by W. Beese.*

9.3.7 Operational Progress

A broad overview of implementation monitoring of operational progress relative to company targets is provided in Chapter 4. We also examined operational progress in retaining habitat elements in combinations of retention types and BEC subzones with at least four sites monitored in one or more of the first three years (1999-2001) plus the fifth or sixth years (2003 and 2004) of the implementation of variable retention. The combinations included mixed retention in CDF and CWHxm, dispersed retention in CWHdm, and group retention in CWHvm+vh. A particular emphasis was placed on three habitat elements that had been identified as weak points from initial monitoring or by the International Scientific Panel reviews: tall snags, deciduous trees, and large-diameter CWD.

Retention of live trees, snags, and, to a lesser extent, CWD rose in operational blocks in all subzones from 1999 (the initial year of VR implementation) to 2000 and 2001. In the CWHdm and CWHvm+vh, levels of some of these main elements stayed the same or continued to increase through 2003-04. In the drier subzones, however, retention levels declined back to or fell below 1999 levels.

Habitat elements specifically identified for operational improvement showed encouraging increases from the first year of implementation (1999) to 2000 or 2001, particularly for live deciduous trees and tall snags (Figure 9.8). Large coarse woody debris showed little improvement (Figure 9.8). However, in most 2003-04 operational blocks, the specifically identified elements were at or below the levels attained in the initial 1999 blocks.

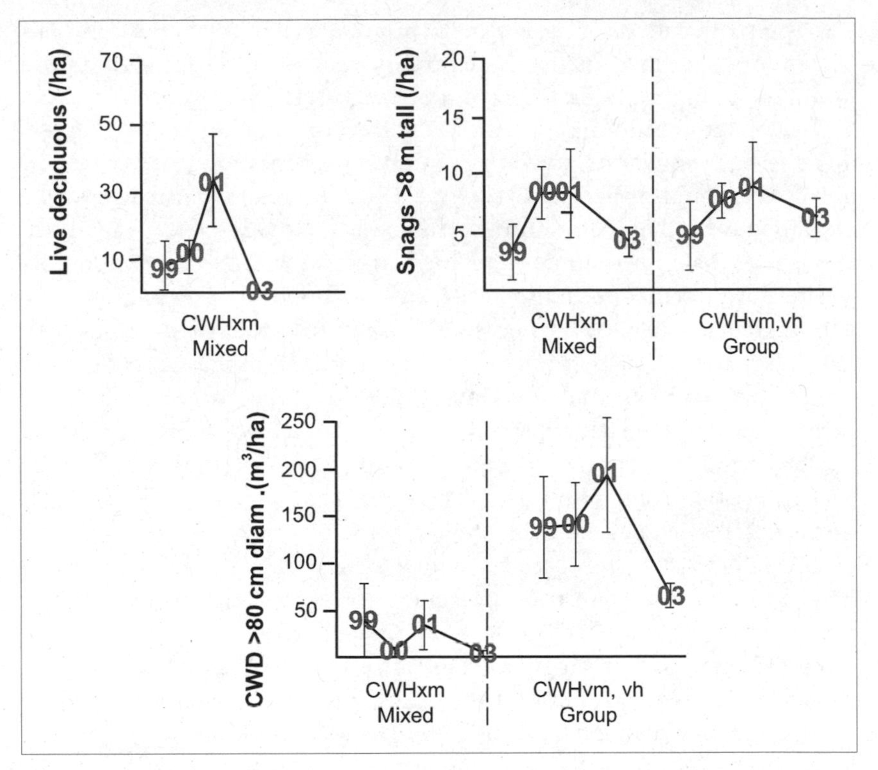

Figure 9.8 Levels of habitat elements initially identified for operational improvements in operational blocks in 1999, 2000, 2001, and 2003-4.
Note: Error bars are ±1 SE.

Although these results suggest that operational retention has not improved in response to initial monitoring and Science Panel recommendations, the results may be misleading because relatively small numbers of operational blocks are monitored every year. Additionally, changing markets and corporate pressures can lead to year-to-year fluctuations in the amount and type of retention or even the predominant subzones harvested. Longer-term monitoring is needed to accurately assess whether retention of habitat elements is improving or declining in operational cutblocks.

9. General Discussion

Monitoring the retention of habitat elements has great potential to provide information to management practices because stand-level elements are directly affected by on-the-ground forest management decisions. There are direct opportunities for rapid feedback. The ease of measuring habitat features allows extensive monitoring, which in turn means that the monitoring can be designed to address many of the priority comparisons identified for the monitoring program (summarized for all indicators in Chapter 12).

The habitat monitoring summarized above led to a number of management recommendations noted in preceding sections. For several reasons, however, monitoring had relatively little direct impact on practices. A major limitation in applying monitoring results was high year-to-year variability in practices that resulted from changing markets as well as changing corporate and government directions. The decline in retention of several important habitat elements in 2003-04 operational retention blocks, for example, may reflect high timber and pulp prices as well as increased emphasis on economic values by government and the company. These fluctuations simply indicate that monitoring of operational progress, and of the effectiveness of practices, is a long-term venture. It is probably naive to expect direct short-term feedback from monitoring results in the face of much stronger economic pressures on forest managers. Ultimately, monitoring may lead to improved practices not by direct feedback but simply by serving as a frequent reminder that particular forest practices are valued for more than their economic contributions.

A second limitation to applying results from monitoring of habitat structures is complexity arising from the large number of potential variables, ecosystem types, and comparisons. Comparing retention results to benchmark sites to identify and focus on the weakest points in retention is a useful approach to simplifying the message for practitioners. That approach created progress in the initial years of implementation, when the operational environment was conducive to improvement. One difficulty with the approach is that it can lose sight of the fact that we want to maintain a diversity of habitat, not just a few focal elements. Relating habitat elements to the needs of a range of organisms through empirical habitat models is one way to incorporate many habitat elements into fairly simple, easily understood indices. So far, relatively little progress has been made on using habitat models within the monitoring program. The primary limitation has been that developing such habitat relationships frequently requires organism monitoring to use study designs and measurements that are different from those needed to make direct comparisons.

A final limitation of applying results from habitat monitoring is the recognition that retention is intended to maintain habitat elements throughout the rotation. Immediate post-harvest monitoring only examines a small part of this long-term objective. Unless long-term trajectories of habitat elements are projected, using simulation models, immediate post-harvest results can always be discounted: high retention levels of some element can be countered by concern that the high levels will not last long into the rotation; low levels can be countered by the hope that the element will be recruited quickly in the developing stand. Projections are limited by uncertainty for many key parameters, particularly because we are concerned with features and harvest systems beyond those that traditionally have been the focus of silvicultural modelling. Effects of retained trees on growth and mortality of regeneration, performance of retained understorey trees, windthrow and other damage to retained trees, and subsequent effects on deadwood and vegetation cover are all relatively poorly known. Some of the required information can be provided by longer-term remeasurements of habitat monitoring plots, but other information needs will have to be met by auxiliary measurements and separate studies (see Chapter 12).

9.5 Summary

Monitoring of habitat structure addressed six comparisons that were ranked as high priority within the adaptive management program.

1 Group retention had higher levels of retention of many habitat elements than did dispersed retention, though dispersed retention did better at retaining large-diameter live trees, large but short snags, and a few other variables. Mixed retention was generally similar to group retention, though variable among different forest types.
2 Compared to benchmark sites, large trees, overall basal areas, and some vegetation layers were retained at lower levels in retention patches, while deciduous elements were more abundant. Differences likely reflect more low productivity sites and wetlands in retention patches.
3 The abundance of habitat elements was proportional to percent retention for many tree and snag elements in group retention but more variable in dispersed retention. Amounts of CWD and most vegetation layers were generally independent of percent retention in both systems.
4 Most habitat elements showed no effect of distance from patch edge in our surveys shortly after harvest. A few variables showed very minor decreases near the edge of retention patches. Only tall snags had enough of a decline near the edge to have any implications for size or shape of retention patches.
5 Patches anchored on wetlands had the same or elevated levels of most habitat elements compared to typical forest, while patches anchored on rock outcrops had substantially reduced levels of many variables, including large trees, snags, CWD, and some cover layers. Patches specifically targeted at wildlife features or habitat structures differed little from other forested patches.
6 Operational blocks harvested in 2000 and 2001 showed substantial increases in retention of many habitat elements, including three specifically

> targeted for improvement, when compared with blocks harvested in 1999, the first year of applying VR. However, most of these variables had declined back to 1999 levels in 2003-04 operational blocks, reflecting good timber and pulp markets and a greater emphasis on economic values.

Most potential benefits from monitoring habitat elements are not accrued in the short term. Monitoring habitat is a long-term venture.

10
Sustaining Forest-Dwelling Species

Laurie L. Kremsater and Fred L. Bunnell

10.1 Rationale

Indicator 3 is the finest assessment of the criterion for maintaining biological diversity and monitors the presence and population trends of organisms. It is informative for species whose requirements we know well or can track over time. There are clear trade-offs between coarse-filter approaches that deal with habitats of several species and fine-filter approaches that address the needs of a few species (Hansen et al. 1999). We chose to allocate effort to both approaches. We believe that the requirements of most species can be met by keeping some unmanaged areas and retaining important habitat structures; they are assessed by Indicators 1 and 2 (ecosystem representation and habitat). Monitoring Indicator 3 (organisms) helps to assess whether assumptions about the provision of habitat structure and representation actually result in persistent populations of species.

There are three reasons to monitor species. The first is to ensure that the generative, adaptive capacity of the forest is sustained (§3.2.1). Sustaining species across their distributions is the simplest and most effective way of sustaining desired genetic variability, a fundamental reason for managing for biological diversity (Bunnell 1998b; Frankel and Soulé 1981; Namkoong 1998; Soltis and Gitzendanner 1999). Evading extreme changes in relative abundance reduces the risk of extinction and loss of genetic variability due to extremely low numbers (Gilpin and Soulé 1986) or possible negative effects on other species due to extremely high numbers. A second reason is to provide credence to Indicators 1 and 2. These indicators are used because it is clearly impossible to monitor the entire 1.1 million hectare study area or more than a small fraction of species. Monitoring species helps to assess the effectiveness of these indicators in representing species. The third reason for monitoring species is that the public views sustenance of species as the ultimate measure of success or failure in sustaining biodiversity.

Two key points guide the monitoring of species. First, monitoring all species is not possible. Second, focusing on individual species can seriously detract from the goal of sustaining species richness by allocating resources to a very small portion of the species present. Monitoring must therefore consider both individual species that are useful in answering specific questions and, more generally groups of species. In §10.2, we consider broad attributes of species that make them informative within the adaptive management program. In §10.3 through §10.8, we review groups of species in terms of these attributes.

10.2 What to Monitor? An Overview

Because monitoring all species is impossible, efforts have focused on selecting a relatively small number to monitor. Much has been written on the use of flagship species, umbrella species, keystone species, indicator species, or focal species in attempts to represent the needs of all species. Each approach for selecting species to monitor has its weaknesses. For example, flagship species are usually those of high public profile but are often not fundamental to the rest of the community and often require prohibitively large resources to monitor (Caro and O'Doherty 1999; Simberloff 1998). Using umbrella species, those whose needs encompass those of many other species, is an attractive concept, but whether other species really fall "under the umbrella" is often more a matter of faith than of evidence (Roberge and Angelstam 2004). Monitoring keystone species, those that play critical roles in a community, also has promise (Simberloff 1998), but keystone species are difficult to identify, and the notion that only a few species are critical to the ecosystem is contrary to other theories of community ecology. The complexity of ecological interactions and ignorance about them argue against the strict application of keystone species as the focus for monitoring (Mills, Soulé, and Doak 1993; but see also Symstad, Willson, and Knops 1998). While the use of indicator species has become a general concept, it is often difficult to know what selected species were intended to indicate. The focal species approach (Lambeck 1997), using a suite of species rather than a single species as indicators, is an improvement over the assumption that a single species can reliably act as a surrogate for a major portion of biodiversity. However, the total number of species still outnumbers feasible focal species by orders of magnitude.

Reviews of species-based monitoring approaches reveal that no single species, nor even a group of species, accurately reflects entire communities. Understanding the response of a single species may not provide reliable predictions about a group of species even when the group is comprised of a few very similar species (Lindenmayer 1999). Typically, the responses of one group of organisms (e.g., vascular plants) does not serve as a comprehensive surrogate for those of other groups (e.g., insects or lichens) (e.g., Chiarucci et al. 2005; Crisp, Dickinson, and Gibbs 1998; Landres, Verner, and Thomas 1988; Oliver, Beattie, and York 1998; Pharo, Beattie, and Binns

1998), and the failure of species-based biodiversity surrogates has occurred across a wide range of environments (Lindenmayer and Fischer 2003).

Despite the weaknesses of any single approach, we still need to select species to monitor. Experience to date reveals attributes that increase the probability that selected species will be useful indicators. Ideally, indicator species should be relevant to the ecological phenomenon of interest, able to differentiate between natural and human-induced stresses, tied to management objectives, sufficiently sensitive to provide early warning, widely applicable geographically, easy and cost effective to measure and interpret, and repeatable (Ferris and Humphrey 1999; Larsson 2001). The preceding attributes are desirable in all indicators. Other characteristics are specific to selecting species. For a species to be a useful indicator, it must have a relatively high population size, well-known taxonomy, good background information (relating to the question of interest), high ecological fidelity, and known statistical properties (Woodward, Jenkins, and Schreiner 1999). Additionally, the species must be sensitive to factors of interest, ideally showing cause and effect rather than merely correlation with the management actions examined (otherwise, species could be increasing or decreasing due to factors not measured). Levels of variability need to be low enough to allow ecologically meaningful data, and ideally, information on the species allows scaling from the site to the landscape or to other geographic areas. When selecting species, it is useful to consider whether they have attributes of umbrella or keystone species.

Because no single indictor possesses all desirable qualities, a set of complementary indicators is often sought. Plants are often chosen as indicators owing to their relationships to edaphic and climatic factors and their role in providing habitat for dependent fauna. Lichens may be useful due to requirements for forest age, close association with particular tree species, and their role as food sources for other organisms. Some invertebrates are sensitive to habitat structure and have large functional significance among a wide array of life history patterns but exhibit high redundancy. Vertebrates are often chosen as umbrella species that need large tracts of habitat (Landres, Verner, and Thomas 1988); some rely on particular structures, and some (e.g., wood peckers) provide habitats for other organisms. Some researchers suggest using exotic species as indicators (Larsson 2001; Noss 1999); others note the value of selecting native species, species with large area requirements, and species that are dependent on specific stand types (Larsson 2001). Choosing species based on specialist versus generalist, body size, area needs, or other features is not as important as ensuring that the species is useful in answering the management questions that focus the monitoring program. Species must be sensitive to the habitat attributes manipulated and measured. Rare species are not usually recommended for monitoring to determine responses to management actions (Broberg 1999). Rarity and population vulnerability to extinction are better used as factors to decide whether they are species to prioritize for conservation (Breininger et al. 1998).

We do not presume that our choice of species to monitor reflects the health of the whole community; rather, we hope to accomplish two goals: first, to select a sufficient range of species to give credence to Indicators 1 and 2; second, to select species that respond to the particular management questions that focus the monitoring program. Meeting both objectives requires that we select a range of species in terms of taxonomic diversity, body size, mobility, and area requirements. Because of the extremely different requirements and life histories of different taxonomic groups, taxonomic diversity is necessary to avoid reinforcing prejudices based on experience with a single taxonomic group. A range of body size, mobility, and area requirements recognizes the wide range of scales affected by forest practices.

To encompass most life forms, we evaluated six broad groups of forest-dwelling organisms: (1) vascular plants, (2) bryophytes, (3) lichens, (4) macro-fungi, (5) terrestrial invertebrates, and (6) terrestrial vertebrates. Our discussion is limited to macroscopic terrestrial organisms. We assume that microscopic organisms are monitored indirectly as other aspects of ecosystems are monitored (e.g., tree and plant growth, organisms higher in the food chains).

Within groups, species selected for monitoring must be chosen carefully. For each group of organisms, we polled experts to identify taxa of concern, then examined those taxa and other potential candidates in terms of four features intended to assess their utility to monitoring:

1 must be forest dwelling;
2 must be sensitive to forest practices employed;
3 must be practical to monitor in terms of sampling, identification, and cost; and
4 must provide information useful in guiding forest practice.

Forests are not the primary habitat of all species occurring on the study tenure. Species that are not primarily forest dwelling require careful design to separate effects of forestry from non-forest-related influences on populations. Some forest-dwelling species are relatively insensitive to changes in forest structure, so they provide no indication of changes that might impact other species dramatically. We want to select organisms that span a range of sensitivities to forest practices, from moderately tolerant to sensitive. This range is useful to explore the different roles of different features of management (e.g., reserve areas, structural retention, landscape features). We want to include species about which there is a range of knowledge, from well-known to poorly known to unknown species. Considering only well-known species creates severe taxonomic restrictions and fails to account for the vast majority of species that are not well known. However, some groups of species (as among the invertebrates) are so poorly known in terms of taxonomy and natural history that they can be neither sampled nor interpreted effectively. Some species that are sensitive to forest practices, and can be sampled, are equally sensitive to other influences, so they cannot be used to guide forest practice without carefully designed comparisons and intensive study information (e.g., separating effects of forestry on salmon from effects of oceanic conditions and fishing is very difficult).

For six broad groups of organisms, we briefly discuss features specific for each group under the headings Rationale, Factors Influencing Monitoring, How to Monitor, and Links to Management. Within each group, we sought species that could inform choices among forest practices and were amenable to sampling and identification. The following sections review how focal taxa were determined from the six broad groups of organisms. We use the term "focal taxa" not in the restricted sense of Lambeck (1997) but in the common English usage of "relating to focus," recognizing our focus as the potential effects of forest practices. Within each group, the selection was made largely on the basis of the species' sensitivity to forest practices, ease of sampling, and utility of information to inform management. We considered rarity in terms of potential influences of forest practice. The resultant list of species must also be evaluated as a whole. An effective package includes species representing a range of mobility, range of turnover rates, specialists and generalists, different trophic levels, large and small home ranges, some with close associations to a range of habitat elements, and some that are known less well. The final list of focal species (§10.10) is an amalgam of potential candidates reviewed in the following sections.

10.3 What to Monitor: Vascular Plants

Vascular plants are obvious, recognizable components of biological diversity, including trees, shrubs, herbs, grasses, and ferns. Because they are long lived and offer unique information, trees are a special case. Within the monitoring program, but outside this indicator, tree growth was tracked as a biomonitor of soil productivity. Estimates of growth, mortality, and decay rates are included in permanent sample plots, and effects of retention on tree growth and decline are examined. Indicator 2 tracks the distribution of tree species relative to VR practices, while Indicator 1 tracks the composition of major overstorey species. Specific growth and yield programs (outside this monitoring framework) look at effects of variable retention on trees.

Indicators 1 and 2 also assess status of some vascular species other than trees. Assessing representation using site series groups[1] (Indicator 1) evaluates habitat provided for plants that are indicators in the BEC classification and an *unknown* number of additional species. Indicator 2 monitors habitat structures left by different forms of retention, recording percent cover for broad groups of plants (e.g., herbs and shrubs). Most vascular plant species, however, are not used to delimit ecosystems or considered as individual elements of habitat structure, so monitoring only Indicators 1 and 2 could overlook impacts on plants.

We addressed three broad features when considering the use of vascular plant species for monitoring: (1) whether they were monitored by other indicators, (2) whether they could reveal impacts of forest practice, and (3) whether they had site specificities that could aid or confuse monitoring.

10.3.1 Factors Influencing Monitoring: Vascular Plants

We considered seven groups of factors to guide the monitoring design and select informative species.

Forest patch size and edge

There is little evidence that vascular plants need large patches of "forest interior" habitat, although some species disperse poorly, so contiguous habitat can govern their presence. However, even small forest fragments can be important for conserving plant species richness (Bratton 1994; Honnay, Hermy, and Coppin 1999; Kremsater and Bunnell 1999; Meier, Bratton, and Duffy 1995; Moola and Vasseur 2004; but see Matlack 1994). Edge effects for vascular plants generally extend only about 25 metres into the forest but farther for some species.[2] In the Pacific Northwest, two species appear to be potentially revealing of forest interior conditions – *Goodyera oblongifolia* and *Trillium ovatum* (Chen, Franklin, and Lowe 1996; Jules 1998; Jules and Rathcke 1999).

Forest age and canopy closure

There is ample evidence that many species do not persist on highly disturbed sites (e.g., Gilfedder and Kirkpatrick 1998; Moola and Vasseur 2004; Trofymow et al. 2003), but we found no literature documenting vascular plants restricted to old forest, even though they may be more abundant there. Species most closely associated with old growth may be those requiring well-developed humus or rotting wood and relying on mycorrhizal fungi in rotting wood for delivery of nutrients (e.g., Leake 1994; Trofymow et al. 2003), such as *Allotropa* and *Hemitomes*. Although not a direct effect of age, clear-cuting could reduce the presence of species associated with partial canopy closure (e.g., devil's club [*Oplopanax horridus*], some *Pyrolacea*, and some *Erythronium*). The ability of variable retention to sustain such species is unknown.

Aspect and elevation

In coastal British Columbia, strong effects of aspect are not apparent, likely reduced by the cool, moist climate conditions. Both species composition and richness of vascular plants vary with elevation. In British Columbia, many plants are associated with, though not limited to, broad elevation bands (e.g., mountain hemlock, *Castilleja rhexifolia, C. elmeri, Valeriana uliginosa*) (Douglas, Meidinger, and Pojar 1999; Douglas, Straley, and Meidinger 1989-94, 1998). Current data suggest that we need to consider only broad elevation bands when monitoring for vascular plants. That can be addressed by Indicator 1.

Soil moisture

Changes in vascular plant composition with soil moisture are well documented. Although many species have clear moisture preferences (e.g., "dryish" or xeric to subxeric sites), none in the study region appears to be restricted to very narrow soil moisture tolerances (i.e., just xeric or just mesic) (Douglas, Meidinger, and Pojar 1999; Douglas, Straley, and Meidinger 1989-94, 1998). We found no plants restricted to moisture gradients finer than the groupings of site series used for analysis by Indicator 1 (§7.3.3). To accommodate moisture, monitoring beyond Indicator 1 appears to be unnecessary. Plants associated with alluvial soil types are not influenced solely by moisture, so riparian areas should be delineated as separate strata when examining effects of forestry on vascular plants.

Soil characteristics

Rare and endemic plants are often restricted to soils of uncommonly low or high pH (e.g., Brunet 1993; Gustafsson 1994). Preferred habitat of the two endemics in the study area confers considerable immunity from forest practices (Figure 10.1). Limestone substrates, which occur on the study tenure, support some rare species (Bunnell and Squires 2005; Lewis and Inselberg 2001). The representation of limestone substrates should be assessed under Indicator 1. Alluvial soils are also uncommon and often support a greater richness of plants, including some rarer species. If rare soil types are not well represented within Indicator 1, then some species showing strong affinities to these soils should be included in species monitoring (see Appendix 1).

Rarity

Monitoring is complicated when plants have narrow habitat tolerances and are rare geographically. Rare species can be geographically widespread but infrequent throughout their distribution or be locally abundant in a very narrow geographical range. Few of the rare plant species listed for the study area[3] are found in sites affected by forest practices, and most of the forested records are from alluvial or riparian areas that will be protected from management activities. Nine rare species are found in forests and should be recorded individually if found in vegetation plots.

Figure 10.1 *Sinosenecio* is common on Haida Gwaii in moist bogs, meadows, rocky slopes, and forests from lowlands to alpine zones. The *Saxifraga* is rare, on moist rock outcrops, talus slopes, and cliffs in the subalpine and alpine zones (on Haida Gwaii and northwestern Vancouver Island). | *Photographs by J. Pojar.*

"Invaders"

Alien or exotic species can become invasive, with detrimental effects on native species (Lonsdale 1999; Mack et al. 2000), and could be encouraged by roads and other activities in the Timber zone (Cadenasso and Pickett 2001; Lonsdale 1999; Pauchard and Alaback 2004). Tracking the distribution of introduced species in recently harvested retention blocks or along newly constructed roads could expose potential invasion.

10.3.2 How to Monitor: Vascular Plants

Habitat specificity of vascular plants suggests that studies of impacts of variable retention should stratify or block by elevation, aspect, soil moisture, and soil type. Indicator 1 provides sufficient guidance for elevation and soil moisture in most instances, but riparian and alluvial sites should be delineated as separate strata of interest when examining effects of forestry. Rare soil types are poorly mapped, suggesting that focal species should include some species restricted to such types.

White trillium and pink fawn

Gnome plant

Figure 10.2 Candidate vascular plant indicator species include: Trillium (*Trillium ovatum*), which may indicate effects of edge or patch size; pink fawn lily (*Erythronium revolutum*), which may indicate sufficient canopy cover within variable retention; gnome plant (*Hemitomes congestum*), which may indicate abundant rotten wood associated with old-forest conditions. | *Photographs by: J. Pojar.*

Species revealing potential impacts of practices are summarized in Appendix 1 in eight groups associated with (A) canopy cover, (B) down wood or rich humus, (C) older forests, (D) forest interior conditions, (E) rarity, (F) stream sides or moist forests, (G) specific forest hosts, and (H) species that are exotic (see Figure 10.2).

Groups A through D help to assess the degree to which forest practices, in this case variable retention, approximate conditions in intact and older forests. Group E includes nine rare species that would be useful to record individually if found in vegetation plots. Some of these species can be selected for their ability to reveal poorly mapped, rare soil types. For the study area, we know of no species requiring conservation measures outside the provincial ecological reserve system. Group F is intended to help assess the degree to which natural species composition in the rich riparian and alluvial areas is sustained. Mistletoe (Group G) was included because a potential and undesirable consequence of widespread variable retention is increased infection rates by mistletoe. Most forest practices have the capacity to increase the numbers of exotic vascular species, so potentially invasive species were noted (Group H). We do not suggest monitoring all species listed in Appendix 1. Rather, representatives from each group should be considered as focal species in the monitoring program.

Monitoring differs among zones. In the Old-Growth zone, it is unlikely that the type and extent of forestry activities (30 percent of the area harvested lightly) will adversely affect vascular plants. Monitoring of vascular plants in the Old-Growth zone should

- periodically (e.g., five-year intervals) check whether known, geographically restricted vascular plants are maintained; and
- include tracking exotics by presence/absence checklists during habitat structure surveys.

Monitoring in the Timber and Habitat zones is focused by the assessment of ecosystem representation under Indicator 1. Issues potentially affecting vascular plants are reduction of old-growth and forest interior habitat and invasion by exotics. Few vascular plants appear to be limited to old-growth or forest interior habitat (see Appendix 1). Unless analysis of data from pilot studies suggests otherwise, monitoring should

- focus on under-represented ecosystem types to assess whether managed blocks are maintaining vascular plants (e.g., low, dry variants);
- select representatives in Appendix 1 as focal species during habitat structure surveys of variable retention, in both patches and openings (perhaps six to ten species in structure plots);
- assess focal species by percent cover or presence/absence for less abundant species;
- establish plots in deer winter ranges and other larger retention patches (e.g., old-growth benchmarks) to assess whether vascular plants associated with old growth or forest interior, or requiring a degree of canopy closure are being maintained;
- establish plots in older clearcuts to assess whether potentially sensitive species recover after traditional harvesting practices (clearcut benchmarks);
- track the persistence of species showing strong affinities for riparian habitat in areas with minimal riparian buffers;[4] and
- periodically track changes in abundance and ranges of exotic species using permanent plots placed in bands of activity: near mainline roads, near secondary logging roads, in retention blocks, and in untreated benchmarks.

We do not suggest an inventory of rare plants, but where they are found, their responses to disturbance should be tracked annually (i.e., included in structure assessment checklists).[5] To help field crews note rare plants when encountered, a small brochure with colour photographs of rare species was prepared (Beese 2003).

The list of species monitored should include a mix of characteristics, among them, pioneer to climax, good to poor dispersers, long versus short persistence (perennial long-lived versus annual), and species showing affinities with particular stand attributes that are changed by forest practices.

During sampling for Indicator 2, potentially sensitive species can be recorded on an individual basis (rather than as percent cover of species groups). Given their relative rarity, presence/absence is likely sufficient. Species believed to be sensitive to forest age should be sampled within larger "set-asides," such as deer winter ranges or other old-growth reserves, as well as retention blocks. Because the sensitivity of most species noted in Appendix 1 is poorly quantified, they should be recorded individually in pre- and post-treatment surveys during active adaptive management.

10.3.3 Links to Management: Vascular Plants

Vascular plants are feasible to monitor, and methods are well established. At question is the value of the information. The most useful information will come from tracking potential expansion of the range and abundance of exotics, changes in range of less common plants, and effects of variable retention for species assumed to be dependent on old growth or forest interior. Habitat surveys may reveal more locations for rare plants that will add to lists supported by the Conservation Data Centre.

Links to management may be specific for each focal species. Effects of variable retention on species potentially associated with older or continuous forest may inform choice of size and shape of retention. Planning the locations of unharvested areas and patches can be guided by the distribution of rare or geographically restricted plants. If representation of these species is strong in the unmanaged land base, then it is less important to create harvesting practices that maintain or allow rapid recovery of these species.

10.4 What to Monitor: Bryophytes

Bryophytes include mosses, liverworts, and hornworts[6] and are a conspicuous component of old-growth forests in the Pacific Northwest (Lesica et al. 1991; McCune 1993; Newmaster et al. 2003; Schofield 1988). Bryophytes reproduce by spores or by vegetative means, as fragments of vegetative parts or by special asexual propagules. Dispersal distances are commonly short (Crum 1972; Khanna 1964; Sillet et al. 2000; Söderström 1987). Most forest species appear to be prone to desiccation because bryophytes lack well-developed systems of water uptake and retention. Narrow microhabitat preferences coupled with short dispersal distances suggest that some could be vulnerable to forest practices. Specifically, species having strong associations with down wood may not survive on wood in exposed openings. Species prone to desiccation also may not survive well in small forest patches where edge effects dominate the microclimate. Conversely, the forest influence of small patches may extend into the opening to create larger areas of favourable habitat. The fate of bryophytes may thus serve as indicators of the response of other poor dispersers that favour sheltered environments or moist down wood.

10.4.1 Factors Influencing Monitoring: Bryophytes

For the liverworts and mosses, we considered features influencing their utility as focal species, including affinity for woody substrate, effects of forest age, and responses to lack of forest interior (patch size or negative edge effects).

Substrate

Many common mosses and liverworts prefer rotting wood as substrate (Table 10.1) (Crites and Dale 1998; La Roi and Stringer 1976; Muhle and LeBlanc 1975; Rambo and Muir 1998; Söderström 1988). Some bryophytes occur primarily as epiphytes (Gustafsson et al. 1992a; Kuusinen 1994a, b; Nadkarni 1984b),[7] some are restricted to particular tree species (Andersson and Hytteborn 1991; Gustafsson et al. 1992a; Hazell et al. 1998), and some prefer specific positions on tree boles (Hazell and Gustafsson 1999; Sillett 1995). Bryophytes are also frequently associated with moist conditions, including riparian areas (Glime and Vitt 1987; Jonsson 1996; Muhle and LeBlanc 1975). Substrate specificity can change with climatic conditions. For example, mosses may do best on specific substrates but can use those substrates in a range of conditions or change substrates under different climatic conditions (Frego and Carleton 1995; Rambo and Muir 1998; Sultan et al. 1998).[8] Because many bryophytes appear to exploit sites at a fine scale and show considerable overlap at larger scales, the choice of focal species for monitoring must be made cautiously.

Forest age

Many studies report that bryophyte richness increases with age, with some species more common in old-growth than in younger forests (e.g., Crites and Dale 1998; Edwards 1986; Lesica et al. 1991; McCune 1993; Newmaster et al. 2003; Rambo and Muir 1998). Studies that report no increase in richness usually report an increase in abundance with stand age (e.g., Boudreault et al. 2002; Sadler 2004). Because of relatively slow rates of colonization and growth, it may take many decades for bryophytes or lichens to colonize and develop appreciable biomass. Some bryophytes disperse poorly, making long-distance colonization events rare. The fact that some species are absent in young stands may simply indicate inadequate time to colonize the stand rather than absence of favourable stand attributes in younger stands. Successful colonization also depends on the presence of appropriate substrate,

Table 10.1

Common liverworts and mosses closely associated with forest substrates in coastal BC forests

Common name	Scientific name	Habitat or substrate
Liverworts		
Maple liverwort	*Barbilophozia lycopodioides*	forest floor humus
Hardscale liverwort	*Mylia taylorii*	logs
Three-toothed whip liverwort	*Bazzania tricrenata*	lower tree trunks/bases of trees
Little hands liverwort	*Lepidozia reptans*	decaying wood/ rotten stumps
Yellow laddle liverwort	*Scapania bolanderi*	tree bark/ stumps/logs
Tree-ruffle liverwort	*Porella navicularis*	tree trunks/branches
Hanging millipede liverwort	*Frullania nisquallensis*	tree trunks and branches
Mosses		
White-toothed peat moss	*Sphagnum girgensohnii*	woodland bogs
Common hairy cap moss	*Polytrichum commune*	terrestrial in moist forests
Magnificent moss	*Plagiomnium venustum*	rotten logs/ lower portions of tree trunks
Badge moss	*Plagiomnium insigne*	humus in moist shaded forests
Menzies' red-mouthed mnium	*Mnium spinulosum*	soil and needles under conifer trees in heavy shade
Clear moss	*Hookeria lucens*	moist humus in rainforests
Douglas' neckera	*Neckera douglasii*	tree trunks and branches
Hanging moss	*Antitrichia curtipendula*	tree trunks and branches
Rough moss	*Claopodium crispifolium*	logs and tree bases
Coiled leaf moss	*Hypnum circinale*	tree trunks and branches
Oregon beaked moss	*Kindbergia oregana*	logs/humus/tree bases
Slender beaked moss	*Kindbergia praelonga*	logs/humus/tree bases
Lanky moss	*Rhytidiadelphus loreus*	humus-rich substrates in lowland forests
Goose-necked moss	*Rhytidiadelphus triquetrus*	humus-rich substrates in montane forests
Step moss	*Hylocomium splendens*	humus-rich soils mainly in cedar-hemlock forests
Wavy-leaved moss	*Plagiothecium undulatum*	logs/humus/tree cotton stumps/soil in shade
Dusky fork moss	*Dicranum fuscescens*	rotting logs
Broom moss	*Dicranum scoparium*	rotting logs and humus
Curly thatch moss	*Dicranoweisia cirrata*	logs/trees/woody substrates (not necessarily in forests)

Source: Summarized from Pojar and MacKinnon (1994).

Figure 10.3 Mosses that are normally terrestrial on humus and down wood (e.g., *Hylocomium splendens, Polytrichum juniperinum*) enter older canopies and gradually migrate upward as the canopy ages. *P. juniperinum* is shown. | *Photograph by: T. McIntosh.*

but the effects of stand age and substrate are rarely separated.[9] Whether bryophytes are useful indicators of old-forest conditions or simply of decaying wood has not been assessed. In some cases, the preference for substrate appears to change as forests age (Figure 10.3) (Pike et al. 1975; Sillett and Neitlich 1996).

Edge and patch size

Because many appear to prefer moist microclimates, liverworts and mosses may be adversely affected by edges between forests and open areas. Findings are variable, with several studies reporting diminished abundance or growth near edges (Harper et al. 2004; Hylander 2005; Moen and Jonsson 2003; Norton 2002; Zartman 2003), some reporting variable responses (Baldwin and Bradfield 2005), and one (in Oregon) reporting that none of the sixty-five epiphytic, non-vascular plants surveyed was seriously declining after two decades of exposure to a clearcut edge (Sillett 1995). Regardless of tree location relative to the edge, two mosses (*Bryum gemmascens* and *Isothecium cristatum*) and one liverwort (*Radula bolanderi*) were more abundant in sheltered portions of tree crowns than in exposed portions. The retention of bryophytes in sheltered positions of the crown suggests that large trees with well-developed crowns may provide refuge and habitat for bryophytes and lichens even if those trees are found in small patches. The evidence for strong effects of edge or patch size is equivocal, particularly for the Pacific Northwest.

Rarity

The BC Conservation Data Centre lists fifty-one moss species as rare or extirpated in the area of the study tenure (Appendix 2). If identification is possible, rare species associated with forest attributes (Table 10.2) should be recorded when encountered and tracked. However, monitoring rare species is unlikely to inform forest practice because their rarity makes them hard to find and hinders comparison of forest practices.

Table 10.2

Rare liverworts and mosses potentially on the study tenure and potentially affected by forest practices

Species	Habitat or substrate
Rare liverworts	
Cephalozia connivens	rotten wood
Chandonanthus hirtellus	humus
Coloeleunea macounii	rock and maple bark
Herbertus sendtneri	humus
Kurzia trichoclados	humus bank
Lophozia ascendens	tree trunk/soil/rock
Tritomaria exsecta	soil/rock/rotten wood
Rare mosses	
Bryhnia hultenii	log
Callicladium haldanianum	rotten log
Daltonia spachnoides	tree trunk
Fabronia pusilla	soil/*Populus* tree trunks
Micromitrium tenerum (extirpated)	humus
Oedipodium griffithianum	humus
Orthotrichum hallii	rock/alder bark
Orthotrichum pylaisii	limestone/base of trees
Orthotrichum rivulare	shrubs/rocks/roots (also in riparian areas)
Orthotrichum tenellum	deciduous trees (*Acer, Populus*)
Tortella humilis	dry log/rock

Source: Rare species extracted from Ryan (1996).

10.4.2 How to Monitor: Bryophytes

Concerns about bryophytes are that they will not persist in retained patches or on down wood after harvest. If they do not persist, then rates of dispersal or recolonization are important. Evaluating rates of colonization is necessarily long term and helpful only if bryophytes disappear quickly. There is little evidence suggesting that species disappear quickly as canopy is removed, so monitoring can focus on persistence and effectively utilize retrospective studies around remnant patches. Properly designed, the monitoring may yield useful information on the importance of substrate in relation to exposure to desiccation and rates of colonization as they are influenced by duration of exposure in openings, forest age, patch size, and edge effects. Monitoring must consider the moisture status of the site (site series or topographic position) because moisture status can confound potential effects of amounts of down wood and other stand variables.

Any monitoring of bryophytes should

- focus on persistence in patches and on down wood in openings;
- consider common mosses selected from Table 10.1 to include species preferring a variety of substrates from decayed wood to live tree boles and branches;
- be designed to distinguish potential effects of amount and condition of substrate from effects of forest or stand age; and
- evaluate potential influences of microclimate conditions by surveying openings, edges, and patch interiors.

The assessment of representation by Indicator 1 can help to focus the choice of bryophyte species on those reported from under-represented ecosystems. A few rare bryophytes have been found on wood or humus (Appendix 2; Table 10.2) and may be affected by forestry practices. If they are encountered during surveys, they should be recorded and their persistence tracked. Rare mosses and liverworts, like any rare species, are infrequently encountered and can be monitored only opportunistically. If key areas of rare mosses are located, then more focused efforts can be considered. Although bryophytes are readily sampled, expertise in identification is more limited than for vascular plants. Consideration of species to receive monitoring effort must reflect available expertise.

10.4.3 Links to Management: Bryophytes

Bryophytes are potentially useful in assessing biological contributions of down wood retained by variable retention. They may also indicate how well small patches, or even dispersed retention, sustain small, relatively immobile organisms that are prone to desiccation. Given the moist coastal environment, bryophytes may benefit significantly from relatively small amounts of retention, but that has not been evaluated. The assessment of patterns of bryophytes across edges in aging retention blocks may evaluate the "life boating" function of retention and suggest effective sizes of retention patches.

10.5 What to Monitor: Lichens

Lichens make diverse contributions in forests: winter forage for ungulates (Bunnell et al. 1978; Hodgman and Bowyer 1985; Rominger and Oldemeyer 1989; Waterhouse, Armleder, and Dawson 1991), nesting material for many birds and small mammals (Hayward and Rosentreter 1994; Richardson and Young 1977), forage for mites, insects, and gastropods (Richardson 1975), and a source of nitrogen.[10] Few areas of British Columbia's forests have been well surveyed for lichens,[11] but current literature suggests that lichens respond to latitude, elevation, tree species, position in the tree crown, and stand age. Terrestrial lichens also respond to soil type and soil disturbance but are not useful candidates for monitoring on the coast (see below). We reviewed the responses of lichens to forest attributes altered by logging and habitat features such as latitude, elevation, and aspect that could affect sampling design.

10.5.1 Factors Influencing Monitoring: Lichens

Forest age

Some lichen species appear to be closely associated with old growth or continuity of forest cover (Nilsson et al. 1994; Rose 1976, 1992; Selva 1994, 2003). The reasons appear to be short dispersal distances, thus slow colonization rates, and sometimes slow growth rates that make continuity of substrate and favourable microclimate important. Older stands in North America and Europe tend to host more rare species than do young stands, possibly because older stands are themselves relatively rare (Holien 1996; Kuusinen and Siitonen 1998; Selva 1994; Thor 1998). Combined evidence suggests that lichen composition is a useful indicator of forest continuity or age.

The lichen group with the strongest affinity to older stands appears to be the calicioids or pin lichens (FEMAT 1993; Holien 1996; Rikkinen 2003a; Selva 1994, 2003). Some of the nitrogen-fixing or cyanolichens also show affinity for older forests (Arsenault and Goward 2000; Goward et al. 1996).[12] In British Columbia, they are most abundant in low-elevation, moist forests (e.g., Coastal Western Hemlock and Interior Cedar Hemlock zones; see §2.1.2 and §7.3.3 and the Glossary) and tend to dominate old-growth canopies of coastal forests (Pike et al. 1975; Sillett and Neitlich 1996). Evidence elsewhere affirms that the cyanolichens *Lobaria oregana* and *L. pulmonaria* are old-forest associates (Lesica et al. 1991; Shirazi, Muir, and McCune 1996; Sillett and McCune 1998). Stands with *L. pulmonaria* in southern Sweden had, on average, nine red-listed lichens, but stands without *L. pulmonaria* had only one such species (Nilsson et al. 1994). These findings suggest that the presence of *Lobaria* species or other cyanolichens represents stand continuity, old-growth conditions, and increased likelihood of colonization by rare lichens. The utility of other species as indicators of older habitat is unclear. Pendent lichens, such as *Alectoria* and *Usnea,* show variable responses (Arseneau, Ouellet, and Sirois 1998; Esseen, Renhorn, and Petersson 1996; Hyvarinen, Halonen, and Kauppi 1992; Lesica et al. 1991). Common lichens occurring on the study tenure and associated with older conifer stands are listed in Table 10.3.

Age effects are also evident in deciduous stands (Gustafsson et al. 1992b). Our summary of larger lichen species associated with old growth (Table 10.3) focused on those primarily inhabiting conifers, but some also occur on older deciduous trees. Indicator 2 tracks the abundance of deciduous trees. Moreover, provided that riparian buffers (where deciduous trees are concentrated) remain intact, lichens preferring older deciduous trees should be sustained. Monitoring lichen species living on conifers can reveal the relative effectiveness of retention in retaining old-growth features. Monitoring should include representatives of both common and rare species (§10.5.2).

Woody substrates

A total of 854 lichen species grow on trees or dead wood in the Pacific Northwest (Spribille et al. 2006). Of these, 510 grow on bark, 46 have been reported only from dead wood, and 298 grow on both bark and dead wood.

Table 10.3

Common lichens on the study tenure that may be closely associated with old growth

Species	Common name	Habitat
*Lobaria oregana**	lettuce lung	trees
*Lobaria pulmonaria**	lungwort	trees
Platismatia herrei	tattered rag	trees
Platismatia glauca	ragbag	trees
Platismatia norvegica	laundered rag	trees
Hypogymnia inactiva	forking bone	conifers
Hypogymnia enteromorpha	beaded bone	conifers
Sphaerophorus globosus	common Christmas tree	conifer trunks/ branches
Usnea wirthii	blood-spattered beard	conifers
Usnea longissima	Methuselah's beard	trees/shrubs
Bryoria fuscescens	speckled horsehair	conifers
Alectoria sarmentosa	common witch's hair	conifers

Note: All these common lichens are found in the main areas of the tenure: Vancouver Island, Haida Gwaii, and the mainland coast. * indicates nitrogen-fixing lichens.
Source: Summarized from Pojar and MacKinnon (1994).

The loss of bark and the stage of decay determine which species are present. Like mosses, lichens can be adapted to very specific microhabitats, such as resin-impregnated beaver scars and conifer wetwood, a form of damaged heartwood (Rikkinen 2003b, c). Because lichens respond to the pH of bark, they are often restricted to specific genera or groups of tree species (conifer bark typically is more acidic than hardwood bark) (Barkman 1958; Goward et al. 1996; Hyvarinen, Halonen, and Kauppi 1992; Kuusinen 1994a, b; Sillett and Neitlich 1996). The variety of substrates used by lichens cannot be encompassed in a general monitoring program but emphasizes the need for monitoring a diversity of habitat components in Indicator 2.

Edge and patch size

Scattered observations suggest that small patches or even single trees can retain arboreal lichens provided that trees are large with well-developed crowns (Bunnell et al. 2007; Coxson, Stevenson, and Campbell 2003; Daniels et al. 1997; Hazell and Gustafsson 1999; Rominger, Allen-Johnson, and Oldemeyer 1994). The distance of edge effects appears to be short and sometimes absent (Gignac and Dale 2005; Kivisto and Kuusinen 2000; Sillett 1995; but see Esseen and Renhorn 1998). However, transplant experiments suggest that lichens require time to adapt to new conditions of exposure (Hazell and Gustafsson 1999; Sillett 1994; see also Shirazi, Muir, and McCune 1996 and Sillett and McCune 1998). Current data suggest that edge and patch size influences are sometimes evident (Hilmo and Holien 2002; Hilmo, Hytteborn, and Holien 2005; Hilmo, Holien, and Hytteborn 2005), so responses of lichens to edge and patch size should be included in the monitoring program.

Latitude

Changes in lichen richness over the latitude spanned by the management area have been documented (Goward and Ahti 1997; Kuusinen 1994a, b). The management area, however, is disjunct, consisting primarily of two large, well-separated groups of islands and the mainland coast (Figure 2.1). The lichen flora are more likely to differ as a function of differential isolation than of latitude. Sampling need not consider latitude specifically.

Elevation and aspect

Given that species composition of lichens can differ between northerly and southerly exposures of tree trunks (Kenkel and Bradfield 1986; Kuusinen 1994a), we expect differences with aspect. On northern Vancouver Island, slope, elevation, and aspect accounted for 82 percent of the variation in *Alectoria* biomass in old-growth forests. Elevation accounted for most of the difference, and biomass was much higher at elevations of 500 m to 730 m than below 500 m, particularly on sunny aspects (Stevenson 1978; Stevenson and Rochelle 1984). Pendent lichens (*Alectoria, Bryoria,* and *Usnea*) showed similar preferences for mid-elevation in Quebec (Arseneau, Quellet, and Sirois 1998). These widely separated studies suggest that pendent, alectorioid lichens do best in a broad mid-elevation belt.

The responses of nitrogen-fixing lichens to elevation differ among genera, but they usually are more abundant at lower elevations (Berryman and McCune 2006; Goward and Ahti 1992; Goward and Arsenault 1997; Sillett and McCune 1998). Because there appear to be broad trends in lichen abundance with elevation, any monitoring of lichens should stratify by elevation.

Rarity

Rare lichens occur in the study area (Table 10.4), but because of their affinity for microsites they are encountered only opportunistically during monitoring. We consider rare lichen species as we do rare bryophytes. If identification

Table 10.4

Rare lichens on the study tenure that may be closely associated with old growth

Lichen	Habitat	Location within the tenure
Heterodermia leucomelos	over seaside conifers; may be restricted to old growth	Ucluelet
Heterodermia sitchensis	seaside conifers; may be restricted to old growth	Tofino, Ucluelet
*Leptogium brebissonii**	over conifers in open forests; may be restricted to old growth	Ucluelet, HG
*Nephroma silvae-veteris**	old-growth conifers	Terrace, Kitimat
Pannaria ahlneri	old-growth conifers	Terrace, Kitimat
Pannaria laceratula	over conifers; may be restricted to old growth	West coast VI, HG
Pannaria rubiginosa	over conifers; may be restricted to old growth	West coast VI, HG (Moresby Island)
Parmotrema chinense	old-growth conifers southern VI	West coast
Parmotrema crinitum	over conifers; may be restricted to old growth	West coast VI, HG (Moresby Island)
*Pseudocyphellaria rainierensis**	over trees; restricted to old growth	Kitimat, Masset, Chilliwack, Nanaimo Lakes, Brooks Peninsula
Leoderma sorediatum	conifers in open forests; may be restricted to old growth	Ucluelet, HG

Note: * indicates nitrogen-fixing lichens; VI = Vancouver Island; HG = Haida Gwaii.
Source: Derived from Goward (1995).

Figure 10.4 Potentially informative lichens include *Lobaria* (left) and *Alectoria* (right), which are indicative of older forests. *Lobaria* appears to be more consistently associated with older stands than does *Alectoria*. | *Photographs by: Lobaria and moss, W. Beese; and Alectoria, M. Preston.*

is possible, then rare species associated with forest attributes (Table 10.4) should be recorded when encountered and tracked. Rare species are not helpful focal species because their rarity makes them hard to find and hinders comparison of responses to forest practices.

Terrestrial lichens

Terrestrial lichens have much different habitat requirements than lichens found on down wood or in trees, but information on their habitat requirements in coastal regions is scarce.[13] In more arid regions, particularly grasslands, they form intimate associations with other soil-inhabiting biota that are vulnerable to disturbances (Belnap 1994; Eldridge 1996; West 1990). They are poor candidates for monitoring in coastal regions because (1) surface disturbance is less pervasive than in grasslands, (2) many preferred substrates are little influenced by forest practices (e.g., pH of rocks), and (3) relationships with forest practices are too obscure to provide guidance. Monitoring should focus on lichens that grow in trees or on wood and are influenced by forest practices.

10.5.2 How to Monitor: Lichens

The responses of lichens to broad features of the environment affirm the basic ecological strata for monitoring suggested by other groups. Difficulties in identification and affinities for specific microsites eliminate many lichen species from a general monitoring program. The same difficulties eliminate lichens as accessible indicators of dead wood in different decay states. It is more effective to monitor those habitat components directly within Indicator 2. However, some readily identified lichens can be used to assess whether environmental features of intact or older forests are sustained under variable retention. They include pendent lichens and the genus *Lobaria*. Common species (Table 10.3) are most useful. If representatives of the rare lichens (Table 10.4) are encountered, then those sites should be monitored periodically (every five years).

Any monitoring of lichens should

- exploit Indicator 1 to assess the likelihood that rare species are encompassed by unmanaged areas (Table 10.4);
- use pendent arboreal lichens to assess the ability of variable retention patches to sustain arboreal lichens;
- include *Lobaria* and similarly responsive cyanolichens as indicators of older forest conditions;
- stratify by elevation and stand age (elevation could confound treatment effects during active adaptive management);
- exploit retrospective studies to examine patches, degree of isolation, and duration of isolation and use multi-year intersampling periods (effects of disturbance do not seem to have immediate impacts on lichen communities);
- consider edge effects across edges in variable retention patches of different sizes (stratified by elevation) and transects through edges of different ages (e.g., old forest to young, old forest to second growth) to examine dispersal and colonization of adjacent forests; and,
- when focal or rare species are found, record details of substrate (e.g., live tree, species, age, down wood) and decay class.

10.5.3 Links to Management: Lichens

The most rapid links to management will come from retrospective studies. Information on lichen occurrence in remnant patches and across edges of various ages can inform the choice of patch size. Monitoring can assess long-term sustainability by evaluating the persistence of lichens in remnant patches of different sizes and when lichens commonly associated with old forests become established in second-growth forests.

Representation analysis can indicate whether particular elevation belts are lacking in the unmanaged forests (Indicator 1) and where, as a result, larger patches may be necessary to sustain lichens in the Timber zone. Similarly, known locations of rare lichens (Table 10.4) should be assessed against results of Indicator 1 to assess risk associated with forest management.

The results of monitoring current practices will accrue slowly. If such monitoring reveals that variable retention provides lifeboats for lichens to recolonize second growth, then effective dispersal distances should be investigated to indicate the arrangement of patches facilitating maintenance of lichens in managed stands. If species believed to be associated with older stands (e.g., Tables 10.3 and 10.4) cannot be retained by VR, they may become limited to the Old-Growth zone or other larger tracts of unmanaged forest. In that case, the distribution of unmanaged or non-harvestable areas merits re-examination.

10.6 What to Monitor: Fungi

Fungi are neither plant nor animal. They perform critical roles as decomposers, parasites, foragers, and symbionts. The most commonly recognized fungi fruit above ground (epigeous fungi); others fruit out of sight below ground (hypogeous fungi) or have fruiting bodies too small to see (microfungi). We consider hypogeous fungi and microfungi impractical for use in a monitoring program assessing the effectiveness of forest practices because of the difficulty in collecting and identifying these groups. We restrict our consideration of potential focal species to epigeous and epixlyic species (growing on wood) that produce fruiting bodies visible to the naked eye – macrofungi. Neither NatureServe nor the BC Conservation Data Centre provides information on the status of fungi, so we cannot consider rarity. Macrofungi may be symbiotic, parasitic, or saprophytic (decomposers).

10.6.1 Factors Influencing Monitoring: Fungi

Symbiotic fungi

Root-inhabiting mycorrhizal[14] fungi form symbiotic relationships with trees and other plants. The fungal hyphae collect phosphorous and other scarce nutrients for the plant; the plant provides energy-rich carbon compounds to the fungi. All of our forest tree species have obligate relations with mycorrhizal fungi. As well as aiding nutrient uptake by vascular plants, these fungi are food for some small mammals, including shrews, deer mice, and squirrels (Hayes, Cross, and McIntire 1986; Lehmkuhl et al. 2004; Maser, Trappe, and Nussbaum 1978). The richness and abundance of forest fungi, including ectomycorrhizal species, have decreased dramatically in Europe (Jonsson, Kruys, and Ranius 2005; Rydin, Diekmann, and Hallinbäck 1997; Siitonen, Lehtinen, and Siitonen 2005).[15] Reasons for the decline are unclear, but loss of forests, especially old forest, is a suspected cause. There is similar concern in the Pacific Northwest that decreasing old growth will decrease (or has already decreased) fungal richness and diversity (FEMAT 1993; Smith et al. 2002). Many ectomycorrhizal fungi do not survive clear-cutting and must depend on dispersal from older stands to colonize young stands, unless rapid planting after logging sustains them (Perry, Molina, and Amaranthus 1987). On Vancouver Island, the numbers of macrofungi genera in regenerating stands were relatively low for both mycorrhizal and saprophytic fungi but returned to pre-harvest levels while stands were still immature (~20 to ~60 years old; mean age 44 years) (Countess, Kendrick, and Trofymow 1998).

Mature (87 and 89 years old) and old-growth stands (288 and 441 years old) had very similar ectomycorrhizal communities (Goodman and Trofymow (1998).[16] Proximity to older stands may be especially important for recolonization by species fruiting below ground that are dispersed mainly by small mammals.[17] The composition of retained tree species in a forest influences the richness and abundance of epigeous ectomycorrhizal fungi, and "partial-cutting systems could allow some timber removal without necessarily reducing ectomycorrhizal mushroom communities" (Kranabetter and Kroeger 2001, 978).

Because mycorrhizal fungi are most active in the upper soil and humus layers, they are sensitive to changes in temperature, compaction, or erosion (Luoma 1989; Luoma, Frankel, and Trappe 1991). Large increases in soil surface temperature can reduce the growth of mycorrhizae (Meijer 1970), as can increased compaction (Bowan 1980). For mycorrhizal and non-mycorrhizal species, however, mushroom production may not decrease with disturbance. In the Pacific Northwest, several edible mycorrhizal mushrooms increased production after partial harvesting (e.g., chanterelles) or after understorey burns (e.g., matsutake mushrooms and morels) (Molina et al. 1993; Pilz et al. 2004). As with other organisms, fungi species respond to disturbances in a variety of ways. Some are positively affected and some negatively.

Down wood is critical for sustaining populations of mycorrhizal and saprophytic fungi (Harvey, Larsen, and Jurgensen 1979; Plochmann 1989), but the recovery of mycorrhizal fungi in logged areas is also affected by distance from standing forest. In British Columbia, as gap size increased, above-ground fruiting bodies decreased exponentially, but ectomycorrhizal richness on seedling roots decreased only slightly (Durall et al. 1999). Maximum richness on both lodgepole pine and western hemlock seedlings was found 7 metres or less from the edge, consistent with other studies (e.g., Jones and Durall 1997; Outerbridge, Trofymow, and Kope 2001; Parsons, Miller, and Knight 1994). Luoma et al. (2004) examined the production of fruiting bodies by epigeous and hypogeous fungi in aggregated and dispersed retention at four different levels of retention (15, 40, 75, and 100 percent). Many fewer fruiting bodies were found in 15 percent retention than in higher levels, but 40 percent dispersed retention did not exhibit significant reductions. Results differed between spring and fall sampling. Although overall richness decreased with increasing gap size and retention level, different mycorrhizal types did not respond the same way. Moreover, the same fungal species can respond differently when on roots of different tree seedlings (Massicotte et al. 1999), and mycorrhizae on root tips and numbers of fruiting bodies (mushrooms) often do not correlate with each other (Allen, Berch, and Berbee 2000). Fruiting bodies permit the dispersal of fungal spores but are subject to poorly understood microclimatic influences and can be highly variable between years and age of forest stand (Allen, Berch, and Berbee

Truncocolumella citrina *Amanita muscaria*

Figure 10.5 Some ectomycorrhizal fungi rarely or never fruit above ground. Others are frequently visible above ground. Left is the ectomycorrhizal system of *Truncocolumella citrina* on Douglas-fir (*Pseudotsuga menziesii*). *Amanita muscaria* (right) is also ectomycorrhizal on Douglas-fir but on hemlock, true fir, spruce, and a variety of hardwood species as well. | *Photographs by: root tip fungi, R. Outerbridge; and Amanita, C. Bunnell.*

2000; Hunt and Trappe 1987; Smith et al. 2002). Richness cannot be determined by sampling only fruiting bodies because many do not fruit in the field and rarely do so in culture (Allen, Berch, and Berbee 2000). Presence on seedling roots indicates an established symbiotic relationship useful to the host plant and shows less interannual variability. We recommend monitoring of mycorrhizal fungi using seedling roots.[18]

Parasitic and saprophytic fungi

Parasitic fungi enter living tissues, causing damage or mortality to plants, including trees. Fungal richness may reduce the impacts of parasitic fungi. For example, *Hypholoma fasciulare* can displace the fungus causing Armillaria root disease from woody substrates (Chapman et al. 2000; Wästerlund and Ingelög 1981). Because they cause damage and death, parasitic fungi have economic impacts. The damage, however, can be critical to other organisms that rely on natural cavities or rotten wood in living or dead trees. Saprophytic fungi decompose organic matter, extend cavities, and play a critical role in nutrient cycling and soil formation, particularly in nutrient-poor or colder habitats. Smaller organisms, such as beetles, feed on both parasitic and saprophytic fungi, breed on them, or rely on environments created by their activities.

There are many parasitic fungi, and for most the life histories are poorly described. It is more reliable to monitor the products of their activities than the organisms themselves. Much of that can be accomplished within Indicator 2 and permanent sampling plots, whose primary function is to estimate growth and yield. Apparent causes of death can be recorded.

Various characteristics of down wood (e.g., size, decay state, distance to edge) influence the abundance and diversity of saprophytic fungi (Dettki et al. 1998; Heilmann-Clausen and Christensen 2003; Siitonen, Lehtinen, and Siitonen 2005). As down wood changes after harvest, so do associated fungal communities – the changes are most evident where forestry has been practised longer. In Swedish forests, a dramatic decrease in dead wood has enforced the decline of now-threatened macrofungi (Rydin, Diekmann, and Hallinbäck 1997). In forests of northern Finland, the distribution of saprophytic basidiomycetes was strongly affected by the patchy occurrence of down wood after logging (Sippola and Renvall 1999). Whereas white rots increased and colonized slash rapidly, some fungi species were not able to colonize logging slash but were restricted to older down wood that had died a natural death. Down wood from pre-logging times hosted many species that were considered old-growth indicators, and the diversity of fungi depended greatly on the amount and distribution of down wood created *before* logging activity. Old-growth forests had more fungi species than younger forests, due in large part to the presence of deciduous down wood (Sippola and Renvall 1999). Even selective cutting can decrease species diversity of fungi (Bader, Jansson, and Jonsson 1995). Although data are too sparse to estimate the relative contributions of different amounts of forest retained, it is clear that retaining trees to permit natural death helps to maintain fungal richness and thus the organisms dependent on them.[19]

Hundreds of saprophytic species decompose wood, yielding natural patterns of succession as wood decays, but the appearance of identifiable fruiting bodies is subject to poorly understood microclimatic and other influences. Again, it is far easier and potentially more reliable to sample the products of their activities rather than the organisms themselves. That is done within Indicator 2 by monitoring amounts and decay classes of dead wood.

10.6.2 How to Monitor: Fungi

We excluded microfungi or fungi fruiting below ground as candidate species for monitoring. They are too difficult to detect. Some mycorrhizal, saprophytic, and parasitic fungi produce readily identifiable fruiting bodies. However, they respond strongly to moisture and weather conditions, producing high interyear variability in the timing and amount of fruiting. Fruiting bodies usually persist for only a short time during appropriate weather conditions. We focused monitoring for adaptive management on effects of variable retention on mycorrhizal fungi; monitoring epixylic macrofungi was secondary.

Monitoring of mycorrhizal fungi should

- consider two key variables: the effect of edges (opening size), and the amounts and distribution of retained down wood;
- use direct monitoring of fungi in root tip samples (e.g., Durall et al. 1999; Outerbridge, Trofymow, and Kope 2001); and
- consider retrospective sampling of older clearcuts to assess the rate of recolonization.

Monitoring of saprophytic and parasitic fungi should

- assess abundance and richness using fruiting bodies, including those of mycorrhizal fungi;
- address whether fungal richness and abundance differ between retention patches and larger areas of older forest;
- compare and track richness in retention patches, edges, and older forest to evaluate the degree and rate at which patches assist the recolonization of openings;
- relate richness and abundance to amounts and decay stage of down wood;
- schedule sampling and collection for the most profitable period (which depends on the species of interest);[20] and
- use paired sampling of specific comparisons (during the same period) to reduce problems caused by the irregular nature of fruiting.

10.6.3 Links to Management: Fungi

There are three ways in which monitoring fungi can help to refine practice. Evaluating effects of edges on mycorrhizal distribution can help to design the best patterns of variable retention to promote mycorrhizal persistence in young forests, thus long-term productivity. Relationships between amounts and distribution of down wood and mycorrhizae abundance and richness can help to guide retention of down wood. Relationships of epixylic fungal richness and abundance to amounts of down wood, patch size, and stand age could reveal whether further attention to epixylic fungi is warranted in the Timber zone.

10.7 What to Monitor: Invertebrates

Invertebrate species represent the vast majority of animal life, and insects alone comprise between 75 percent and 85 percent of species in the animal kingdom (Gaston 1991; Wilson 1992). Estimates of species richness per square metre of soil in western coniferous forests range from 200 to 550 species of soil arthropods (Marshall 2001; Parsons et al. 1991), and many more species occur above ground. Given the exceptional abundance and high diversity of invertebrates, it is safe to assume that there are few ecological processes in which they do not play a significant role.

Amid all this richness, an obvious question is: are there specific groups or species that indicate the persistence of ecosystem functions or specific forest habitats? Invertebrates may be stronger indicators of ecological processes than of broad habitat types. For example, only a few species are "shredders" that initiate the process of nutrient recycling. The presence or absence of shredders alters humus types, overall rates of nutrient mineralization, and the growth of tree seedlings (Anderson et al. 1985; Moldenke, Pajutee, and Ingham 1996). Although some invertebrates are associated with particular habitat elements, using invertebrates to monitor forest habitats does not appear to be promising (Koen and Crowe 1987; Kremen 1992), primarily because of sampling issues associated with their use of microsites (Grove 2002b; Niemelä, Haila, and Punttila 1996; Speight 1989). However, precisely because invertebrates respond to different scales than other organisms, they are relevant to a monitoring program (Niemelä, Haila, and Punttila 1996; Oliver, Beattie, and York 1998).

10.7. Factors Influencing Monitoring: Invertebrates

We found no examples of invertebrates successfully used as indicators in monitoring programs. Their richness and abundance ensure that they play significant roles in ecosystem function but thwart focus on a few key species. We evaluated four broad features for forest invertebrates: relations with forest structure, sensitivity to forest practices (specifically stand age and patch size), site associations, and rare species. Most literature addresses groups easiest to study: ground beetles (carabids), spiders, and lepidoterans (moths and butterflies). Because invertebrates are so diverse, general relations are uncommon.

Relations with forest structure

Reviews of invertebrates find that many are closely associated with the same habitat elements generally used to describe habitat for vertebrates (e.g., Berg et al. 1994; Niemelä, Haila, and Punttila 1996; Sippola, Siitonen, and Punttila 2002). This association is both real and illusory. It is real because invertebrates can also be limited by amounts of dead wood in particular decay classes or of deciduous species. It is illusory because the microsites sought by invertebrates often are much narrower than is usually described for vertebrates. Nonetheless, the broad associations often hold, even if they do not reveal causal relationships well. Two elements of forest stands appear to be particularly influential – dead wood and tree species composition.

Many authors note close associations of invertebrates with specific forms of dead wood and note pronounced effects of logging (e.g., Grove 2002a; Marshall, Setala, and Trofymow 1998; Niemelä 1997; Nilsson and Baranowski 1997; Okland et al. 1996) or with mycorrhizal fungi (Sippola, Siitonen, and Punttila 2002), which are often associated with dead wood. Within the study area, fewer species of collembolans (which often graze fungi) were found in stumps in early-seral than in older-seral stages (Marshall, Setala, and Trofymow 1998),[21] perhaps due to differences in fungal colonization. Associations with particular tree species or groups of species are also commonly reported (Ås 1993; Hammon and Miller 1998; Magura, Tóthmérész, and Elek 2003; Niemelä 1997).[22] Given our limited knowledge of invertebrate habitat requirements, direct monitoring of dead wood or tree species composition (Indicator 2) will be more useful than monitoring associated invertebrate groups.

Forest age

Given the enormous variety of invertebrates, we expect, and find, a wide range of responses to forest age. We grouped observations from the literature into three categories: openings and recent cutovers, plantations and young stands, and old stands.

Openings and recent cutovers: Studies of invertebrates, including those in early-seral forests, typically report that many species are distributed relatively evenly across age classes, and a few are restricted to young openings (Berg et al. 1995; Greenberg and Forrest 2003; Niemelä 1997; Niemelä, Haila, and Punttila 1996; Niemelä, Langor, and Spence 1993; Okland et al. 1996). Within the study area, Brumwell, Craig, and Scudder (1998) examined four age classes of Douglas-fir forest: regeneration 4-6 years, immature 32-43 years, mature 77-99 years, and old growth >280 years. They reported that litter spiders and most carabid beetles were more abundant in the regeneration seral stage (one species of twenty-eight avoided the regeneration stage). There were no significant differences among the other three age classes.

Plantations and young forests: About 90 percent of the species found in forests in Alberta were in the growing forest by the time it was twenty years old (Spence et al. 1996). Similar results were obtained in Finland: within twenty years after logging, populations of all taxa had recovered from decline following logging and were similar to those in old growth (Niemelä, Haila, and Punttila 1996). Older stands showed little further increase in abundance or richness. Rapid recovery following logging is not always the case. Canopy arthropods, particularly predators and detritivores, appear to recover more slowly (Schowalter 1995; Schowalter, Stafford, and Slagle 1988). Slow recovery has been noted in other parts of the world as well; for example, effects of harvesting on ants in Japan persisted, so that forty to seventy years after logging, ant assemblages were still different from those of old-growth forests (Mato and Sato 2004).

Thinning seems to have little effect on most forest-dwelling carabids (e.g., Koivula 2002; Villa-Castillo and Wagner 2002), but the effects of partial cutting on invertebrates have not been well studied. Selection harvesting (30 percent removal, thus similar to thinning) had little effect on arthropod communities: clear-cutting and gap felling had the clearest impacts (Siira-Pierikainen, Haimi, and Siitonen 2003). Some studies suggest increasing richness with increasing canopy heterogeneity (Grove 2002a; Hamer et al. 1997; Willott 1999), but others showed no differences between unevenly managed forests and mature, even-aged forests, suggesting that complex canopy did not affect those species (Brouat, Meusnier, and Rasplus 2004).

Older forests: Relatively few invertebrates appear to be restricted to old stands, but some are strongly associated with old growth, and continuous forest cover may be important for many others (Grove 2002a; Niemelä 1997; Niemelä, Haila, and Punttila 1996; Schowalter 1995; Schowalter, Stafford, and Slagle 1988). More rare invertebrate species occur in old forests than in younger ones (Berg et al. 1995; McLean and Speight 1993; Niemelä 1997). Generally, where forestry has been practised longer, rare species are concentrated in older forests. Species strongly or exclusively associated with old growth usually inhabit well-decayed wood, suggesting that variable retention would help to maintain them (Hammond, Langor, and Spence 2004; Martikainen et al. 2000; Vanha-Majamaa and Jalonen 2001). Potential candidate species or groups strongly associated with old growth include the ground beetle *Scaphinotus marginatus* (Spence et al. 1996), mycetophilid flies,[23] and some canopy arthopods (Schowalter 1995), though no species was unique to old growth.

Patch size

Variable retention maintains some old-growth conditions in scattered, small patches. It is thus useful to assess whether some invertebrates, particularly those more abundant in older stands, reveal effects of patch size or isolation. There is ample evidence of patch size or isolation effects in non-forested habitats (De Vrie and Den Boer 1990; Hopkins and Webb 1984; Kindvall and Ahlen 1992; Klein 1989; Schultz 1998; Thomas, Thomas, and Warren 1992; Tscharntke 1992; Turin and Den Boer 1988), but evidence of patch size effects in forests is sparse. Most studies on insects and spiders in forests report little response to patch size or isolation (Abensperg-Traun et al. 1996; Ås 1993; Davies and Margules 1998; Halme and Niemelä 1993; Niemelä 1997; Okland 1996; Pajunen et al. 1995; Usher, Field, and Bedford 1993). Small fragments of old forest contained nearly as many red-listed wood beetle species as did larger fragments in southern Sweden (Nilsson and Baranowski 1997).

Particular species or groups may be sensitive to patch size. In Finland, abundance and richness of carabid beetles were lowest in contiguous forest and highest in fragmented forests (patches of 0.5 ha to 21.5 ha), but three flightless species were found only in contiguous forest (Halme and Niemelä 1993). Species richness of mycetophilids in Norway was increased by forest continuity (Spence et al. 1996).

Some dung and carrion beetles of Brazilian rainforests would not cross even narrow clearcut barriers (Klein 1989). Although there was no effect for patch size, researchers in Finland found that spider communities differed in composition between old forest interior and the surrounding managed forest and attributed it to edge effects (Pajunen et al. 1995). Edge effects have been reported (positive for butterflies and negative for grasshoppers) (Bieringer and Zulka 2003; Kremen et al. 1993; Thomas 1991). Some spiders, carabids, or gastropods might reveal effects of retained patch size and forest edges and so be useful indicators of the value of retained patches to sustain populations of poorly dispersing species long enough to recolonize the harvested area (Bonham, Mesibov, and Bashford 2002; Okland 1996). However, gastropods appear to recover quickly from disturbance (e.g., timber harvesting and vegetation management), and year-to-year variations in weather can have effects on gastropods of similar magnitude to effects of forest management (Prezio et al. 1999).

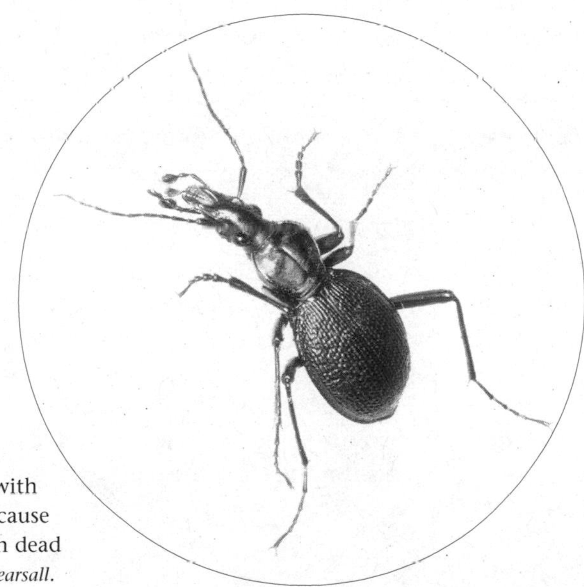

Figure 10.6 Scaphinotus marginatus is associated with older stands, possibly because its larvae feed on fungi in dead wood. | *Photograph by: I. Pearsall.*

Site associations

Invertebrates can have very specific associations with microsites and substrates. Some have narrow microclimatic niches, and the growth rates of many populations are sensitive to temperature, humidity, and rainfall (Kremen et al. 1993). Soil invertebrates respond to numerous characteristics that change over fine scales, including soil particle size and type, the amount and flow of water and gases in the soil, the amount and type of organic matter, and chemical factors (Wall and Moore 1999). Although a square metre of forest soil may contain fifty to seventy-five species of turtle mites (Oribatida or Cryptostigmata; Moldenke 1989), individual species may be limited to foraging on only a few fungal species (Stefaniak and Seniczak 1978). Mutualism affects species composition and distribution (e.g., bacteria, fungi, and mites interact to decompose complex organic compounds and create soil). Different species prefer different temperature and moisture conditions. Given the underlying complexity, it is no surprise that studies examining effects of site have found that invertebrates respond to factors at a variety of scales (Niemelä et al. 1992).

Because the fine-scaled response of invertebrates to soil elements greatly hinders their value as focal species, we considered relations with plant species. Many insects are flightless and restricted to small areas or specific host fungi, plants, or plant groups (Crisp, Dickinson, and Gibbs 1998; Loertscher, Erhardt, and Zettl 1995; Ødergaard et al. 2002; Scott 1997; Wahlberg, Moilanen, and Hanski 1996). In coastal forests, there are undoubtedly many invertebrates with specific preferences for soil characteristics or reliant on particular plant hosts. Generally, we do not know what the preferences or host plants are. The site series used in Indicator 1 reflect both soil conditions and plant composition. Measures of Indicator 2 include decay classes. We believe that there is no practical way of monitoring the wide range of sites used by invertebrates other than through Indicators 1 and 2. Moreover, the range encompassed by these indicators is broader than could be attained with a few focal species.

Rarity

In British Columbia, the Conservation Data Centre lists rare invertebrates but is hampered by lack of data. Whenever more intensive or more extensive sampling has been undertaken, the number of species new to science is disconcertingly large and difficult to place into context (e.g., Humble, Winchester, and Ring 2001; Kondla, Crispin, and Shepard 2001). In short, there is no adequate reference source from which to evaluate rare species, and no rare species in the study area is sufficiently well known to make it a useful indicator.

10.7.2 How to Monitor: Invertebrates

The richness of invertebrates frustrates any search for a few indicator species. Moreover, identification of most species is difficult and expensive, often requiring experts from far afield and associated shipping costs.[24] As well, dispersal capabilities and habitat requirements or preferences usually are poorly known. This lack of basic knowledge hinders their utility in guiding forest practice but emphasizes that, because we cannot predict their responses from changes in habitat structure, direct monitoring may be necessary to instill confidence that they are being sustained. One difficulty in monitoring any specific group of invertebrates is that we do not know which, if any, other species respond similarly. However, because of the degree of mutualism shown among some species, selected species may represent larger assemblages. On the positive side, studies on carabid beetles identify some temperate forest species that are closely associated with older forests and relatively easy to identify. Furthermore, sampling for invertebrates often costs less than for vertebrates but more than for vascular plants (Oliver, Beattie, and York 1998), so pilot studies to evaluate useful invertebrate indicators are not impractical.

For the study area, the literature provides little guidance to an effective way of using invertebrates in monitoring. Recommendations for monitoring can only be general.

- Evaluate invertebrates as a means of revealing fine-scale effects.
- Exploit more than one scale because sampling larger areas masks small-scale heterogeneity (Chust et al. 2003; Niemelä, Haila, and Punttila 1996; Spence et al. 2003). Observing clumped occurrences may reveal the heterogeneity as perceived by the organism.
- Use pilot studies to evaluate the occurrence of species strongly associated with older forests that could evaluate the efficacy of retention patches and potential edge effects. Carabid beetles and gastropods are potential candidates.[25]

10.7.3 Links to Management: Invertebrates

Although relations among invertebrate groups are poorly understood, invertebrate surveys in retention patches of different sizes may help to reveal patch sizes that would maintain invertebrate assemblages. Species associated with older forest conditions may reveal the capacity for retention to maintain such species in managed forests. Species associated with moist forested conditions (some gastropods) may reveal whether retention patches provide microclimatic conditions that allow these species to persist in patches until the open areas become more hospitable. Conversely, site moisture may dominate any effect of forest cover.

If species respond to edges, then they could reveal the extent of edge effects within retention blocks and, if surveyed at repeated intervals as the matrix ages, could document duration of edge effects and dispersal from retention patches or surrounding forests into the harvested area.

10.8 What to Monitor: Vertebrates

There are several reasons to monitor terrestrial, forest-dwelling vertebrates.

- *Vertebrates are readily monitored.* Like vascular plants, vertebrates are well known, easily identified, and large enough to lend themselves to sampling.
- *Vertebrates include wide-ranging species.* Using vertebrates to assess success in sustaining biological diversity considers larger areas than are necessary for many other organisms.
- *The responses of many vertebrates to forest structure are known.* Although many vertebrates are quite plastic in their habitat requirements, several have well-understood associations with particular forest structures that can be used to guide choices among forest practices.[26]
- *Many other organisms are linked to the structural features of forests to which vertebrates respond.* Monitoring vertebrates may provide evidence on the potential health of many other forest-dwelling organisms.[27]
- *Vertebrates connect directly with public concerns.* The public shows more concern for vertebrates than for the other groups that we have evaluated for monitoring.

10.8.1 Factors Influencing Monitoring: Vertebrates

Determining whether vertebrate species are forest dwelling is sometimes more problematic than it is with many other organisms because vertebrates are sufficiently visible that more of their flexibility is exposed. For example, although the peregrine falcon nests predominantly in cliffs, where it is unaffected by forest practices, it sometimes nests in trees and was considered forest dwelling. Because amphibians spend a considerable portion of their lives on land, we included them among terrestrial forest-dwelling species even though several breed in water. We estimated that 175 terrestrial vertebrate species potentially breed in coastal forests of the study tenure.[28] Several unconfirmed species (e.g., sharptail snake and bobcat) are included in summary tables (e.g., Table 10.5). More species are present than are summarized in the tables (e.g., mallard, cinnamon teal, marsh wren, American goldfinch, muskrat), but they are largely limited to non-forest vegetation.

More is known about vertebrate response to forest practices than for all other groups of organisms. In fact, part of the rationale for adopting variable retention is to permit sustained provision of structures common in older forest that are known to influence vertebrate abundance. We considered

Table 10.5

Habitat use by native forest-dwelling amphibian, reptile, bird, and mammal species potentially breeding on the study tenure

		Percent restricted to or favouring						
Class	Total spp	Cavities	Downed wood	Shrubs[1]	Hardwoods[2]	Riparian	Early Seral[3]	Late Seral[4]
Amphibians	10	0	50.0	20.0	20.0	70.0	0	40.0
Reptiles	6	0	33.3	16.7	33.3	66.7	50.0	0
Birds	108	27.8	4.6	27.8	30.6	50.9	14.8	31.5
Mammals	51	33.3	31.4	13.7	23.5	47.1	13.7	27.5

1 Birds may nest low in trees (<3 m) surrounded by understorey.
2 At least two-thirds of reported bird nests are in hardwoods.
3 Breed predominantly in stages 1 and 2 of Figure 8.1a or at forest edges.
4 Breed predominantly in stages 5 and 6 of Figure 8.1a.

vertebrate responses to specific forest structures, landscape features (including forest age), patch size (including edge), and rarity of individual species.

Forest structure

Table 10.5 summarizes associations of vertebrate classes with different habitat structures or elements and some broader landscape features. Data of Table 10.5 reveal habitat elements and landscape features altered by forest practices to which species are likely to be sensitive.

When stewardship zones were applied, they were not applied uniformly across the landscape. Analysis of Indicator 1 (ecosystem representation) reveals that much of the non-harvestable, thus undisturbed, forest occurs in the least productive BEC zones, such as the Mountain Hemlock zone (Figure 7.2). The distribution of vertebrate species across BEC zones is summarized in Table 10.6.

Landscape features

It is apparent that affinity for riparian areas in the study area is high, particularly in the driest forest type, Coastal Douglas-Fir (Table 10.6). As expected, given the natural disturbance regime (§2.1.2), only about 15 percent of the forest-dwelling vertebrates show strong associations with early-seral habitats, while about 30 percent show associations with late-seral or old-forest habitat. No vertebrate species appears to be restricted to old growth by stand age itself. Rather, late-seral associates appear to respond to habitat features, such as large snags or downed wood, that are more common within old growth (Bunnell, Kremsater, and Wind 1999; Hansen et al. 1991; Mazurek and Zielinski 2004). That phenomenon illustrates the potential for variable retention to sustain species and emphasizes the importance of monitoring habitat variables as well as species.

Table 10.6

Habitat use by native forest-dwelling vertebrates in the three forested biogeoclimatic zones of the study tenure

		Percent restricted to or favouring						
BEC zone[1]	Total spp	Cavities	Downed wood	Shrubs[2]	Hardwoods[3]	Riparian	Early Seral[4]	Late Seral[5]
CWH	169	26.6	16.0	23.1	28.4	50.9	14.2	30.8
MH	109	26.6	18.3	22.0	21.1	52.3	14.7	33.9
CDF	139	30.9	13.0	22.3	30.9	54.0	12.2	32.4
ALL	175	26.9	16.0	22.9	28.0	50.9	14.9	29.7

1 Biogeoclimatic zones of the area are described by Green and Klinka (1994); CWH = Coastal Western Hemlock; MH = Mountain Hemlock; CDF = Coastal Douglas-Fir.
2 Birds may nest low in trees (<3 m) surrounded by understorey.
3 At least two-thirds of reported nests in hardwoods.
4 Breed predominantly in stages 1 and 2 of Figure 8.1a or at forest edges.
5 Breed predominantly in stages 5 and 6 of Figure 8.1a.

Patch size and edge

Responses to edge are summarized in Table 10.7.

Negative responses to edge are designated "forest interior" in Table 10.7. We included any species for which at least one publication argued for a negative response to edge. For positive responses to edge, we included species for which a significant statistical response had been documented (designated "tested" in Table 10.7) and species for which data implied a positive response to edge but did not permit a statistical test (designated "not tested"). For example, forty-six bird species apparently respond positively to edge, but we could test the response in only twenty-three. The proportion of species showing responses to edge is relatively high, 57 of 108 bird species (53 percent). Given that VR increases amounts of edge, edge effects must be evaluated. Many species showing a positive response to edge also show strong affinities for riparian areas (thirty of seventy-one species; 42 percent), suggesting responses to particular forms of edge.

Table 10.7

Number of native forest-dwelling vertebrate species responding to forest edge within the study tenure

	Positive edge effects			
Class	Tested	Not tested	Riparian	Forest interior
Amphibians	3	0	2	1
Reptiles	5	1	4	0
Birds	23	23	19	11
Mammals	10	6	5	2

Rarity

We evaluated red and blue lists for species in the study area, including trends for those species for which there were data (primarily Breeding Bird Surveys). Reasons for exclusion or inclusion of individual rare species or species "of concern" suggested by experts are summarized elsewhere (Appendix 4.5 of Bunnell et al. 2003).

Focal species must be forest dwelling, sensitive to forest practices, practical to monitor in terms of sampling, identification, and cost, and provide information useful in guiding forest practices (§10.2). Forest-dwelling species are summarized in Table 10.5. The flexibility exhibited by many vertebrates makes determining their sensitivity to forest practices challenging. Some species are relatively insensitive to changes in forest cover or structure (e.g., long-toed salamander, common garter snake, barred owl, deer mouse), so they provide no indication of changes that might impact other species dramatically. Inferring sensitivity to forest practices is further complicated because past data cannot be confidently extrapolated to the new forest practices being implemented on the study area. Our initial estimates of sensitivity were based on associations with particular habitat elements or landscape features (summarized in Tables 10.5 and 10.7).

Among the forest-dwelling vertebrates that are potentially sensitive to forest practices, a few cannot be easily identified (e.g., Keen's long-eared myotis), and some would be very costly and so not practical to monitor (e.g., gray wolf, fisher). Other species are clearly forest dwelling but are equally or more responsive to changes other than forest practices and thus provide no useful guidance to forest practices (e.g., pine siskins or other irruptive species).

As with other groups of organisms, we followed the two-step procedure of polling local experts or management agencies as well as relying on literature and available data to select focal species. The response differed from other groups in that more "favourite" species, not amenable to monitoring, were named during polling.

10.8.2 What to Monitor: Vertebrates

We sought a list of candidate focal species that would cover the range of expected response patterns, with some species indicating suitability at the stand level and some over broader areas (Figure 10.7). For the stand level, we recognized the five standard habitat elements to which groups of vertebrates show strong responses (Bunnell, Kremsater, and Boyland 1998; Bunnell, Kremsater, and Wind 1999; Thomas 1979). These elements included (1) dying and dead trees, (2) down wood, (3) shrubs, (4) deciduous trees, and (5) large, live trees, both conifer and hardwood. The habitat elements guided selection of a suite of taxa useful primarily for assessing conditions at the stand or treatment level (Table 10.8).

Figure 10.7 Among terrestrial vertebrates, potential indicator species were closely associated with forest elements or features altered by forest practices. The black-capped chickadee (*Poecile atricapillus*) is associated with dying or dead trees, while the blue grouse (*Dendragapus obscurus*) is associated with downed wood. | *Photographs by M. Preston.*

Table 10.8

Potential terrestrial vertebrate focal species monitored to assess effects primarily evident at the scale of stands or treatment units

Dying trees and snags (cavity sites)
- Red-breasted sapsucker (BS)
- Downy woodpecker (BS)
- Hairy woodpecker (BS)
- Northern flicker (BS)
- Pileated woodpecker (BS)
- Chestnut-backed chickadee (BS)
- Red-breasted nuthatch (BS)
- Brown creeper (BS)
- American marten (WT)

Down wood
- Blue grouse (BS)
- Winter wren (BS)
- Common shrew (PF)
- Dusky shrew (PF)
- Southern red-backed vole (PF)
- American marten (WT)

Shrubby undergrowth and berries
- Cedar waxwing (BS)
- Willow flycatcher (BS)
- Hutton's vireo (BS)
- Swainson's thrush (BS)

Shrubby undergrowth and berries
- Hermit thrush (BS)
- Black-throated gray warbler (BS)
- MacGillivray's warbler (BS)
- Spotted towhee (BS)
- Song sparrow (BS)
- Black-headed grosbeak (BS)

Deciduous trees
- Red-breasted sapsucker (BS)
- Swainson's thrush (BS)
- American robin (BS)
- Cassin's vireo (BS)
- Warbling vireo (BS)
- Chestnut-backed chickadee (BS)
- Black-throated gray warbler (BS)

Large live trees
- Great blue heron (FS)
- Bald eagle (FS)
- Red-tailed hawk (nests)
- Merlin (BS)
- Brown creeper (FS)

Note: Survey methods are indicated in brackets: FS = focal surveys (as around known nests or colonies); PF = pitfalls; BS = bird surveys; WT = winter tracking.

To recognize the different scales over which species integrate habitat, we used six other habitat features to create the range of species useful for monitoring (Table 10.9). The six additional features are intended to represent the broad scale of impacts of forest practices and include (1) riparian habitat, (2) early-seral stages, (3) late-seral stages, (4) desirable edges, (5) forest interior, and (6) a range of scales of habitat use.

We know enough about many vertebrate species that the lists of species summarized in Tables 10.8 and 10.9 can be refined to meet local conditions. For example, we do not expect to find all species listed throughout the tenure. Black-throated gray warblers and spotted towhees will be encountered only on the southern coast. Similarly, additional species could be listed for specific locations on the tenure. Bewick's wren is associated with dying trees and

Table 10.9

Potential terrestrial vertebrate focal species monitored to assess effects primarily evident at the broad landscape or forest scale

Riparian
- Pacific treefrog (BS)
- Hooded merganser (BS)
- Common merganser (BS)
- Bald eagle (BS)
- Spotted sandpiper (BS)
- Willow flycatcher (BS)
- Cedar waxwing (BS)
- Song sparrow (BS)

Early-seral stages
- Blue grouse (*if burned,* BS)
- MacGillvray's warbler (*if shrubs,* BS)
- White-crowned sparrow (BS)
- Long-tailed vole (PF)
- Townsend's vole (PF)

Late-seral stages
- Northern goshawk (BS)
- Chestnut-backed chickadee (BS)
- Brown creeper (BS)
- Golden-crowned kinglet (BS)
- Varied thrush (BS)
- Townsend's warbler (BS)
- Western tanager (BS)
- Southern red-backed vole (PF)
- American marten (WT)

Edges
- American kestrel (BS)
- Rufous hummingbird (BS)
- Olive-sided flycatcher (BS)
- American robin (BS)
- Orange-crowned warbler (BS)
- Western tanager (BS)

Forest interior
- Pileated woodpecker (BS)
- Brown creeper (BS)
- Varied thrush (BS)
- Southern red-backed vole (PF)

Note: Survey methods are indicated in brackets: PF = pitfalls; S = bird surveys; WT = winter tracking.

snags, but only on the southernmost coast, varied thrush with down wood on Haida Gwaii, and alligator lizard with down wood on the Lower Mainland coast (Figure 2.1). Hermit and Swainson's thrushes are useful indicators of desirable edges primarily on Haida Gwaii. Similarly, some potential focal species are evident or readily sampled only during limited times of the year. For example, western screech owl and northern pygmy owl indicate the presence of dying trees and snags but must be sampled in February and March. Band-tailed pigeons indicate the presence of deciduous cover but primarily during July and August. Some species indicating sustained provision of large live trees are not commonly encountered (e.g., osprey, Cooper's hawk, great horned owl).

Our review of vertebrates on the tenure suggests the following considerations when selecting candidate species for monitoring.

- Include species responding to both specific habitat elements and broader landscape features (Tables 10.8 and 10.9).
- Acknowledge regional variation in the fauna.
- Reduce the potential for equating population presence with population health by (1) repeated survey routes to establish trends and (2) recording evidence of productivity during surveys (e.g., nesting, young of the year).
- Exploit the large literature on appropriate sampling methods (e.g., Heyer et al. 1994; Ralph, Sauer, and Droege 1995; Skalski and Robson 1992; Thompson 2002). General techniques are noted in brackets in Tables 10.8 and 10.9. For all groups of terrestrial vertebrates except reptiles, we used pilot studies to evaluate specific methods for monitoring.

10.8.3 Links to Management: Vertebrates

Studies comparing relative abundance or trends of vertebrates in retention treatments can indicate types and levels of retention that benefit various focal species. Tracking long-term trends helps to focus any detailed analysis of species with declining (or markedly increasing) trends. Where declines can be linked to habitat features, forest practices can be appropriately adjusted – either through adjusting the mix of retention types or by targeting certain habitat elements in retention areas (e.g., large snags, down wood). Monitoring of the unmanaged land base, including riparian areas, will indicate which species are likely accommodated in unmanaged areas and which must be accommodated elsewhere. Most effectiveness monitoring will not be detailed enough to expose mechanisms underlying habitat associations, but when combined with existing knowledge it can aid the development of species-habitat models (§12.8.2). These models can then project habitat quality over large areas and long time periods. Testing such models refines our knowledge of species requirements and subsequent impacts of forest management.

10.9 Overall Feedback to Management

Information on species can help management decisions in five broad ways.

1 *Species information can reveal broad-scale effects.* The occurrence (presence or absence) of species can be compared to known range maps to examine reductions or expansions in species' ranges. The expansion of any group of species over other groups indicates potential problems. For example, the expansion of exotic or pioneering species, combined with reductions in ranges of species associated with older stands, could indicate reduced productivity within the latter group. Conversely, constancy or expansion in ranges of species commonly associated with older forests may indicate the success of variable retention in providing habitat structures for species associated with older forests. Management responses could be to reduce the spread of exotics (e.g., closing roads) or to retain more of specific structural attributes (where associations are relatively well known).

2 *Species information can provide an early warning system.* Trends in population can be useful over both the short and the long terms. When associations of species with particular habitat elements are known and species are not wide ranging, reasons can likely be deduced for downward trends in populations and steps undertaken to increase needed habitat elements. Even where associations with particular habitat elements are poorly known, pronounced, long-term trends (either upward or downward) should trigger closer scrutiny of potential causes. Trends in some organisms (e.g., mycorrhizal fungi, arboreal lichens) can help to determine whether variable retention is providing microclimatic conditions suitable to sustain species associated with older forests.

3 *Species information can guide specific practices.* For some species, relationships with particular forest elements or features are sufficiently well documented that early warning of declines can, at least initially, be attributed to lack of that element or structure. Management actions to increase supply of those elements (e.g., down wood, dead trees) or landscape features (e.g., forest interior) can be implemented. Those actions could include changing patterns of variable retention from dispersed to patches or from small patches to large patches, or vice versa, or changing patterns of harvested blocks on the landscape to retain larger contiguous blocks of forest if that were indicated. Confidence in such actions is strengthened if the monitoring design evaluates relations to forest structure.

4 *Species information can refine modelling efforts.* Information on species' associations helps to refine relationships to allow projection over long time periods and large areas. Use/availability studies also can help to refine model relationships. For example, several fungi, bryophytes, invertebrates, and vertebrates are associated with down wood, and information on the species and decay classes they use would improve habitat models.

As the models increase in their predictive ability, they are better able to guide improvements in practices.

5 *Species information can focus fine-scale monitoring.* Occurrence information or population trend information may indicate productivity problems for some species, providing reasons to investigate their natural history more closely. Studies on dispersal or productivity may be warranted. Findings of these studies may suggest ways to improve limiting factors (e.g., add dispersed retention or keep down wood to allow dispersal of bryophytes or amphibians).

These broad links to management are applicable to all species. However, different groups of species are appropriate for asking different questions or examining different scales. Trends in songbird populations may be good for long-term trends but less suitable for detecting differences in benefits of different patterns of variable retention. Specific comments for species and groups of species are offered in §10.3 through §10.8.

10.10 Summary

To be at all feasible and cost effective, the list of focal species used for monitoring cannot be lengthy. That is particularly true when efforts are made across taxa to encompass a breadth of species. Different taxa require different techniques and skills. The preceding sections provided the rationale for selecting potential candidate species from a broad range of taxa. The summary list is too long to be practical and must be distilled further while considering two broad features: (1) the habitat element, landscape feature, or forest practice that would be revealed by the species' response, and (2) inclusion of a range of taxa.

Candidate taxa should include those integrating their habitat responses over a range of scales. In particular, it is important to assess both stand-level effects (about 0.25 ha to 100 ha) and forest effects (500 ha to 10,000 ha). The list of resultant focal species is summarized in Table 10.10, along with the particular habitat element or phenomenon that they are intended to represent.

Table 10.10

Focal species and the habitat element or phenomenon they are intended to represent or index

Organism group	Scientific names	Common names
Canopy closure		
Vascular plants	Listed in Appendix 1	Listed in Appendix 1
Mosses	*Mnium spinulosum, Plagiomnium insigne, Hookeria lucens, Polytrichum commune*	
Late-seral /large trees		
Vascular plants	*Poa laxiflora*	Lax-flowered bluegrass
Liverworts	*Bassania tricrenata, Apometzgeria pubescens, Conocephalum conicum, Frullania nisquallensis, Metzgeria conjugata, Radula complanata, Riccardia multifida, Porella cordeana*	
Mosses	*Antitrichia curtipendula, Claopodium crispfolium, Heterocladium procurrens, Hypnum circinale, Metaneckera menziersii, Neckera douglasii, Thamnobryum neckeroides*	
Lichens	*Hypogymnia enteromorpha, Usnea wirthii, Sphaerophorus globosus, Alectoria sarmentosa, Hypogymnia inactiva, Platismatia norvegica, Lobaria oregana, Lobaria pulmonaria, Usnea longissima, Platismatia glauca, Bryoria fuscescens, Platismatia herrei*	
Invertebrates	*Pterostichus crenicollis, Scaphinotus angulatus, Scaphinotus angusticollis, Zacotus mattewsii*	
Vertebrates	*Ardea herodias, Falco columbianus, Buteo jamaicensis, Regulus satrapa, Accipiter gentilis, Dendroica townsendi, Poecile rufescens, Certhia americana, Piranga ludoviciana, Ixoreus naevius, Haliaeetus leucocephalus, Martes americana, Clethrionomys gapperi*	Great blue heron, pigeon hawk, red-tailed hawk, golden-crowned kinglet, goshawk, Townsend's warbler, chestnut-backed chickadee, brown creeper, western tanager, varied thrush, bald eagle, marten, southern red-backed vole

►

◄ *Table 10.10*

Organism group	Scientific names	Common names
Snags		
Vascular plants	None	None
Mosses	*Plagiomnium venustum, Plagiothecium undulatum*	
Liverworts	*Porella navicularis, Scapnia bolanderi*	
Vertebrates	*Poecile rufescens, Certhia americana, Picoides pubescens, Picoides villosus, Colaptes auratus, Sphyrapicus rubber, Sitta canadensis, Martes americana*	Chestnut-backed chickadee, brown creeper, downy woodpecker, hairy woodpecker, flicker, red-breasted sapsucker, red-breasted nuthatch, marten
Down wood		
Vascular plants	*Athyrium filix-femina, Polypodium glycyrrhiza, Allotropa virgata, Menziesia ferruginea, Hemitomes congestum, Vaccinium parvifolium, Monotropa uniflora, Chimaphila menziesii, Pterospora andromedea, Hypopitys monotropa, Chimaphila umbellata, Gaultheria shallon, Streptopus streptopoides, Corallorhiza striata, Corallorhiza maculata*	Maiden hair fern, licorice fern, candystick, false azalea, gnome plant, huckleberry, Indian pipe, Menzie's pipsissewa, pinedrops, pinesap, prince's pine, salal, small twisted stalk, striped coral root, western coral root
Mosses	*Dicranum fuscescens, Dicranum scoparium, Kindbergia oregano, Kindbergia praelonga*	
Liverworts	*Mylia taylorii, Lepidozia reptans, Porella navicularis, Scapania bolanderi*	
Vertebrates	Plethodontid salamanders, *Troglodytes troglodytes, Dandragopus obscurus, Sorex cinerus, Sorex monticolus, Martes americana, Clethrionomys gapperi*	Lungless salamanders, winter wren, blue grouse, masked shrew, dusky shrew, marten, southern red-backed vole
Humus		
Vascular plants	*Athyrium filix-femina, Polypodium glycyrrhiza, Allotropa virgata, Menziesia ferruginea, Hemitomes congestum, Vaccinium parvifolium, Monotropa uniflora, Chimaphila menziesii, Pterospora andromedea,*	Maiden hair fern, licorice fern, candystick, false azalea, gnome plant, huckleberry, Indian pipe, Menzie's pipsissewa,

►

◄ *Table 10.10*

Organism group	Scientific names	Common names
Vascular plants (cont'd.)	*Hypopitys monotropa, Chimaphila umbellata, Gaultheria shallon, Streptopous streptopoides, Corallorhiza striata, Corallorhiza maculata*	pinedrops, pinesap, prince's pine, salal, small twisted stalk, striped coral root, western coral root
Mosses	*Plagiothecium undulatum, Dicranum scoparium, Kindbergia oregana, Kindbergia praelonga, Rhytidiadelphus triquetrus, Rhytidiadelphus loreus, Hylocomnium splendens, Mnium spinulosum, Plagiomnium insigne, Hookeri lucens, Polytrichum commune*	
Liverworts	*Barbilophozia lycopodioides, Herbertus aduncus*	
Vertebrates	None	None
Shrubs		
Vertebrates	*Pheucticus melanocephalus, Catharus guttatus, Catharus ustulatus, Vireo huttoni, Pipilo erythrophthalmus, Dendroica nigrescens, Oporornis tolmiei, Bombycilla cedorum, Melospiza melodia, Empidonix traillii*	Black-headed grosbeak, hermit thrush, Swainson's thrush, Hutton's vireo, rufous-sided towhee, black-throated gray warbler, MacGillivray's warbler, cedar waxwing, song sparrow, willow flycatcher
Deciduous trees		
Vertebrates	*Ensatina eschscholtzii, Poecile rufescens, Sphyrapicus rubber, Piranga ludoviciana, Dendroica nigrescens, Catharus ustulatus, Turdus migratorius, Vireo cassinii, Vireo gilvus*	Ensatina, chestnut-backed chickadee, red-breasted sapsucker, western tanager, black-throated gray warbler, Swainson's thrush, American robin, Cassin's vireo, warbling vireo
Edge		
Vertebrates	*Falco sparverius, Selasphorus rufus, Contopus borealis, Piranga ludoviciana, Turdus migratorius*	Sparrowhawk, rufous hummingbird, olive-sided flycatcher, western tanager, American robin

►

◄ *Table 10.10*

Organism group	Scientific names	Common names
Fungi	Mycorrhizal	
Mosses	*Plagiothecium undulatum, Dicranum scoparium, Kindbergia oregana, Kindbergia praelonga, Rhytidiadelphus triquetrus, Rhytidiadelphus loreus, Hylocomnium splendens, Mnium spinulosum, Plagiomnium insigne, Hookeria lucens, Polytrichum commune*	
Invertebrates	*Pterostichus crenicollis, Scaphinotus angulatus, Scaphinotus angusticollis, Zacotus mattewsii*	
Interior/patch size		
Vascular plants	*Goodyera oblongifolia, Trillum ovatum, Taxus brevifolia*	Rattlesnake plantain, trillium, western yew
Mosses	*Plagiomnium insigne, Hookeria lucens, Polytrichum commune*	
Vertebrates	*Poecile rufescens, Certhia americana, Ixoreus naevius, Clethrionomys gapperi*	Chestnut-backed chickadee, brown creeper, varied thrush, southern red-backed vole
Invertebrates	*Pterostichus crenicollis, Scaphinotus angulatus, Scaphinotus angusticollis, Zacotus mattewsii*	
Riparian		
Vascular plants	*Taxus brevifolia, Barbarea orthoceras, Senecio pauperculus, Petasites frigidus* var *palmatus, Corydalis scouleri*	Western yew, American bittercress, Canadian butter weed, palmate coltsfoot, Scoular's corydalis
Vertebrates	*Taricha granulosa, Hyla regilla, Lophodytes cucullatus, Mergus merganser, Actitis macularia, Oporornis tolmiei, Ixoreus naevius, Turdus migratorius, Bombycilla cedorum, Melospiza melodia*	Roughskinned newt, Pacific tree frog, hooded merganser, common merganser, spotted sandpiper, MacGillivray's warbler, varied thrush, American robin, cedar wax-wing, song sparrow
Exotics		
• Vascular plants	Listed in Appendix 1	Listed in Appendix 1

11 Learning from Organisms

David J. Huggard and Laurie L. Kremsater

11.1 Context

During its pilot phase, the program for monitoring organisms was intended to select candidates for monitoring from the review of different taxonomic groups (Chapter 10), test our ability to monitor each group during the first five years, then use these results to outline the long-term program of organism monitoring. The pilot study phase was meant to ensure that any components chosen for the long-term program would be feasible, statistically meaningful, and efficient. In reality, the organism monitoring assumed a more opportunistic, evolutionary pathway, and the studies fulfilled their intended role as pilot studies to different extents. In most cases, organisms were selected based on the review; then researchers selected areas of interest within the broad framework of the adaptive management program and addressed those, often in quite different locations throughout the tenure. A few studies focused on high-priority questions, but many were on medium-priority topics that were attractive to researchers and funding agencies. While many studies piloted methods, few piloted them in a way that would allow improved choices of statistically efficient study designs. To date, there has been no definitive statement describing the long-term program for organisms, and there may never be one: the program may need to evolve continually in the face of frequent changes in company ownership, priorities, and external funding sources.

11.2 Intended Roles of the Pilot Study Phase

Pilot studies were intended to provide five types of information to develop the long-term monitoring program (see below). In the initial five years of monitoring, different projects provided this pilot information to different degrees.

11.2.1 Assess Sensitivity to Forest Practices

The review of possible indicator organisms (Chapter 10) identified a number of taxonomic groups containing species sensitive to forestry. However, for lesser-known groups, we were less sure that there would be sensitive species or that these species could be detected frequently enough to be useful for monitoring.

Two of the larger initial monitoring projects, carabid beetles and gastropods, included simple comparisons of harvested and unharvested areas in their first years. Most other studies of these groups have found species affected by forestry, but many were conducted in other forest types, often in Europe. The comparison of harvested versus unharvested areas was thus considered worthwhile to repeat on Vancouver Island, where local conditions might make harvested areas less hostile for sensitive species in these two groups – abundant post-harvest debris for carabid beetles, and generally wet conditions for gastropods. Both studies found species that were much reduced in harvested areas and were abundant enough to provide useful monitoring information. For carabids, they included *Scaphinotus angusticollis,* a predator of snails that is commonly considered a forest species, and *Zacotus mattewsii,* another apparent forest dweller (Pearsall 2005). Among gastropods, robust lancetooth (*Haplotrema vancouverense*), western flatwhorl (*Planigyra clappi*), and tightcoil snails (genus *Pristiloma*) were common species reduced in harvested areas (Ovaska and Sopuck 2005). Introduced gastropods were very rare and thus not useful as disturbance indicators.

In general, any moderately diverse group of organisms inhabiting forests is likely to contain some sensitive species. Identification of these species in initial studies is most useful if it leads to focused, more efficient sampling for future monitoring. This has not been the case so far with the carabid and gastropod monitoring, which use general sampling techniques that catch a wide range of species. Focusing on a few sensitive species could be helpful in reducing time spent on identification, which can be a major cost in the monitoring of species-rich groups. For example, carabid species in the genus *Pterostichus* were typically generalists; these specimens would not need to be identified to species. Researchers, however, prefer to identify each specimen to the extent possible in case rare species are missed. The potential cost savings from selecting sensitive focal species have not been realized with these projects. One benefit could be simplifying reporting by focusing on the sensitive species instead of indices of species richness, which usually are uninformative.

Another aspect of sensitivity examined by several projects was how quickly species recovered from logging, using a chronosequence of disturbed stands of different ages. Some of this work was done as part of an initial examination of "life boating" – how quickly forest species can recolonize regenerating stands from retained patches (§2.3.2). Variable retention is expected to make regenerating stands habitable continuously or earlier than regenerating clearcuts, but for many species, we do not know at what age a clearcut becomes suitable. If some species quickly recovered in clearcuts, then those species would likely do well in any harvested landscape and would be less useful as indicators for monitoring retention.

A pilot study of red squirrels on Vancouver Island suggested that they were as abundant in twenty- to thirty-year-old hemlock stands as in older stands, although this finding was based only on sign surveys (Huggard and Herbers 2000). Carabid beetles were sampled in a range of stands from recently harvested to approximately 100 years old, as well as in old-growth stands, in northern and southern Vancouver Island. Total carabid captures either levelled off by fifty years of age or declined in the oldest stands, but individual sensitive species showed continual increases in abundance with stand age. *Scaphinotus angusticollis* and *Zacotus mattewsii* were rare in stands <30 years old and at about 50-60 percent of their old-forest levels in fifty-year-old stands. Vascular plants were more abundant overall in recent harvest blocks, with only one genus, *Chimaphila*, largely restricted to mature forests. *Lonicera ciliosa* and *Vaccinium* species occurred in mature forests, persisted in recently harvested blocks, but disappeared in regenerating stands. More non-vascular plants appeared to be sensitive to stand age. Three bryophyte species were largely restricted to mature forest >90 years (*Dicranum fuscescens, Ptilidium californicum,* and *Scapanium bolanderi*), three occurred primarily in mature and fifty-five-year-old stands (*Hylocomium splendens, Hypnum circinale,* and *Radula complanata*), and two were found in stands >20 years (*Kindbergia oregana* and *Isothecium stoloniferum*) (Sadler 2004). Alecterioid lichens increased fairly rapidly with the age of regenerating stands, while other epiphytic lichens continued increasing in abundance through to old stands (Stanger 2004).

The patterns of moderate sensitivity to the age of a regenerating stand suggest that some carabid and non-vascular plant species would be good indicators of effects of variable retention – the species are not so dependent on old forest that they would disappear from a managed stand of any age, nor are they indifferent to harvesting. If variable retention makes younger stands more hospitable, then such retention would make a substantial contribution to the overall occurrence of such species in extensively managed landscapes.

11.2.2 Define Ecological Strata

Because the tenure is ecologically diverse (§2.1), a critical aspect of designing the long-term monitoring program is stratification. Stratification includes determining which parts of the tenure require separate sampling to allow generalization and which parts can be ignored. With the mountainous terrain, elevation is an obvious potential gradient for stratification, examined by pilot studies on birds, carabid beetles, and gastropods. All three studies found lower abundance and richness of species at the higher-elevation sites. High-elevation species were typically a subset of the low-elevation species, although there were a few species in all three groups that were found only in higher-elevation forests. This pattern with elevation, along with the greater amounts of non-harvestable forest at higher elevations (Chapter 7), led to an emphasis on lower elevations for subsequent organism monitoring.

Pilot studies that examined animal groups in more than one region of Vancouver Island (birds, carabids, gastropods, amphibians) typically found substantial differences in abundance but mainly the same species or pairs of similar species within a genus. Pilot studies on vascular plants, bryophytes, lichens, and fungi did not make regional comparisons. Animal communities differed somewhat among Vancouver Island, the mainland, and the isolated islands of Haida Gwaii, although the most abundant species were often found in all areas. The community patterns found in pilot studies left unresolved the more relevant question of whether results of adaptive management comparisons – such as the effects of dispersed versus group retention or edge effects – differed among the geographic regions or BEC subzones. This question is important for determining how to allocate effort for operational comparisons and especially for assessing the potential value of the experimental sites, where replicates of treatments are in widely different forest types.

Seasonality was another potential stratification variable assessed by some of the animal monitoring. Variation in peak singing dates of different bird species is well known; bird monitoring is therefore typically repeated several times through the breeding season (or standardized to a fixed short period for annual trend information, as in Breeding Bird Surveys). Different species of carabid beetles were similarly found to be more abundant at different periods from spring through autumn, requiring several trapping sessions for a full community sample. Sampling songbirds or carabids are instances where focusing on a few key species might reduce costs by allowing fewer sampling sessions. For gastropods, variation between trapping sessions was very high. It is unclear, with the extreme variability, whether there were species-specific differences in seasonality or simply overall changes in catchability (due to weather events). A key unresolved issue for all groups that show strong seasonal variation is whether the seasonality of different species is itself affected by harvest treatments. If that is the case, then repeated sampling throughout the possible season is required to avoid confounding timing and treatment effects.

11.2.3 Define Appropriate Sampling Methods

For some organisms, field protocols have become standardized through long trial and error or simple tradition (e.g., point counts and Breeding Bird Survey routes for songbirds, pitfall trapping arrays for carabid beetles). For these species, using standard methods has the benefit of allowing more direct comparisons with results from other work. Pilot studies can then be used to optimize some details of the sampling, such as plot size, number of traps, or number of repeat surveys. This quantitative comparison of methodologies was often neglected in the pilot studies. For other groups, such as aquatic amphibians, gastropods, and epiphytic lichens, field methods are much less fixed. A substantial part of the pilot phase was used to evaluate different methods. Formal analyses of the costs of different methods versus the value of their information, or of biases in different methods, were not carried out. Most of the learning about methods for these groups simply entailed rejecting techniques that were infeasible or inefficient.

11.2.4 Guide Optimization of Sampling

As well as to evaluate alternative methods, one goal of pilot studies was to document natural variability well enough to provide quantitative measures of the expected precision of estimates with different efforts (costs) and to determine the optimal sampling design. Optimization ensures that the long-term monitoring program is using the most efficient design. Precision estimates allow the program to be designed to provide the most adequate answers within a given budget. Such analyses avoid the rare problem of excessive effort being spent providing unnecessarily precise answers for too few questions, or the more common problem of too little effort being spent trying to answer too many questions, with all results imprecise. Formal analyses of precision and optimization were conducted for four studies in the monitoring program. Most other studies continued to refine methods and provide ad hoc comparisons and were not designed in a way that allowed these analyses (e.g., using nested sampling designs and recording time or cost components).

The expected precision of trends derived from Breeding Bird Surveys was calculated based on results of the first three years of these surveys on the study tenure. Expected precision increased sharply as the number of years of annual surveys increased. However, even with the relatively large number of survey routes on the tenure, only extreme trends are likely to be detected reliably within five years for common species and within ten years for species at moderate abundance. A trend of -2 percent/year is highly likely to be detected only after eight to ten years for the most common species and after fifteen to twenty years for the moderately abundant forest-dwelling species that are more suitable indicator species. With a high component of random variation between individual survey points, alternative sampling schemes of equal effort, such as fewer stops on more routes, or two surveys per year on half as many routes, had little effect on predicted precision. Even a substantial increase in effort each year would have relatively little effect compared with simply conducting the surveys for a few more years. This result emphasizes the need for a long-term commitment to conducting these annual surveys, which are currently the only tenure-wide organism monitoring.

A pilot study on red squirrels showed low densities in all forest age classes on eastern Vancouver Island, and very low capture rates, making precise mark-recapture estimates impossible within a reasonable level of effort. Usefully precise estimates are more feasible with transect surveys of feeding sign, but biological interpretations of such an indirect index are questionable. These sampling difficulties, together with the finding that squirrels in the study were not closely associated with older forests but were also quite abundant in young stands (twenty to thirty years old), reduced interest in red squirrels as an indicator organism.

Formal precision analyses of pilot study results were conducted for two other groups, aquatic-breeding amphibians and gastropods. In both cases, species showed an extremely aggregated distribution at the level of individual plots. In many cases, over half the total captures of a species were at a single location, while the large majority of plots had none of the species. Under those conditions, it is virtually impossible to produce precise estimates with any amount of "bulk" sampling. This result led to a complete redesign of the aquatic amphibian study, using an approach of experimentally managed ponds with stratification based on pre-treatment surveys of amphibians. The amphibian project was expanded to monitor vegetation and physical attributes to ensure some usable information if amphibian results remained too variable. The high variability of gastropods might have been addressed by restricting sampling to particularly suitable microsites, such as seeps. Instead, the project focused on using the experimental VR sites, where pre-treatment sampling was possible. That strategy assumes that the extreme variability among plots remains the same from year to year, which is untested.

11.2.5 Illustrate Ways to Generalize

Direct monitoring can only examine a small proportion of all possible management options and in the near term is limited to recently harvested stands. Developing ways to predict long-term consequences of different options, and options not directly monitored, must therefore be part of the long-term monitoring program. Several pilot studies began to explore the relationships of organisms to habitat elements that can be projected in managed stands (§12.7). Some groups of snails were most closely correlated to herb and fern cover, while slugs were associated with moss. Although much of the variation of these groups is unexplained, these initial relationships suggest a more detailed examination of ground-level vegetation, including more effort in projecting these layers. Stepwise multiple regression with songbird abundance found a variety of statistically significant relationships with habitat elements. Weak relationships were also found with landscape composition variables in surrounding neighbourhoods of different sizes. The predictive ability of these models with independent data is not known. The pilot study on bryophytes compared different substrate types in clearcuts of different ages as a possible way to project effects on this group, finding a clear pattern of increasing bryophyte abundance with increasing decay state of logs. Overall, these initial examinations of habitat relationships have given some direction for future work but are not definitive, mainly because study designs to develop good predictive relationships are different from the designs for comparisons of treatments or the designs for collecting quantitative pilot information.

11.2.6 Summary of the Pilot Phase

We anticipated that work conducted during the pilot stage would establish which groups contained species that were sensitive as indicators and feasible to work with. A significant contribution of the pilot phase was to identify and abandon projects that faced huge variability or seemed to be logistically impossible. We also anticipated that pilot phase work would provide information on sampling strata, optimal sampling design, and expected precision for a given monitoring effort. The quantitative design information would be especially important in deciding which projects to include in the long-term monitoring program, how much effort to allocate to making different comparisons with the different indicator groups, and how long we could expect to wait for definitive results. Knowing when usefully precise information on important management comparisons would be available was a key request from reviewers of the monitoring program, including scientists, environmental groups, and operational personnel of the company. Unfortunately, while much interesting information has been provided by the organism monitoring to date, relatively little quantitative contribution was made to designing and justifying the long-term monitoring program.

Several factors explain our limited ability to execute the pilot phase as originally intended.

- There was inadequate communication of the goals of the pilot phase among the scientists designing the monitoring program, the company scientists running the project, and the researchers conducting the actual monitoring work.
- External funding encouraged specific interests of individual researchers and funding agencies rather than research designed and rationalized explicitly within the context of long-term monitoring.
- Exploiting external funding also meant that the focus on formal piloting was thwarted by an expectation that each project would produce a "stand-alone" product every year.
- The many possible issues that need to be addressed to set up biodiversity monitoring in a complex environment diffused focus.

- The relative novelty of monitoring some of the lesser-known groups captured researchers' interest but directed focus away from common key questions.

We expand on these factors in Chapter 13.

11.3 Individual Monitoring Projects

Because of the numerous threads of research on different organisms and questions, usually carried out in different areas and by different methods, there is no overall "organism program" that can be summarized simply. Even summarizing individual projects undertaken over the five-year pilot phase is not straightforward – questions, locations, and methods often changed from year to year. A further impediment is data ownership, particularly when academic researchers participating in the program are funded by public agencies. Here we briefly outline the activities of the various projects that were part of the initial organism monitoring, some reasons for their changes, and initial findings.

11.3.1 Breeding Bird Surveys

Breeding bird surveys are the only currently feasible way to track tenure-wide trends in a group of organisms, and they contribute to the broader North American Breeding Bird Survey. In 1999, an initial set of routes on Vancouver Island was surveyed. Some of them were modified for logistical reasons, and additional routes were added in 2000. Some routes were also surveyed twice to estimate temporal variation as part of the pilot study. These twenty-eight routes plus an additional six new routes on Haida Gwaii and sixteen new routes on the coastal mainland were surveyed annually from 2001 to the present. Precision analysis (expected uncertainty in trend estimates for different sampling intensity and duration) was conducted for the Vancouver Island routes after the third year. All survey points have been located with GPS and landscape characteristics within different radii of each point assessed with GPS to estimate neighbourhood models for the survey points (Preston and Campbell 2005; Vernier 2004).

11.3.2 Songbirds

Two songbird projects were undertaken as part of the pilot phase in monitoring organisms, initially with relatively little co-ordination. Lack of co-ordination reflected different external funding sources and academic objectives for the two projects. The projects recently have become better co-ordinated to provide complementary data. Two projects working independently on the same group of species within one monitoring program appear to be inefficient but have the advantage of providing alternative perspectives, especially for a well-known group such as songbirds.

One project began in 2000 with point counts in southern Vancouver Island, using operational retention blocks plus four uncut benchmark stands. Most of the blocks from the initial year of harvesting in this area were dispersed retention with low retention levels. In 2001, more group and mixed retention blocks were added to balance the comparison of retention types. Dispersed retention sites with higher retention levels were sought out to expand the range of retention levels and avoid confounding retention type and level. Pre-treatment surveys were also conducted at one of the experimental sites, and habitat structure plots were measured at the survey points. The operational blocks were surveyed again in 2002, along with post-harvest surveys at two experimental sites and pre-harvest surveys at a third.

In 2003, six uncut benchmark sites were monitored, and post-harvest surveys continued or began at the three experimental sites. Modelling of habitat and neighbourhood relationships was conducted using the operational survey results (Chan-McLeod and Vernier 2005). This project has spent considerable effort at the experimental sites, which hopefully will produce useful information when comparable replicate sites have been completed.

The second songbird project also began by using point counts at operational retention blocks in 2000 and 2001 but on northern Vancouver Island examining only the patches in fifteen group retention blocks and comparing them to eight uncut benchmark sites (Figure 11.2). The researchers' original objective was an examination of patch size effects. The surveys included low- and higher-elevation sites. In 2002, the design was revised to focus on the more species-rich lower-elevation forest types and to sample the complete cutblock (including harvested areas) with a layout similar to the first songbird project. In total, twelve clearcut sites, twelve group retention sites, and twelve benchmark sites were used. The group retention sites were chosen to cover the range of operational retention levels. Habitat sampling was also conducted at the point count stations. The same blocks were surveyed in 2002 and 2003. In 2004, an additional study component was added, using fine-scale mapping of a few focal species to more closely examine the use of group retention blocks and retained habitat elements. Group retention supported similar species in similar abundance as benchmark forests. The likelihood of a given species occurring in a group retention block appeared to be more dependent on the sizes of patches retained than on the total percentage of trees retained within the cutblock (Preston and Harestad 2007).

11.3.3 Owls

A feasibility study of owl monitoring was conducted using road-based playback surveys in 2000. However, responses were few and patchy, making reliable estimates of relative abundance or trends almost impossible with any affordable effort. Additionally, it was difficult to relate the results of playback

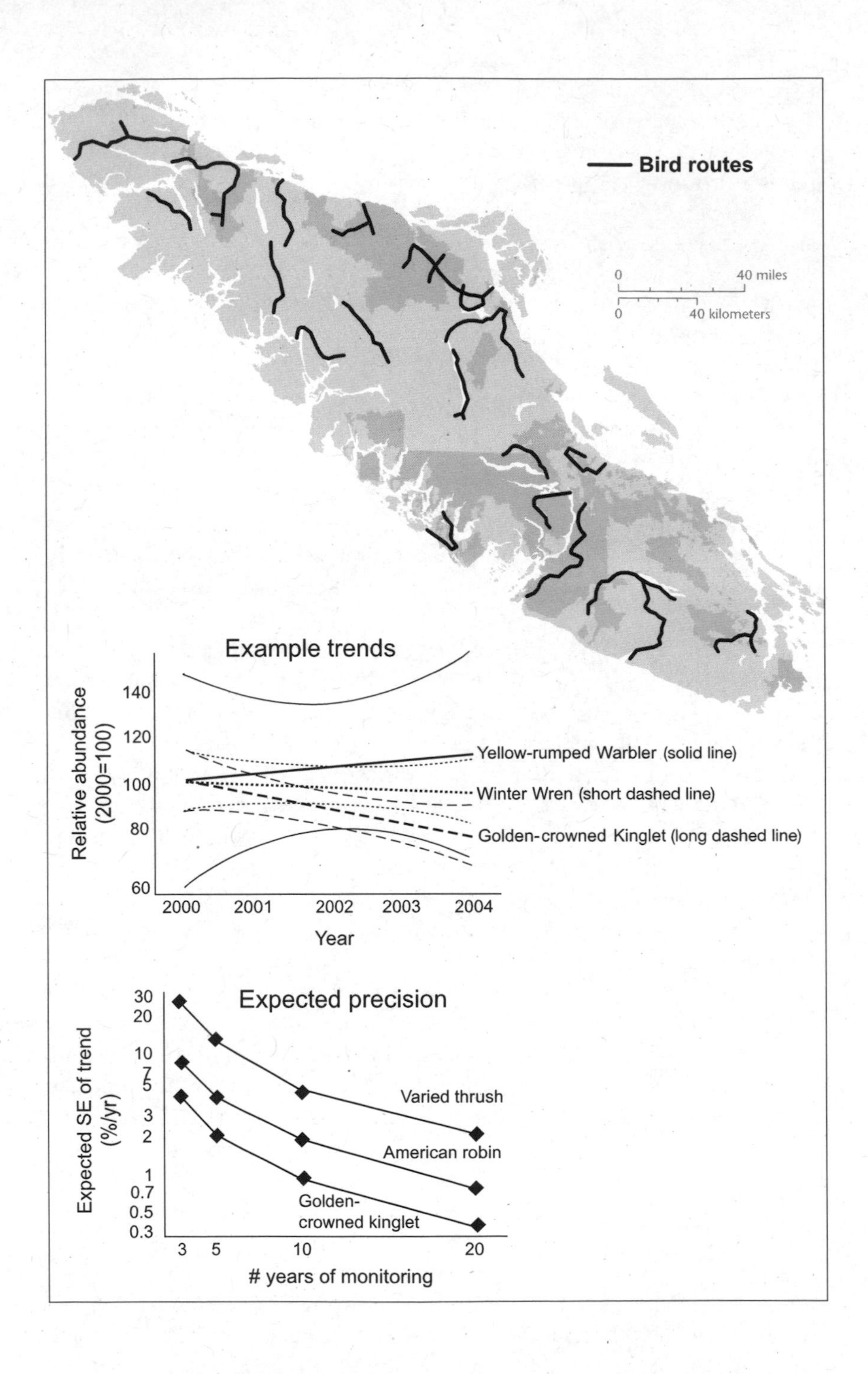

Bird routes
0
40 miles
0
40 kilometers
Example trends
Relative abundance (2000=100)
140
120
100
80
60
2000
2001
2002
2003
2004
Year
Yellow-rumped Warbler (solid line)
Winter Wren (short dashed line)
Golden-crowned Kinglet (long dashed line)
Expected precision
Expected SE of trend (%/yr)
30
20
10
7
5
3
2
1
0.7
0.5
0.3
3
5
10
20
years of monitoring
Varied thrush
American robin
Golden-crowned kinglet

◄ *Figure 11.1* Breeding bird survey routes and trends on the study tenure. Most initial trends (upper) have been relatively flat (e.g., winter wren), though uncertainty is still high for most species (e.g., confidence limits for yellow-rumped warbler). Three relatively common species have shown apparent declines (e.g., golden-crowned kinglet). Precision analyses (lower) suggest that eight to ten years are necessary for reasonable confidence that a common species is declining at 2 percent per year, increasing to fifteen to twenty years for species at moderate abundances.
Source: Summarized from data provided by M. Preston. | *Map: MacMillan Bloedel Limited.*

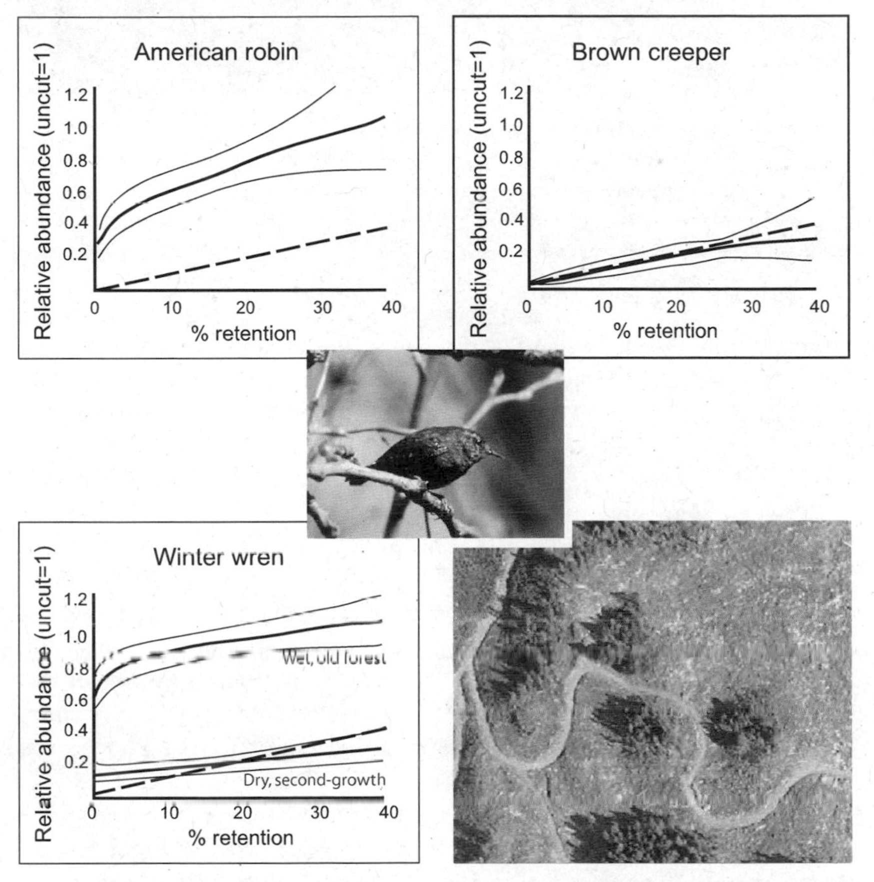

Figure 11.2 Forest-dwelling bird species' responses to variable retention. Many birds were retained in moderate densities in variable retention. The brown creeper was severely reduced in VR and abundance was nearly proportional to percentage of forest removal (the dotted line shows a 1:1 relationship between abundance and percent retention.) Winter wrens showed little response to harvesting in wet ecosystems but exhibited strong declines in dry forest. Detailed spot-mapping (bottom right) can be used for a more detailed examination of effects of habitat features, edges, and patch sizes. *Source:* Results summarized from data provided by M. Preston and A. Chan-McLeod; spot-mapping figure from M. Preston. | *Photographs by: winter wren, F. Bunnell; and map, MacMillan Bloedel Limited.*

Figure 11.3 Although they are potentially sensitive to forest practices and appeal to many people, squirrels and owls proved to be unsuitable for further monitoring. Detection rates were low, and indirect survey methods were dubious. Intensive radio-telemetry work is needed to provide meaningful new information on these animals. | *Photographs by: M. Preston.*

surveys of highly mobile species to stand types or even local landscape characteristics. Thus, it would be difficult to use owls to address any of our comparisons without intensive, expensive, radio-telemetry of individuals. Monitoring of this group was therefore not pursued (Preston and Campbell 2001).

11.3.4 Red Squirrels

A pilot study on red squirrels in 2000 examined mark-recapture live trapping, radio-telemetry, and transect surveys, primarily of feeding sign. Live trapping was inefficient because of low squirrel densities and almost immediate loss of trapping bait to high populations of deer mice. Although these problems could be overcome for a research project, they would make the species prohibitively expensive for broad monitoring. Radio-telemetry provided information on the use of retention blocks by squirrels, but limited home ranges and the frequent use of particular habitat features meant that a large number of squirrels would need to be followed for results to have general relevance. That again would be expensive. Transect surveys detected few squirrels directly but found considerable amounts of feeding sign (middens and consumed cones), including in young regenerating stands. Usefully precise estimates of feeding sign could be obtained with reasonable effort, but biological interpretation of such an indirect index is dubious. For these reasons, monitoring of squirrels was not pursued.

11.3.5 Carabid (Ground) Beetles

Monitoring of carabid beetles began in 2001 by comparing old growth, young forest, and clearcuts to see whether there were species affected by logging

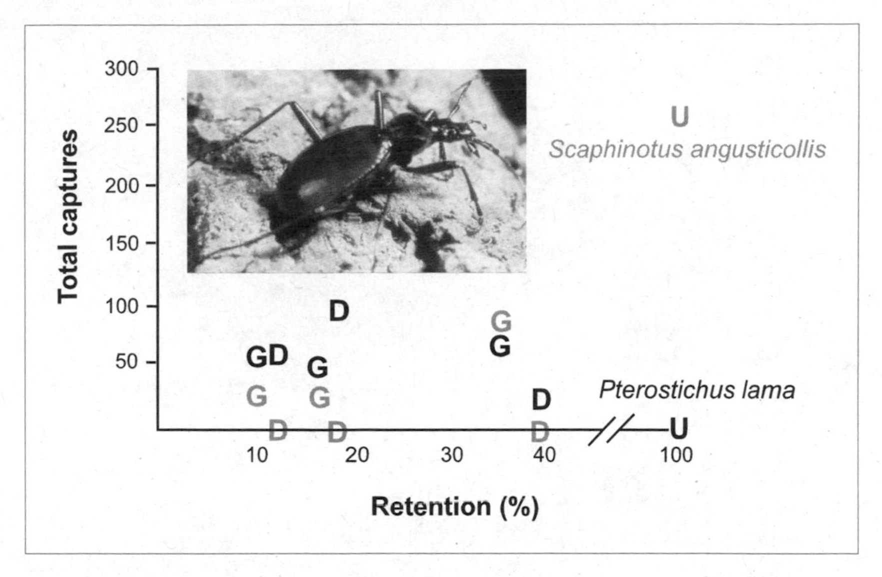

Figure 11.4 Forest-dwelling carabid beetles, such as *Scaphinotus angusticollis*, appeared to persist in group retention patches (G) immediately after harvest but at lower abundances than in uncut forest (U). Dispersed retention (D) supported fewer or none of these species. Early-seral species, such as *Pterostichus lama*, arrived in the harvest treatments, both dispersed and group retention. *Source:* Results summarized from data provided by I. Pearsall. | *Photograph by I. Pearsall.*

and how abundant they were. High- and low-elevation sites were also compared. In 2002, the study moved to two post-harvest experimental sites, including sampling across edges of patches at the group retention experiment. Carabid communities in the patches of the group retention site were similar to those in the uncut control site. Without replication, little could be said from that year's results. In 2003, the project sampled across edges of different ages of clearcuts on northern and southern Vancouver Island and into the adjacent uncut forest. This work was intended to assess whether some species would be suitable for monitoring "life boating" and was partly a response to pressures to produce short-term interesting results. In 2004, the project again shifted, to operational retention sites four to six years old on western Vancouver Island. The carabid project is an example of how changing annual priorities within the monitoring program and from external funding agencies led to interesting but scattered and non-definitive results, without having firmly established how well the group might be monitored in the long term.

11.3.6 Gastropods

The gastropod project began in 1999 and continued through 2001 by comparing retention patches, harvested areas, and uncut sites to see whether

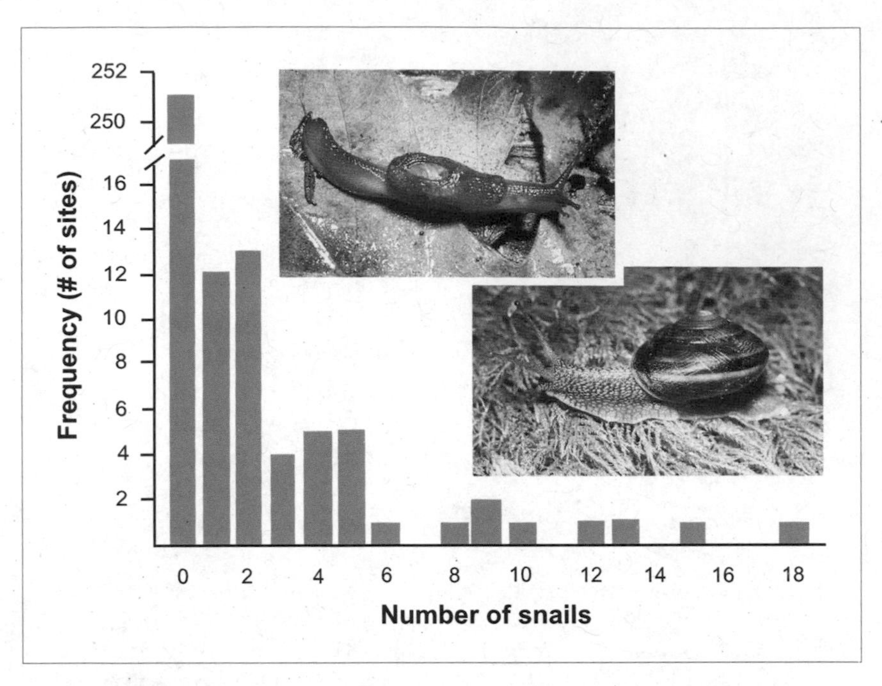

Figure 11.5 Surveys of gastropods on Vancouver Island and Haida Gwaii found sixteen snail and four slug species, including the snail *Monadenia fidelis* and a new species of jumping slug (upper inset). Monitoring this group is challenging because of low numbers of large species and great variability among individual sampling stations for smaller snails; most samples had no individuals, while a few had large numbers.

Source: Results summarized from data provided by L. Sopuck and K. Ovaska. | *Photographs by: L. Sopuck and K. Ovaska.*

any species of this poorly known group was affected by logging and how frequently they were captured. One operational site on northern Vancouver Island and one on southern Vancouver Island were sampled intensively to evaluate methods. Active searching of plots, artificial cover boards, and litter extraction were compared and suggested that well-designed cover boards (complemented by litter extractions for small species) best surveyed the species present. High- and low-elevation sites were also compared. Recognizing the extreme variability among plots, in 2002 the study switched to making pre-treatment measurements at the experimental sites, with the hope that pre- versus post-treatment values would reduce the high variability.

11.3.7 Bryophytes and Vascular Plants

The initial study of bryophytes focused on remnant patches that had been retained during past logging because of physical inoperability (Baldwin 2000). The concept was that they might act as possible benchmark sites for

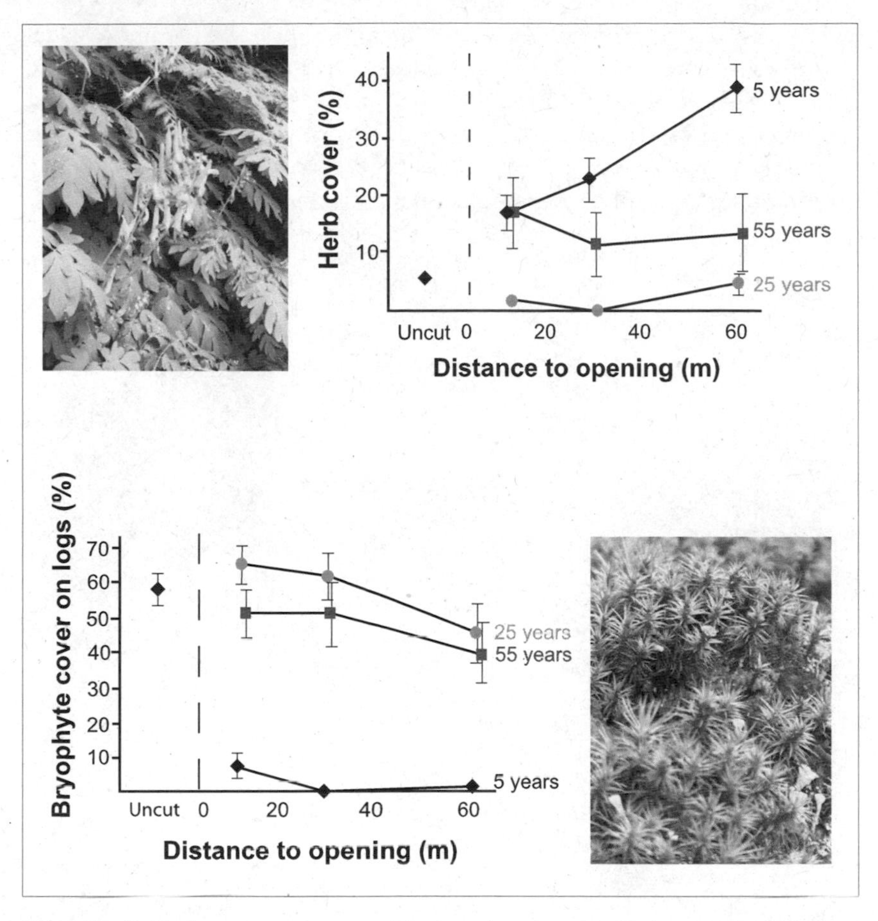

Figure 11.6 Understorey response with time since harvest. Herb cover declined by year twenty-five before recovering by year fifty-five. Bryophytes, particularly those on logs, were more sensitive to harvesting than were vascular plants. Several bryophyte species increased with age in older clearcuts and were slightly more abundant near adjacent forest. Such relationships determine how successful VR will be at promoting recolonization of harvested areas.

Source: Results summarized from data provided by K. Sadler. | *Photographs by: Scouler's corydalis, W. Beese; and moss, T. McIntosh.*

comparing retention with older surrounding matrix conditions and would help to reveal how well patches would retain species while the surrounding forest regenerated. This approach was later abandoned because the remnants were highly anomalous, low-productivity sites unlikely to reflect future conditions in current retained patches. However, within these anomalous sites, the study found that interior habitats of patches had both greater abundance of old-growth associated groups and total bryophyte cover than occurred near patch edges (Baldwin and Bradfield 2005).

A second study of vegetation species, including bryophytes and vascular plants, took place in 2003 in the same old-clearcut sites used by the carabid study. These old clearcuts surrounded mature and old-growth patches, and the intention was that information would contribute to an understanding of whether retained patches encouraged the recolonization of surrounding areas for a variety of organism groups. As well as transects into the cutblocks

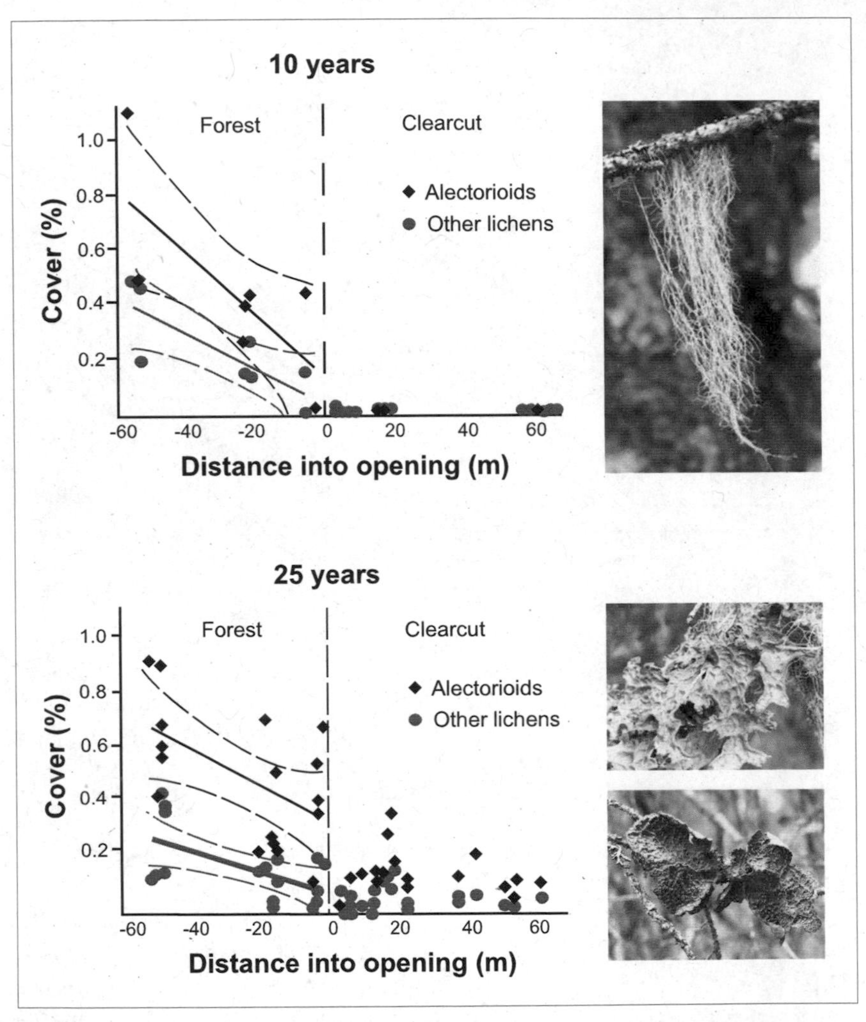

Figure 11.7 Epiphytic lichens are a physical challenge to sample, and their vertical and horizontal stratification in tree crowns adds statistical difficulties. Nonetheless, some species are potentially useful for monitoring programs because of their sensitivity to cutblock age, persistent negative edge effects in retained forest, and limited dispersal ability. *Source:* Results summarized from data provided by N. Stanger. | *Photographs by: I. Houde.*

and reference sites in the adjacent mature forest, the bryophyte study compared bryophytes on different substrates – logs of different decay classes, coniferous and deciduous tree bases, and forest floor.

11.3.8 Epiphytic Lichens

A pilot study on epiphytic lichens was conducted in 2002 and 2003. The main goals in the first year were to determine how to sample retained trees safely and to test strata for sampling epiphytes within the trees. In 2003, the project monitored across the edges of older clearcuts, although the difficult sampling reduced the number of sites.

11.3.9 Ectomycorrhizal Fungi

Monitoring of ectomycorrhizal fungi began in 2000 with sampling at different distances into areas recently harvested by variable retention, in three

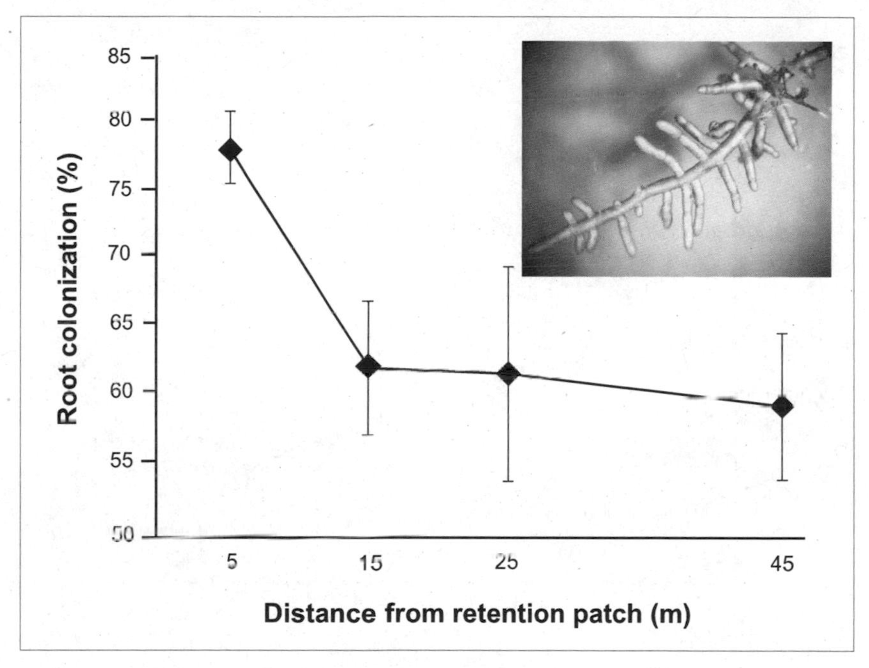

Figure 11.8 Ectomycorrhizal fungi declined in openings. Establishing reference collections is an initial requirement for monitoring diverse and difficult-to-identify ectomycorrhizal fungi, such as this *Lactarius rubrilacteus* on Douglas-fir (*Pseudotsuga menziesii*). Abundance and diversity of these species decline in openings beyond the spread of the roots of trees retained in adjacent patches. Current research is examining whether dispersed trees retain fungi well enough to maintain ectomycorrhizae across the harvested area.
Source: Summarized from Outerbridge and Trofymow (2004). | *Photographs by: R. Outerbridge and J.A. Trofymow.*

second-growth and three old-growth sites. Roots of sample seedlings were assessed in 2002. This part of the project also established a reference collection of ectomycorrhizal species and morphotypes. In 2004, the project put transects of sample seedlings into dispersed retention at different distances from the cutblock edge and 1 metre and 10 metres from individual dispersed trees to evaluate whether single trees are as useful sources of ectomyccorhizal fungi as VR patches.

11.3.10 Aquatic-Breeding Amphibians

This project initially examined aquatic-breeding frogs and salamanders on land using pitfall trapping in 2000 and 2001. However, variability was extremely high, with most captures of a species usually occurring in a single location. This made informative comparisons unlikely without immense sampling effort (Wind 2005). In 2002, the project switched to the aquatic habitat, piloting methods to examine the effects of different levels and types

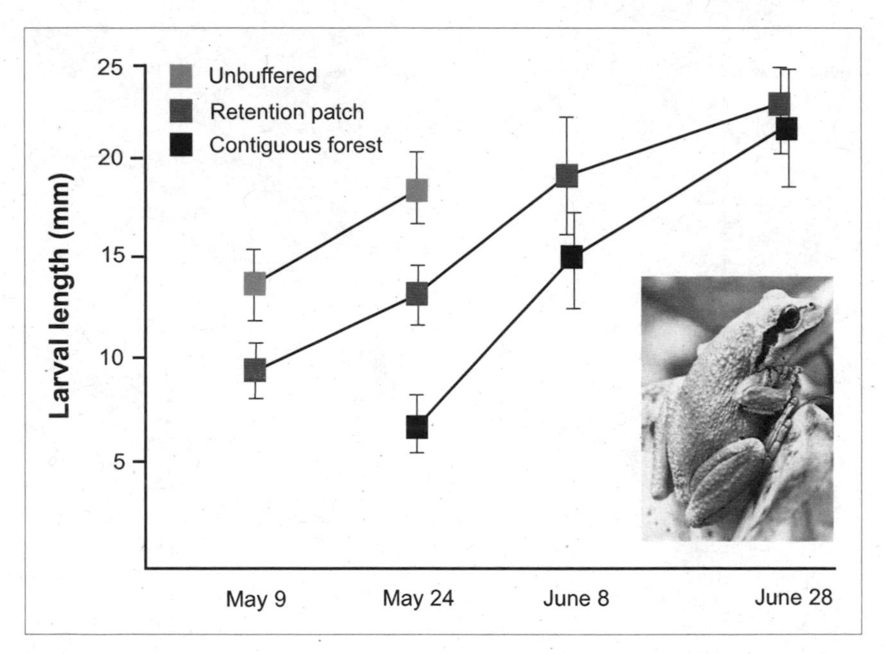

Figure 11.9 An experimental design, with treatment allocation based on initial surveys of wetland characteristics and amphibian occurrence, helped to overcome problems caused by the patchy distribution of species such as this Pacific chorus frog (*Pseudacris regilla*). Larval frogs grew faster in unbuffered small wetlands but died before metamorphosis when these ponds dried out. Larvae in wetlands in retention patches hatched earlier than in contiguous forest, but sizes had converged by the time of metamorphosis.
Source: Data provided by E. Wind.
Photograph by: E. Wind.

of canopy retention on small wetlands. Given the variability in wetland types and operational variability in treatment of these wetlands, the project changed to a designed experiment implemented in operational cutblocks on southern Vancouver Island, where dry summer conditions created the most concern for wetland environments. In 2003, a large number of small wetlands in operational forests were classified and buffer treatments designed (no buffer, typical narrow operational buffer, and wide buffer). In 2004, pre-harvest physical and vegetation measurements were made of the ponds, along with sampling of larval amphibians, and were used to assign comparable wetlands to each treatment. Given the high variability, this blocking appeared to be the only way to ensure useful results with moderate sample sizes. The physical and vegetation measurements should provide more certain results even if amphibian responses are variable.

11.4 Summary

As anticipated, monitoring organisms proved to be the most challenging of our three indicators. Some pilot studies provided guidance around the five points that they were expected to address.

- Sensitivity to total forest cover was documented for species within groups poorly studied in the region (gastropods, carabid beetles) and to forest age for species among lichens, bryophytes, vascular plants, and carabid beetles.
- Attention to lower elevations (where variable retention was most widely practised) during monitoring was affirmed by reduced abundance and richness among songbirds, gastropods, and beetles in response to harvest.
- Sampling methods found to be infeasible or highly variable were discarded.
- Studies of songbirds, red squirrels, amphibians, and gastropods provided sufficient information to estimate precision under modified sampling regimes and to guide more optimal sampling.
- Responses to habitat variables of bryophytes, snails, and songbirds provided information on how more refined, predictive models could be developed.

However, we encountered significant difficulties in both the initial pilot study phase and the transition to implementing the full monitoring design, though the causes and possible solutions to these problems differ. They were noted in §11.2.6. They are reviewed in detail along with potential solutions among "lessons learned" in Chapter 13.

Part 3: Summary

Part 1 provides context for the material summarized in this book. Part 2 considers the three major indicators of success in sustaining biological diversity and information derived from each of them. Part 3 provides a summary in two chapters. Chapter 12 illustrates how the individual parts of the monitoring program, many discussed in Part 2, were integrated into an overall monitoring design, and it summarizes lessons learned during the design. Chapter 13 provides an overview of the progress made and lessons learned from attempting to implement an adaptive land management program.

During the course of this project, the company went through three ownership changes: from MacMillan Bloedel Limited to Weyerhaeuser Limited to Cascadia Forest Products Limited to Western Forest Products Incorporated. With each change in ownership, personnel within the company also changed. Those changes hindered continuity and are partially responsible for some of the failings in attaining the goals of the adaptive management program. They were not the only reasons. Within any resource industry subject to changes in market, government policy, and other social and economic influences, adaptive management is difficult to implement. The initial question that managers ask – where are we going? – represents a moving target. As well, it is difficult to encourage individuals subject to different risks and rewards to move in a consistent direction. The two chapters of Part 3 confront a difficult balance – to review what was intended at the outset of the project, a kind of ideal, and to juxtapose that with what actually happened, less than ideal.

Chapter 12 is a summary of the most intellectually challenging portion of an adaptive management program – the design for effectiveness monitoring. It is one thing to conceive of the ideal design and quite another to implement that design within the complexity of nature and the vagaries of economic emphasis. The chapter has three broad goals: (1) to document how features noted for each broad indicator are integrated into a single monitoring design, (2) to review the principles that must underlie a monitoring design (e.g., blocking principles to permit extension of findings, pros and cons of experimental versus operational sites), and (3) to present our solutions when confronted with real-world complexity or caprice. It attempts to illustrate both what could be and what actually happened.

Chapter 13 provides a review of the progress made on the six management questions formulated at the beginning of the program and reviews lessons learned from attempting to implement adaptive management. These lessons are aggregated under three broad headings: organizational structure, design, and feedback. Feedback is the final step of the four that we used to describe an effective adaptive management process (§1.4) – linking the results of monitoring to management decisions. We report our inability to consistently close the adaptive management loop and evaluate why that happened. We close with summary thoughts.

12
Designing a Monitoring Program
David J. Huggard, Laurie L. Kremsater, and Fred L. Bunnell

12.1 Context

The primary role of monitoring is to reveal what has been achieved, or will likely be achieved, from particular management options. A secondary role, embedded in the monitoring design, is to make subsequent monitoring more cost effective by revealing useful refinements. We tried to design effectiveness monitoring that included both roles. We noted (Chapter 3) that the following steps facilitate effectiveness monitoring.

- *Determine major issues.* Our issues were to implement a management system that sustained native species richness and to reserve more old growth (Chapter 2).
- *Clearly define objectives* and associated indicators of success by focusing monitoring on appropriate variables. We used a single criterion and three broad indicators of success (§3.2.2).
- *Identify the management plan and practices.* Stewardship zones and variable retention were chosen to address the major issues (§2.3). The approach to planning and practice became hypotheses to be evaluated by the monitoring process (Davis et al. 2001).
- *Bound the problem.* Establish the physical, functional, and conceptual boundaries. The largest issues, spatial extent of the program and conceptual extent of biodiversity, were addressed in §3.4.1. We also attempted to select specific, interpretable indicators that did not introduce extraneous influences over which the manager had little control. Chapters 6, 8, and 10 summarize the search for effective indicators.
- *Identify the major monitoring questions.* Ideally, a monitoring design asks questions in a fashion that reduces statistical uncertainty, properly estimates parameters from noisy data, and assigns probabilities to alternative hypotheses (Mulder et al. 1999; Noss 1990, 1999; Noss and Cooperrider 1994). It also attempts to ask questions that are most relevant to management actions. The major questions of the program were selected relatively early (§3.4.2).

- *Identify data needs.* An assessment of current conditions indicates broad knowledge gaps, but specific needs must be identified in terms of data best suited to answer the major monitoring questions. The reviews of Chapters 6, 8, and 10 attempted to do that. Results of pilot efforts indicate variable success in addressing those data needs (Chapters 7, 9, and 11).
- *Rank the questions and data needs.* Available resources for monitoring are always limited. Ranking is necessary because it focuses on questions that represent greater risk, thus data needs. We summarize here our process of refining questions and ranking them.

The steps above address *what* to monitor. This chapter reviews what we learned about *how* to learn effectively from the monitoring program and summarizes the overall design of the monitoring program. The design provides structure to the monitoring to allow learning. At the outset (§1.1), we noted that assessing how best to manage for biodiversity fell into the class of "wicked" problems. Features of wicked problems, particularly the lack of a single right or true test and the absence of simple true or false solutions, had to be confronted within the monitoring design.

12.2 How to Ask Questions

Simply measuring is not enough. We learn from a monitoring program by making comparisons and by understanding mechanisms.

12.2.1 Comparisons and Mechanisms

Knowing that we have 7.2 (±1.3) large snags per hectare does not teach us anything or help to guide management. The measurement becomes useful for learning only if we compare it to some scale or standard or use it to clarify a mechanism. We could compare measured snag density to results from an alternative harvest method, a natural benchmark, or a regulatory target. Or we may have designed the study to clarify mechanisms creating that level of snag retention – for example, the rule of thumb that fallers use to decide which snags to leave or to fall when logging. It is helpful to consider the relative contributions that making comparisons and exposing explanatory mechanisms make to effectiveness monitoring.

Most simply, direct comparisons are easier, more precise, and often cheaper than explanatory studies, but application of the results is limited to the specific cases compared. Explanatory studies generally involve more thought and understanding of the system, are more complicated, and require more expensive measurements and more assumptions, but they are potentially applicable to a wider range of situations. We can compare large, old snag retention in the same amounts of dispersed and group retention with a rather simple design yielding a precise comparison. The results allow a recommendation of which retention method to favour to retain large, old snags at that retention level or which method of retention better meets predetermined target values. However, the comparison tells us nothing about the two retention methods at other retention levels or about the consequences of other modifications to variable retention. Conversely, a study of how fallers' decisions led to the observed level of snag retention would be more difficult, involving several field methods as well as work to validate the proposed explanations. Predictions of retained snags in a particular harvest type would require assumptions and be relatively imprecise. However, a good explanatory model could predict snag retention under a wide variety of real and proposed management options. It might also help to predict additional variables that were not measured directly in the comparative approach (e.g., short snags or non-merchantable live trees). The point is that both comparisons and mechanistic explanations should be considered in the monitoring design, even though the latter are often considered research, not monitoring. In fact, the two approaches are necessary components of meeting monitoring goals of efficient precision and generality of application.

Because evaluating mechanisms is costly and the consequences of variable retention invoked many mechanisms, we emphasized making comparisons.

12.2.2 Types of Comparisons

Two types of comparisons are useful in adaptive management: direct comparisons of management alternatives, and comparisons to targets or benchmarks. Direct comparisons may involve only current operational activities ("passive" adaptive management), former activities, novel practices being tried operationally, or controlled experimental treatments ("active" adaptive management). Targets or benchmarks can be derived externally, as with regulatory requirements, or they may require field measurements employing natural benchmarks.

Approaches comparing alternatives and comparing to targets are not fully separable because of the need for scale or context. "Targets" should represent a useful scale, so that we know, for example, whether the level of snag retention in either dispersed or group retention is adequate or inadequate to meet objectives. We found that regulatory requirements rarely had been evaluated in terms of their effectiveness, so we often chose natural benchmarks as our standard. Even when targets are agreed on, comparisons to targets alone may not help to improve practices. For example, if the retained snags are too few in a particular harvest type, what should be done instead, and which alternatives would be more successful?

12.3 Stand-Level Comparisons

For stands, we derived twenty-five informative comparisons that potentially could address our major questions (§3.4.2). Some of them represented different aspects of the comparison. There were five broad groups of comparisons:

1 among retention types: retention type and amount;
2 within retention blocks (retention may be implemented differently within the same type and amount of retention): edge effects, patch size, patch "anchor points," patch arrangements, harvest method;
3 to other operational harvesting: clearcuts (recent and old), other silvicultural practices (with and without retention), existing regulations, certified blocks, old-forest restoration;
4 over time: changes in practice or retention, changes within retention stands, chronosequences; and
5 to targets or apparent thresholds: older forest, external targets, naturally disturbed stands, species' requirements, non-harvestable forest.

Allocating effort to any specific comparison must be made in the context of the overall program as well as priorities for the major questions. Here we review how we guided the monitoring effort by assigning priority to each comparison. Priorities were based on direct or potential relevance to management decisions and the feasibility of collecting information that could guide those decisions. Priorities below are current and will change over time as information accumulates or new issues arise. In some comparisons (e.g., with clearcuts), different aspects of a comparison received different priorities. Assigning priorities at the outset permits thoughtful responses when budgets change.

Clearcuts and uncut forest provide the "book ends" between which variable retention fits. There is a wealth of data for many organisms within clearcuts, so the highest priorities for monitoring were assigned to uncut forests and those harvested by VR.

12.3.1 Very High-Priority Comparisons

Four comparisons designated very high priority were the main focus of the adaptive management monitoring program and received the most effort. Direct comparisons of how retention is implemented provide the most immediate feedback to management practices at the stand level. The main variables are the type and amount of retention.

Types of retention (dispersed, group, and mixed)

Dispersed retention and group retention are distinct treatment types (not a gradient of retention patch size) because they involve different operational processes (mapped reserves versus faller selection) and different ways of evaluating retention (area versus volume). Mixed retention also differs because it can involve different processes for locating retained trees than does either group or dispersed retention. Thus, dispersed, group, and mixed retention were compared as three distinct types of variable retention.

Amounts (percent) of retention

Designated retention levels are based only on retained area or volume of live trees, so relationships of biodiversity indicators with retention levels are not necessarily directly proportional. Snag retention, for example, could drop off more quickly than live tree retention because of edge effects and safety concerns, or it could remain higher than live tree retention because snags were favoured during retention. Relationships between retention levels and forest-dependent organisms are even more complex and less certain.

The two variables, type and amount of retention, permit direct comparisons among current operational practices. Designs for such comparisons must include interactions between the variables. There is no reason to expect a change in percent retention within dispersed retention to have the same biological effects as the same change in group or mixed retention. Examining interactions is difficult because dispersed retention tends to retain fewer trees than group retention. This is a clear instance in which refinement monitoring (§3.4.3) should seek out rarer combinations (e.g., high-percent retention in dispersed retention), whereas monitoring of current performance should seek samples more representative of operational practices.

The very high priority reflects the extensive use of variable retention. Finding better ways to implement retention operationally was an immediate need. Comparison was possible with operational cutblocks, which also contributed to implementation monitoring, and with experimental blocks, which provided a greater range of treatments and other advantages for learning.

Edge effects within blocks

Measuring edge effects was a very high priority because edge effects directly affect two of the main motivations for using VR: maintaining forest influence over most of the cutblock, and providing "lifeboats" of forest structure that maintain forest organisms in retention areas when most forest cover is removed. Moreover, measurements of edge effects permit calculating effects of patch size and shape by simple geometry. Stratifying by edge distance is, in any case, necessary for most monitoring that samples heterogeneous retention blocks.

Design for measuring edge effects is complicated by the fact that, in many operational retention patches, distance from edge is confounded with changes in habitat types, because patches are anchored to features such as wetlands or outcrops near their centres. Monitoring edge effects in retention patches as they currently are implemented should sample in randomly selected or typical retained patches. However, monitoring of edge effects to learn how to improve effectiveness through patch design should try to avoid confounding distance effects and habitat changes by seeking out retention patches that are in single habitat types.

Targets from nature – older forest benchmarks

The ecosystem management paradigm, which influences much thinking about forest management, specifies management targets by reference to natural conditions (Galindo-Leal and Bunnell 1995; Grumbine 1994). These targets can either be derived from an imaginary static or average natural condition or be bounded by an estimated range of natural variability. The general implication for monitoring design is the need for information from unmanaged forests to establish natural targets. Many ecological issues make implementation of the ecosystem management paradigm unreliably complex, including the appropriate period for definition of the targets, the relevant scale, natural long-term changes in ecosystems, and pervasive but uncertain human effects altering even unlogged forests. These uncertainties hinder comparisons to natural targets. A more immediate difficulty for any monitoring study design is determining the appropriate unmanaged areas to use as benchmarks. "Natural" is sometimes thought to equate to "old growth," but this ignores the role of natural disturbance and succession in forests. The most credible approach to using natural forests as benchmarks would measure attributes in stands ranging in age from immediately following natural disturbances to old forest. The "target" would then be the array of attributes found in natural stands of different ages when the stands were considered in proportion to their occurrence under natural disturbance regimes, thus producing a range of conditions. Attributes across the managed landscape would then be compared to this range of natural landscapes. Beyond the amount of work involved, practical difficulties include finding naturally disturbed stands of a variety of ages (particularly in areas that have experienced human disturbance) and determining the natural disturbance regime.

Because "natural" is so often selected as a desirable target, we assigned this comparison a very high priority. We measured natural benchmarks in old forests in wetter ecosystem types that were naturally dominated by old forest. We established benchmarks in old-growth stands or the oldest class of stands in fire-dominated ecosystem types without representative old growth (provided they were >80 years old). Emphasis was on using benchmarks in stand types comparable to those being managed rather than on using the oldest stands if they were on atypical sites.

12.3.2 High-Priority Comparisons

High-priority comparisons were considered important for a complete monitoring program but were assigned less intensive effort than very high-priority comparisons. We initially recognized nine high-priority comparisons.

Patch size (beyond or instead of edge effects)
Retention can be implemented differently within the same type and amount of retention, providing opportunities for management. For example, it is cheaper to change retention patterns at a given level of retention than it is to increase retention levels. Patch size can have effects beyond those that can be calculated directly from edge effects. Examples are reduced, or increased, use of small patches by mobile organisms compared to expectations from edge effects or reduced longer-term persistence of populations of small organisms in small patches. Patch sizes should be compared directly where it is not possible to measure edge effects directly. For example, point counts of songbirds provide insufficient resolution to measure edge effects. To avoid confounding patch size and amount retained, monitoring should be done in blocks with similar average levels of percent retention or at two or more levels that bracket the range of operational retention.

Patch anchor points
Within group and mixed retention, a major influence on many ecological variables appears to be the feature used for establishing or "anchoring" retention patches (e.g., rock outcrop, wet area, snag patch). Effective sampling within blocks, contrasting patches having different types of anchor points, could contribute to recommendations about which anchor points to favour (§9.3.6).

Recent clearcuts
Clear-cutting without retention is no longer used in the study area. In the short term, comparisons between clearcuts and variable retention are trivial for certain habitat structures (e.g., snags), but expected differences are less obvious in the longer term and for many organisms. The comparison was nonetheless a low priority for fieldwork because the social mandate to avoid clearcutting means that it would have little influence on management, regardless of outcome. However, collating available published data on organisms in a fashion that permits comparisons was a high priority.

Silvicultural systems during retention
Comparisons of retention types and amounts refer to the pattern of permanent retention in the cutblock. That retention can be created within a variety of silvicultural systems (e.g., clear-cutting, shelterwood, group selection). Different silvicultural systems in retention settings are a potential source of variability, providing options for maintaining different ecological components in a stand. Comparing other silvicultural systems was a high priority for learning about effects at the stand level, but there were few operational blocks that allowed these comparisons. Monitoring them would contribute little to monitoring current performance. Moreover, existing examples were

often in high-elevation, heli-logged sites, which were not a priority for monitoring because they were well represented in the non-harvestable land base. Some experimental treatments involve multi-pass systems, providing opportunities for comparisons in lower-elevation blocks.

Old-forest restoration

We determined that assessing the effectiveness of active restoration of old-forest characteristics would be a high priority, provided that restoration activities become more than a pilot study. Some stands manipulated to enhance old-forest characteristics for deer winter ranges twenty years ago (Bunnell 1985; Nyberg et al. 1986; Nyberg et al. 1989) provided an opportunity to foresee possible effects of restoration practices and were considered a high priority for monitoring. Because no stands were designated for restoration, the opportunity was not used.

Changes in practice of variable retention over time

Continual improvement in practices, or the results of practices, is a primary measure of the success of adaptive management and an important indicator under some forest certification systems. The only way to assess such improvement is to collect unbiased measurements of overall operational performance over time (Chapter 4; §9.3.7). A problem in measuring ecological improvement in practices is that measurements need to focus on recently harvested blocks to represent current practices, yet the ecologically meaningful effects occur over the entire rotation. The only solution is to couple direct measurements of recent practices with reliable projection models that can answer the question: are the future effects of current practices better than the future effects of past practices? Developing and documenting effective sampling methods and archiving data to allow comparisons in the future were high priorities. Once baselines were established, the priority became low because there has been little time for VR practice to evolve. Resampling every five years probably will be adequate to indicate major changes in practice. Initial sampling should be a random and representative sample of operational blocks, not blocks selected to show the best operations.

Changes within retention stands over time

The direct way to measure change in variable retention stands over time is to resample, in the future, stands that were sampled immediately after harvest. That is a slow process. Moreover, the oldest variable retention stands at a given time may be atypical because they were created by early harvesting when practices were changing rapidly. Simultaneous comparisons of VR stands of different ages, rather than resampling through time, would have the same confounding of aging and changing practices as well as additional variation from comparing ages across different blocks. The short-term implication for study design is to establish permanent plots for a wide variety of indicators in current variable retention blocks that are thought to be representative of how retention will be practised in the future. Resampling every five years after harvesting should reveal details of successional changes and provide reasonable estimates of rates of windthrow, snag fall, wood decay, and other dynamic processes.

In the short term, estimates of changes in stands with time must rely on simulation modelling (§12.8). Developing reliable projection models was a high priority. We therefore ensured that the monitoring design would provide the information that future monitoring will require to track changes in retention stands through time. That included establishing permanent sampling sites in current retention blocks and permanent benchmarks, carefully documenting and archiving data and metadata and committing to appropriate remeasurement schedules.

External (non-empirical) targets

Targets may derive from a number of sources that do not require field measurements. They include regulatory requirements, targets specified by forest certification systems, internal company objectives, perceived needs of the public or environmental organizations, or expert opinion about biologically significant differences between practices. The only implication for the monitoring design is to determine existing, relevant, external targets and to ensure that appropriate variables are measured. Such collation was a high priority because it is cheaper than field monitoring, yet provides context for evaluating the results of field monitoring. It became a lower priority as it became clear that standards for forest practice in British Columbia were in considerable flux and of questionable ecological relevance.

Targets from species' requirements

Targets for habitat attributes within stands could be derived from the requirements of the species intended to be maintained by the attributes. The requirements are broadly known for some vertebrate species (reviews in Bunnell, Kremsater, and Boyland 1998, Bunnell, Kremsater, and Wind 1999; Thomas 1979) but would have to be determined in the field for many other organisms. Generating habitat models for indicator or focal organisms requires a different study design than simply monitoring organisms in different harvest types. Sampling sites need to be found that help to separate the commonly confounded occurrence of some habitat elements (e.g., low canopy closure and large amounts of woody debris). Sites providing broad ranges of habitat elements are needed to develop models with general applicability. Further sampling is needed to test habitat models once they are developed. The benefit of all this additional work is a credible basis for evaluating levels of structural retention in VR that are directly tied to the goal of maintaining species. However, the expense of reliably documenting such habitat requirements leads to focusing on very few species, thereby ignoring many others. We gave a high priority to compiling existing information on species' requirements for habitat structures and a low priority to costly new studies. However, where adding relatively inexpensive habitat measurements to field sampling of species could augment existing information, that was considered a high priority.

12.3.3 Moderate-Priority Comparisons

Comparisons of moderate priority were recognized as useful contributions if sufficient funds became available that higher priorities were not compromised. Three potential comparisons were considered of moderate priority.

Older stands of clearcut origin

Monitoring across the edges of older clearcuts (>20 years) permits testing the ability of organisms to recolonize different distances into harvested areas after moderate periods of time. Such tests help to evaluate the concept of "life boating," which underlies the choice of variable retention. If fewer mobile organisms have recolonized the centres of larger, old clearcuts, then their presence suggests that distances between retention patches need not be small. The focus then would shift to ensuring that the organisms can persist in the patches until the harvested matrix is suitable. Conversely, if organisms have recolonized only a limited distance into young stands following clear-cutting, this finding could help to inform the spatial planning of VR blocks.

Harvest method

The harvest method (e.g., ground based, cable, helicopter) has substantial effects on initial conditions in the harvested part of variable retention blocks and on the features retained, so it should be a high priority. However, harvest type corresponds closely to forest type, so direct comparisons that are not confounded by forest type are nearly impossible. Moderate priority reflects the limited opportunities for finding operational sites where a choice in harvest method is available. Harvest method was recognized as a blocking factor when generalizing monitoring results (§12.6.2).

Non-harvestable land base ("scrub")

Establishing benchmarks for comparisons with older forest recognizes and samples one part of the non-harvestable land base – productive stands that are constrained from harvesting for some reason (e.g., ungulate winter ranges and riparian reserves). Scrub forest, which was non-harvestable because of low standing volumes in old forest (<210 m3/ha), represents a large component of the non-harvestable land base. Monitoring the scrub part of the non-harvestable land base has two purposes: (1) to allow integration of monitoring results across the whole tenure, and (2) to test whether non-harvestable areas assessed by Indicator 1 (ecosystem representation) are representative of habitat structures and organisms across the tenure. Ecosystem types where scrub was a large proportion of the non-harvestable forest were sampled to assess the degree to which this scrub area contributes to ecosystem representation. The purpose of this monitoring was to help landscape or tenure-wide monitoring, not to contribute directly to guiding the management of retention stands. For comparability, we used the same form of monitoring plots as for other stand-level monitoring. Although such monitoring can contribute usefully to evaluating Indicator 1, its priority was moderate because non-harvestable areas are physically difficult to access (and therefore expensive), and there were few opportunities for efficiencies from co-ordinating with other stand-level sampling.

12.3.4 Low-Priority Comparisons

When assessed, low-priority comparisons currently did not merit extra effort but were retained in case funding and design permitted their addition to higher priorities. Initially, we believed that comparisons of variable retention to other types of operational harvesting would allow us to learn what ecological contribution variable retention was making relative to those alternatives. Several such comparisons were considered, but most eventually ranked low primarily because the company had committed to retention harvesting; the alternatives were not available on the tenure. There were nine such comparisons: some were methodological differences in comparisons discussed earlier.

Changes in variable retention over time – chronosequences

A chronosequence approach of finding existing older stands that were similar to VR stands when they originated might provide immediate information about the possible future of VR stands. However, rare old stands with permanent retention are in atypical locations, and the habitat retained was retained for different reasons than with current retention (generally lack of access or value). Such old blocks do not aid this comparison and may be more useful for examining potential changes in edge effects over time. Attempting a chronosequence was a low priority but could be revealing where older retention patches are not atypical.

Existing regulations

Under the legislation existing when changes in practice were made,[1] clearcuts with wildlife tree patches were the norm for operational harvesting in most of the province. Comparisons would be possible if standard blocks with wildlife tree patches were being implemented operationally in similar environments close to variable retention. The main difference between patches under group retention and wildlife tree patch blocks is fewer, larger reserves and often less overall retention in the latter. Wildlife tree patches may also use different anchor points than VR patches. Provided that monitoring of the effects of variable retention patch sizes and anchor point types is done well, inferences about wildlife tree patch blocks can be made without direct comparisons, making this a low priority. Inferences would be most reliable if retention patches as large as wildlife tree patches were sampled.

Silvicultural systems without permanent retention

At the onset of the study, even-aged management with permanent reserves was the dominant silvicultural system on the BC coast. It is likely that uneven-aged management with individual or group selection or strip cuts will receive increasing attention for ecological and aesthetic reasons or to spread risks. This comparison was a low priority because comparisons with such alternative silvicultural systems were not possible in most ecosystem types, and the company intended to include permanent retention in any silvicultural system that it used.

Certified blocks

Comparisons of VR sites with cutblocks formally identified as ecologically sustainable might help to evaluate the effectiveness of certification. Such comparisons could be local or international (relying on monitoring information collected by others in the certified blocks). The objective would be to compare variable retention to other certified blocks, rather than to certification targets, because the question is how well variable retention and other certified blocks maintain components of biological diversity when they are implemented. This became a low priority because certification standards are variably applied and apparent targets change frequently. Additionally, effectiveness monitoring is rarely conducted in certified cutblocks, so the relevant information for comparisons is not available.

Patch arrangements

The arrangements of patches within a block – whether they are spread out or clustered together and how far they are from each other – likely affect their ecological contributions but were ranked low. The low ranking was assigned because we believed that effects were minor compared to retention level, edge effects, and patch anchor type. Moreover, opportunities for unconfounded comparisons of different patch arrangements were unlikely.

Targets from nature – naturally disturbed stands

Patterns of natural disturbance provide two broad forms of guidance to management: background rate, and the causal agents of disturbance. Neither form of guidance provides a clear target or requires detailed analyses. Initial guidance was provided by the approximate background rate of natural disturbance that will occur regardless of management practices. The rate can be specified only crudely, and the locations at which the disturbance occurs cannot be predicted accurately. Estimating this rate crudely, however, is an important contribution. For example, any calculation of sustainable yield must recognize that it will be determined not only by growth rates but also by losses imposed by the natural disturbance regime (recognizing that some losses are incorporated into estimated growth rates). Similarly, efforts to sustain older structures must recognize that these structures will gradually be depleted and replenished by the natural disturbance regime. A commitment to sustainable forest management must consider the way in which these structures are recruited and depleted naturally as well as by management practices. Rates of natural disturbance had been estimated (Dunsworth et al. 1997).

Knowledge of agents of natural disturbance can also guide forest practice because different regimes create different mixtures of habitat elements. At the broadest level, at least some forest-dwelling organisms are correlated to elements of the disturbance regime (Bunnell 1995). Knowing the relative rates of different natural agents of mortality (e.g., relative proportions of snags generated by different agents) helps in understanding the kinds of forest structure to which local flora and fauna are adapted. For example, logs from trees that died a natural death may have heart rot and become hollow (forming dens for black bears, martens, and fishers), whereas logs from trees felled by wind are unlikely to become hollow.

Despite broad value, developing a target from the natural disturbance regime was considered a low priority because

1 we monitored more stable old-growth stands as natural benchmarks;
2 documenting natural disturbance patterns well is very complicated and expensive (§12.3.1);
3 the rate of natural disturbance was crudely known and estimated to be low (Dunsworth et al. 1997; Lertzman et al. 1996; MacKinnon and Trofymow 1998);
4 most stands in many ecosystem types would naturally be mature or old (which received very high priority for monitoring); and
5 patterns of agents of natural disturbance are approximately known from analyses of permanent sample plots (Dunsworth et al. 1997).

12.3.5 Summary of Comparisons

Table 12.1 summarizes stand-level comparisons by priority class. Most comparisons requiring fieldwork that were ranked as "very high" or "high" focus on retention blocks or sampling within retention blocks. Several could be addressed by planning subsampling to address within-retention questions while providing a good stratified sample for comparisons of different retention types and amounts. Separate field sampling was required to establish older forest benchmarks.

The other high priorities for monitoring, including compiling external targets and literature on species' requirements, currently involve only thoughtful use of existing information. Three moderate-priority topics would

Table 12.1

Priority rankings for the possible design comparisons made primarily at the stand level

Very high
- Type of retention
- Amount of retention
- Edge effects within blocks
- Targets from nature – older forests

High
- Patch size
- Patch anchor types
- Recent clearcuts (from the literature)*
- Silvicultural system used during retention (but limited opportunities)
- Old-forest restoration (assessing old deer wintering range stands)
- Changes in practice of retention over time
- Changes within retention blocks over time (simulation)
- External targets*
- Species' requirements (from the literature* or add-on measurements in the field)

Moderate
- Older stands of clearcut origin (evaluate "life boating")
- Harvest method
- Non-harvestable forest (scrub)

Low
- Recent clearcuts (field sampling)
- Species' requirements as separately designed studies
- Changes in practices over time (by field sampling)
- Changes in VR stands over time (by chronosequences)
- Blocks under existing regulations
- Silvicultural systems without VR
- Certified blocks
- Patch arrangements
- Naturally disturbed seral stands

Note: Comparisons marked * do not require fieldwork.

be included if adequate funding became available. The low-priority topics require substantial additional field sampling or appear to be uninformative.

The preceding review exposes a critical trade-off in all monitoring programs: examine many sites and few factors or few sites and many factors or variables. Even with efficiencies from overlapping sampling designs, monitoring all the identified high- and very high-priority comparisons requires extensive sampling (many sites). As a result, the range of indicator variables that can be examined for each comparison is limited. The next two sections briefly review stand-level variables of the indicators of biological diversity, then provide a matrix of indicator variables to be used with each major monitoring comparison.

12.4 Selecting Indicator Variables

Specific indicator variables appropriate for stand-level monitoring were suggested by our literature review (Chapters 8 and 10), used in previous models, recommended by the Scientific Review Panel, or included in pilot studies. Variables for each indicator were discussed at a workshop in November 2001 (Appendix 5.3 of Bunnell et al. 2003). We did not rank each indicator on its individual merit but grouped and organized the indicators to reduce their complexity.

1. Some potential indicator variables were excluded outright because it seemed to be unlikely that we could learn enough about them to guide management because of logistical difficulties in sampling, because of extreme rarity, or because we already knew enough about them (e.g., deer and elk).
2. Some variables were combined into a single group because they could all be sampled together (e.g., vocalizing squirrels combined with songbirds, various habitat elements). Related indicators that could not be sampled well with the same method were kept separate (e.g., owls separate from songbirds, large snags from standard habitat sampling).
3. Some variables were recognized as "add-ons" to other monitoring (e.g., vascular plant species added on to standard habitat plots, terrestrial arthropod by-catch added on to carabid beetle sampling).
4. Some potentially useful variables were designated as "needing a pilot study" (e.g., winter carnivore tracks or resident birds in winter) because we did not know whether they could be monitored efficiently and with useful precision.

Beyond Chapters 8 and 10, more detail on the rationale for inclusion or omission of specific stand-level indicators is provided in Bunnell et al. (2003). Stand-level indicators used in assessing habitat are summarized in Table 8.2; those proposed for organisms are summarized in Table 10.10.

12.5 Matching Indicators with Comparisons

The monitoring portion of the adaptive management program can be summarized within a "what and how" matrix. Most of the preceding text of this chapter deals with "how" to ask questions and learn – direct comparisons or comparisons to targets. The "what" aligns specific forest elements or organisms believed to reflect success in sustaining biological diversity with the various "hows" or comparisons. The matrix summarizes the basic framework of the stand-level monitoring program (Table 12.2).

All the stand-level comparisons noted in §12.3 have relevance to informing current and future management. Similarly, many features or organisms have been suggested for monitoring but combined are only a tiny fraction of resident biodiversity. The ideal monitoring design would make all comparisons with all suggested features and organisms. That would require an immense amount of work, or each part would be done so superficially that it would provide no useful information.

One solution for making most comparisons and using a wide range of biodiversity indicator variables is a "cross-design," in which a few critical variables are used to make all selected comparisons, while the highest-priority comparisons are made with a more extensive array of variables. Additional comparisons can be made for indicators requiring only minor additions to the sampling or where there is a pressing need. A broad array of indicators for major comparisons facilitates a more complete picture of the effects of the main management variables (e.g., Table 12.2).

Benefits afforded by the cross-design include the following.

- It avoids reducing the huge complexity of biodiversity to a few surrogate indicators.
- It bases the most revealing comparisons on a more complete representation of biodiversity.
- It provides tests of how well the few extensive indicators correspond to many other biodiversity variables.
- It allows more comprehensive simulation modelling of indicators across the tenure through time.
- It encourages synergisms among results that help to provide program cohesion and efficiency by focusing a number of projects on the same basic questions at the same sites.

The concept of a cross-design, combined with the priorities assigned to the comparisons and the reviews of potential variables for use as indicators, was used to fill out the matrix of "which indicators for which comparisons" for the stand-level monitoring program (Table 12.2). Low-priority comparisons, comparisons not requiring fieldwork, and excluded indicator variables are not included in the table. The cells of Table 12.2 show a ranking for use

Table 12.2

Priority indicators for monitoring within each comparison for selected stand-level monitoring comparisons of moderate to very-high priority

		Retention*	Silvicultural system	Edge effects	Patch size	Retention anchors	Harvest method	Older clearcut edges	Nature – older	Non-harvestable scrub
Stand-level indicators' priority		VH	H	VH	H	H	M	M	VH	M
II. STRUCTURE										
Habitat elements										
Standard elements + heterogeneity		1	1	1		1	1	3	1	1
Rare elements (e.g., large snags)		3	3	3		3	3		2	2
Non-standard elements (leaning snags)		3	3	3		3	3		3	2
Surface/soil structure		4	4				4			
Habitat integrators										
Special site types		3	3	3		3			3	2
Site series		1+	1+						1	1
Microclimate		3	3	3					4	
III. ORGANISMS										
Vascular plants										
Selected subsets of sensitive + exotic		1+	1+	1+		1+		3	1+	
Bryophytes										
Forest substrate associates		2	2	2		2		1	3	
Lichens										
Old forest associates (epiphytes)		2	2	2				2	3	
Fungi										
Ectomycorrhizal				2				2	4	
Conks/bracket fungi		3+	3+						3+	
Invertebrates										
Gastropods (+ amphibian by-catch)		2	2	2	3	2		2	3	
Carabid beetles		2	2	2	3			2	3	
Terrestrial arthropod by-catch		3	3						4	
Vertebrates										
Amphibians	• terrestrial breeding	4	4						4	
	• aquatic breeding	4	4		2				4	
Birds	• Common songbirds (+ calling squirrels)	2	2	4	1				3	
	• Woodpeckers and owls mentioned in text	3	3							
Mammals	• squirrels (telemetry)	3	3		3					
	• shrews (as carabid by-catch)	2+	2+	2+	3+			2+		

Notes: + = add-on to some higher-priority indicator for that comparison. No priority number means much lower priority.

* Retention type and amount

of that indicator variable to make that comparison: that is, the ranks are estimated *separately within each comparison.* These rankings can be used to guide which monitoring projects are implemented, depending on funding levels. The intention was to ensure that priority 1 and 2 variables were monitored for the very high- and high-priority comparisons over the next five years. Additional, less certain funding could be used to expand the monitoring to the highest-priority variables for moderate-priority comparisons and to the lower-ranked variables for the very high- and high-priority comparisons. Sometimes lower-priority combinations can be added on to other monitoring with little additional cost.

12.6 Answering Questions Well

Selecting comparisons and indicator variables to ask the questions is fundamental to the monitoring design. To ask the questions well includes choosing between operational or experimental comparisons, selection of blocking factors, and use of pre-treatment measurements.

12.6.1 Operational versus Experimental Comparisons

Adaptive management can, as in this program, include replicated, experimental treatment units. The design must then allocate effort to monitoring the experiments versus monitoring operational blocks. Some advantages gained from each approach are summarized below.

Scientific and statistical advantages derived from experimental units include the following.

- There is a greater range of treatments than is typical of operational units. The major retention variables are extended further (e.g., higher-percent retention), and other approaches to retention (including uneven-aged systems) can be considered. A greater range of treatment types facilitates comparisons within variable retention and with alternative systems, creates a wider range of conditions for generating and testing explanatory models, and provides a broader range of experience to help guide future management.
- Controlled experiments reduce the confounding of variables that occurs in operational settings, such as the association of dispersed retention with lower retention levels, or larger patches associated with higher overall retention levels. Reduced confounding helps to assess the best approach to variable retention and to develop explanatory models.
- Uncut controls are immediately adjacent to harvest treatments and are randomized instead of being absent, distant, or anomalous (as with many operationally harvested stands). The potential value of pre-treatment measurements is discussed in §12.6.3.

- With adequate sample sizes, the randomized location of blocks reduces concerns about confounding of harvest types and underlying conditions, such as stand type or topography.

Advantages gained from operational units include the following.

- Measuring representative operational blocks is the only way to monitor the implementation of operational guidelines.
- Although operational blocks are not strictly randomized replicates, there are many more of them to monitor and use when comparing types of variable retention. High site-to-site variability often requires large sample sizes for moderately precise estimates of ecological variables.
- Because operational blocks are numerous and widespread, blocks can be found that include rare features absent from experimental units (e.g., particular wetland types, certain rare species).
- Some useful comparisons cannot be implemented in the limited number of experimental units, such as different spatial configurations of reserve patches. The desired comparison may be found among the more numerous operational blocks.
- Operational retention blocks have the context commonly found in managed stands, such as riparian areas, adjacent harvesting of different ages, et cetera. The context may have important influences on the ecological responses within the block.

In addition to these differing scientific values, other factors can affect the allocation of effort to experimental versus operational units. Experimental sites allow easier local logistics with adjacent units, have more appeal to scientific reviewers, and may attract additional scientific collaborators. They can also serve as demonstration sites, communicate corporate commitment to research, help to reduce conflicts between management operations and monitoring operations, and help to foster collaboration among researchers. Conversely, operational sites are immediately available for pilot studies, fostering communication between managers and scientists and helping to make adaptive management an integral part of operations rather than a separate activity for research. The allocation of monitoring effort to experimental versus operational sites depends on the relative importance given to different aspects of monitoring. Generally, experimental sites are probably the best choice for the direct comparisons that they cover, while operational sites need to be monitored for evaluating operational performance, for additional comparisons, and for any monitoring that requires large sample sizes.

12.6.2 Blocking Factors

We do not expect monitoring results from a particular site to apply across all of the management area, but we do expect some generality beyond the sampled sites. A "block," in the strict sense of study design, is the population of sites from which actual sampling sites are chosen randomly. More loosely, it is the set of conditions that we assume a particular sample represents. We used four related variables to define blocks to which particular results apply.

Ecological units

We considered variants of the BEC system as the most generally applicable blocking units for biological sampling (Chapter 7). Blocking should include a range of units – at least a dry unit, a unit with moderate precipitation, and a wetter unit. The wettest and highest-elevation units are lower priorities because they occur extensively in non-harvestable forest (Chapter 7). Monitoring could include other units, especially at the pilot study stage, to help determine empirically suitable strata for future monitoring.

Harvest method (ground, cable, helicopter)

Different harvesting systems can have substantially different effects on residual habitat structures, patterns of retention patches, and resulting species inhabiting the stand. Generally, only one or two harvesting techniques are dominant in each ecological unit or BEC variant. Where one harvest method is clearly dominant within an ecological unit, monitoring should focus on that method. Comparisons of harvest method should be restricted to units for which more than one method is frequently used. The exception might be unusual techniques that may become important in the future. They can be treated as a direct comparison with standard techniques, using similar blocks harvested by the two methods.

Age of stand harvested

Harvested stands are either old growth or second growth derived from harvesting or stand-replacing disturbances. Stand age is linked closely to ecological units, with second-growth stands in the drier units and more remaining old growth in the wetter units. Monitoring should focus on the dominant age class in the ecological unit being monitored. It can be useful to target second-growth stands being harvested in units composed mostly of old growth. Monitoring such stands helps to reveal what post-harvest stands may look like in future rotations in those units.

Company divisions

Companies are often organized by divisions. There is no ecological reason to spread sampling across divisions except that their operating areas tend to correspond to different ecological units. However, divisions can be an important blocking factor for implementation monitoring (Chapter 4). Monitoring designed for direct comparisons should not be spread across divisions unless this is necessary to create a geographic range of ecologically important variables (or institutional reasons outweigh sampling efficiency). Replicates in monitoring using experimental treatments should be chosen for the comparisons that they allow and not to allocate monitoring effort equally across corporate divisions. Credible results from well-designed sampling, with appropriate ecological blocking, will ultimately help individual divisions more than simply monitoring something in each division. Encouraging the active participation of divisions in the adaptive management program is necessary to build cohesion between managers and researchers but is best done using a combination of operational, experimental, and implementation monitoring rather than forcing equal allocation of experimental sites.

12.6.3 Pre-Treatment Measurement

Pre-treatment measurements can be very useful prior to assigning experimental treatments to replicate units. These measurements can be used to block the units, creating a similar suite of initial site conditions for every treatment. Blocking based on site conditions helps to avoid the possibility of a major driving variable being confounded with a particular treatment by chance alone, which can happen with full randomization and limited replicates. Such blocking can be done only for one or two dominant variables, such as elevation or ecological unit, which are known to affect many other variables. There is no way of ensuring good blocking for many different ecological variables with only a few replicates.

If particular treatments have been assigned to specific locations, then pre-treatment measures may still be useful. They are most useful for variables that change slowly over time (e.g., habitat structural elements) and whose spatial variability is high or whose relative abundances do not change much even if overall abundance fluctuates (some organisms). In such cases, pre-treatment information can reduce the need for large numbers of spatial controls.

If the variables of interest change rapidly over time, but vary little through space, then pre-treatment measurements are of little use, and simultaneous spatial controls are more helpful. This is the case for many animal populations and for erratic floral and fungal species. Effort is then better spent on increasing replication of treatments and/or simultaneous spatial controls.

Pre-treatment measures should receive the same sampling effort as post-treatment measures. That is particularly true if they are going to be used directly – for example, by calculating differences or ratios between pre- and post-treatment values. If pre-treatment measurements receive less sampling effort, then the precision gained in the more intensive post-treatment measurements is wasted when included with imprecise pre-treatment estimates.

12.7 Monitoring over Larger Areas

We monitor landscapes or larger areas because we believe that the ecological contribution of a larger area of forest is different from the sum of the values of the individual stands. If there were no landscape effects, then we could simply add up the stand-level values for each indicator under current conditions or under projections of current practices and alternative scenarios. Monitoring would focus entirely on determining the values in different stand types, and the rest would be simple arithmetic. However, the same composition of stand types can provide different ecological values in a landscape depending on the spatial distribution of the stands because

1 adjacent stands affect each other (e.g., edge effects);
2 spatial arrangement of stand types affects ecological processes, including dispersal, foraging of mobile organisms, spread of natural disturbances, and physical flows; and
3 the marginal value curve for amount of a stand type is non-linear, including asymptotes, thresholds, and minimum requirements (e.g., the change from 10 suitable stands to 0 is more important than the change from 110 suitable stands to 100).

Such emergent properties of landscapes require more than stand-level monitoring. We emphasize monitoring at the stand level because knowing the values of constituent stand types is necessary (if not sufficient) to evaluate landscapes. Moreover, over a moderate range of landscape conditions, many species appear to respond largely to the sum of stand types (Cushman and McGarigal 2004; Fahrig 1999; Schmiegelow and Mönkkönen 2002).

Two scales merit distinction when monitoring larger areas.

1 *Overall tenure:* Monitoring trends in organisms across the tenure is a direct requirement of the goal of maintaining all native species that occur on the tenure. This monitoring is the fundamental check that the sum of stand-level and landscape-level practices is actually meeting that goal. Direct monitoring of all three broad indicators across the tenure is therefore a high priority. Direct comparisons cannot be made at the overall tenure scale (there is no alternative to compare). Instead, we monitor to indicate the current status of the indicators and, by monitoring over time, any changes or trends. Few indicators can be monitored at this scale, and we must be careful that any monitoring program across the tenure has sufficient precision to detect substantial trends in the indicators.

2 *Smaller landscapes:* Landscape effects are any effects beyond the stand scale, including the interaction of two adjacent stands along an edge. However, by landscape monitoring, we usually mean larger areas such as officially designated landscape units or watersheds, which is the scale at which longer-term, forest landscape planning occurs. Unlike the overall tenure, direct monitoring of several smaller landscapes could allow comparisons of different landscape patterns that might directly inform landscape management. These comparisons could be "passive," relying on existing differences, or "active," involving planned manipulations at the landscape scale. Although direct comparisons of smaller landscapes are theoretically possible and would parallel the informative direct comparisons at the stand level (§12.3), both passive and active comparisons at the landscape level face severe difficulties. Passive comparisons based on differences in existing landscapes are almost certain to be confounded by differences in variables with known strong ecological effects, such as stand types or ecosystem types. For example, landscapes with high amounts of edge almost always have much young forest. Finding existing landscapes with distinctly different edges but the same amount of young forest is unlikely (let alone finding replicates). Different patterns in existing landscapes are likely to reflect different ecosystem types that guided the historical choice of practice. Naive comparisons of landscape features that ignore these confounded differences can be misleading. Active landscape manipulations are logistically very difficult. Finding comparable initial conditions for replicates may be impossible, and landscape treatments require a long time before differences become apparent.

Initial stages of direct, large-scale monitoring therefore should focus on developing a program to monitor conditions and trends of the three major indicators across the *overall tenure*. Comparisons of different planning options for smaller landscapes rely on projection modelling rather than on direct measurements of measures acquired in the field. Below we summarize features of each of the three major indicators that are suitable for monitoring over large areas. Our emphasis is on the overall tenure, but we note instances where smaller areas may be informative.

12.7.1 Indicator 1: Representation of Ecosystem Types in Non-Harvestable Areas

Ecosystem representation as a basic broad-scale indicator is discussed in Chapters 6 and 7. Monitoring representation reports the current status. The appropriate design to track changes through time is simple: update the analysis whenever substantial changes are made in the non-harvestable land base. Such changes could result from repositioning Old-Growth stewardship zones or changing their amount, moving discretionary reserves, or responding to policy, technology, and market changes that make formerly non-harvestable areas accessible. Representation analysis should also be updated as finer ecological mapping becomes available. Stand-level monitoring results can be used to test the assumption that ecosystem types are biologically appropriate. Although additional monitoring specifically to test this assumption was considered a low priority, we recorded site series, elevation, and topography during other monitoring to permit preliminary analysis.

12.7.2 Indicator 2: Stand and Landscape Features

The finer scale of stands was discussed in Chapter 9. Specific recommendations for monitoring landscape features were offered in §8.2.4, along with rationales, details on how to measure them, and reasons for excluding some GIS-based landscape indices. Landscape features recommended for monitoring include

- age and patch size distributions (recognizing different retention types);
- edge-contrast lengths and effective interior area;
- roads – linear density, distance-to-road distributions, bisection of land base, and interceptions of riparian areas; and
- a range of idealized "organisms" used to integrate organism-specific effects of changing stand quality with age and management and organism-specific spatial effects such as patch size, fragmentation, and connectivity.

The first three points are simple GIS analyses based on forest cover information and can easily be done across the tenure, with summaries by whatever units are considered relevant (ecosystem type, division, landscape unit). Trends can be tracked by updating the analyses every three or five years. Hypothetical or idealized "organisms" (§8.2.5) require more elaborate analyses, limited to a subset of smaller areas. Trends can be estimated by conducting the same analyses on the same areas in future years. The subset of areas therefore should be representative of the range of ecosystem types, geographic areas, and tenure types (private, public) and be experiencing common practices. Because non-harvestable areas are important parts of the landscape, the subset of areas should include typical distributions of unmanaged land for the ecosystem type.

The biggest initial challenge in direct monitoring of landscape structures is creating the targets needed to assess current conditions and give perspective to changes in the future – that is, to estimate targets well enough to know whether an observed change is large enough to warrant changing management practices or even whether the change is good or bad. Targets for some landscape structures may be available from external policies (e.g., regulatory targets for seral stage and interior forest). Others may be derived from landscape-level studies of organisms in the literature. However, the relevance of specific organism-based landscape targets is dubious when applied to landscapes with different stand and ecosystem types hosting many species. Desirable targets for landscape-level features could also be derived from natural disturbance regimes or our best guesses at them where large unmanaged areas no longer exist. Targets based on natural disturbance regimes probably are most useful for indicating which landscapes and which landscape structures differ most from the presumed natural conditions. Such monitoring highlights areas of most concern for current landscape features, helps to improve stand-level practices, and provides focus to future monitoring.

We considered that trying to derive targets from natural disturbance regimes at the landscape level was a low priority (§12.3.4). These regimes are likely to be well outside the range of socially and economically acceptable choices for managed landscapes (e.g., widespread small-gap dynamics, huge but very erratic hurricane blow-down events, large stand-replacing fires in the heavily developed eastern side of Vancouver Island). Greater priority was given to developing the ability to evaluate projected landscapes under a range of planning options, using predicted landscape indices and real or hypothetical organisms.

12.7.3 Indicator 3: Organisms

The kinds of organisms that could indicate landscape-level effects and that could be monitored directly and practically at broad scales include the following (listed in approximate order of priority for establishing tenure-wide monitoring programs).

- *Breeding songbirds and calling squirrels:* They are currently being monitored with standard Breeding Bird Survey routes modified to assess habitat.
- *Plants:* Invasive species are candidates for large-scale monitoring, along with sensitive native species. The objective is to monitor changes in range distribution by presence/absence in permanent plots.
- *Calling amphibians:* Broad-scale amphibian surveys are conducted in some parts of eastern Canada and the United States and are being developed by the BC Frogwatch Program.
- *Hunted species or species of concern:* They are candidates for large-scale monitoring because widespread surveys are often employed for inventory by management agencies.
- *Specific organisms* suspected of being sensitive to issues at broader scales are noted in Chapter 10 (e.g., species responsive to patch size or forest interior or using a variety of patch types).

To represent biodiversity, several different groups of organisms should be monitored. The monitoring design needs to include sampling across the tenure or at least in a representative range of ecosystem types and geographic locations. The need for extensive, representative surveys of several groups of organisms creates a design conflict. Extensive sampling requires easy access (e.g., the road-based Breeding Bird Surveys), but easily accessed areas are generally biased away from the unmanaged areas that are an important part of many landscapes. Within a limited budget, possible solutions are to monitor representative areas away from roads less frequently or to intentionally choose road routes that pass through unmanaged areas. Surveys that are intended to indicate trends in species' abundances (e.g., Breeding Bird Surveys, amphibian surveys) ideally are conducted annually so that actual trends can be detected more quickly, given natural annual variability. Surveys intended to show range contractions or expansions (e.g., of introduced plants) can be conducted less often. Using survey techniques that are conducted across a larger area, such as the Breeding Bird Surveys, has the advantage of theoretically being able to separate out effects of management on the tenure from overall regional or global changes in species (when local trends differ from regional or global trends).[2] The power of extensive surveys to detect meaningful changes in species in a reasonable time was analyzed for pilot study results before they were adopted for tenure-level monitoring. Analyses suggested that many extensive surveys have low power to detect trends without considerable effort. It is particularly important to start with an efficient and powerful design for tenure-wide monitoring of organisms to minimize the need for future refinements to the design that reduce the value of data previously collected. Unlike representation or landscape structure, which can use archived maps, there is no way of redoing past monitoring of organisms.

12.7.4 Other Possible Ecological Variables for Broad-Scale Monitoring

There are other variables that integrate ecological effects over areas larger than individual stands and are relatively easy to monitor. Among them are mass wasting events, distribution of pathogens (root rots, mistletoe, insects), windthrow occurrence and distribution, and stream hydrology, sediment loads, and nutrient content. These indicators are included in several programs for monitoring sustainable forestry at larger scales (e.g., Canadian Council of Forest Ministers 1997) but do not fit readily into the three indicators chosen to reflect changes in terrestrial biological diversity. They are, however, directly relevant to other criteria in forest certification programs, within which the company is certified. Although monitored for other reasons, these variables have direct effects on many organisms and are thus useful in assessing success in meeting more than one criterion.

12.8 The Role of Models

The monitoring program outlined above assesses immediate, or short-term, consequences of forest practices, but that is insufficient. The consequences of forest management decisions lead to changes over long periods of time as stands develop. Moreover, most comparisons noted provide information only on the retention types sampled, with limited extrapolation to similar types. Applying monitoring results to unsampled options, to alternatives that have not been implemented, and especially to future conditions requires a modelling framework to project probable outcomes. The ability to extrapolate monitoring results to large scales is essential because we measure success at meeting goals for biological diversity at the large scale but have limited ability to monitor large scales directly (§12.7). Modelling is thus an integral part of the monitoring program, intended to aid management by projecting future and cumulative consequences of management actions. In turn, adaptive management is intended to challenge and improve the models over time. Here we focus on the use of models in adaptive management, particularly the modelling framework needed to project the indicators of biological diversity. Addressing specific questions often requires a specific form of model (e.g., Bunnell and Boyland 2002). Details of model architecture, assumptions, variables, and outcomes of runs are described in separate documents.[3]

Models of phenomena at local scales typically are more mechanistic and detailed than those used to project consequences over large areas or long periods. Local models may attempt to ensure that each variable included can be related directly to something that can be assessed in the field. Models projecting over large areas often incorporate a more general relation based on an initial guess or derived from analyses of a local model. For example, a local model of dead wood might use growth, mortality, and decay projections of individual trees. That is impossible to implement over a large area, and a more integrative estimate must be used, such as the average volume of downed wood in different stand types.

12.8.1 General Modelling Approach

The modelling program for biological diversity parallels the direct monitoring program (Figure 12.1) because we want to project future consequences of alternative management options for each indicator. It is also the only way in which we can extrapolate the necessarily local monitoring of many elements to the larger landscape or tenure-wide scales.

The overall framework employs five steps:

1 projecting stand-level habitat features, such as live trees, dead wood, or shrub cover, for stands managed under the various management options or non-harvestable stands in different ecosystem types;

2 projecting the types of stands that will occur across the landscape and their ages, given the management plan and effects of natural disturbances;
3 populating each stand in these projected landscapes with the projected habitat features for that type and age of stand;
4 using relationships between indicator organisms and the stand- or landscape-level habitat features to predict habitat suitability of the projected stands or landscapes; and
5 "scaling up" by combining the predicted habitat suitability with spatial characteristics of organisms in the landscape (e.g., dispersal ability and edge effects) and relationships with nearby stands (neighbourhood models) to project where the organism will occur across the landscape.

This five-step approach allows projections of each major indicator (ecosystem representation, stand elements and landscape features, and indicator organisms) into the future and across large areas. The results of these steps of the modelling program should be an accounting system showing where across the tenure we expect to maintain each of the indicator components of biodiversity. Until we actually get to that future, projection modelling (Figure 12.1) is our best guarantee that current management decisions will meet the goal of maintaining species across the tenure.

12.8.2 Specific Forms of Modelling

The five forms of modelling noted in Figure 12.1 are discussed separately below.

Stand projection models: Models to project the standard habitat elements in stands (§8.1.1) are driven primarily by the dynamics of live trees. Live trees themselves are habitat attributes for many organisms and comprise canopy cover and much horizontal and vertical heterogeneity (habitat structure, §8.1.2). Live trees die to produce deadwood elements. Fall rates, fragmentation rates, and decay rates of snags determine the population of snags in the stand. Down wood is generated by falling live trees, snags, and parts of snags, with decay rates determining abundance and state of logs. Growth and mortality of trees, particularly the canopy closure that they create, determine shrub and herb understories in conjunction with successional pathways of different stand types. The litter layer can be modelled based on inputs from canopy trees and understorey plants and litter decay rates. Tree cover and its effects on understorey vegetation mediate the microclimate, which may be critical for many small and immobile organisms. Edges between stands are also defined by differences in live tree height. Live trees thus directly or indirectly determine the abundance and distribution of other habitat elements within stands.

Live tree growth can be modelled with considerable precision for simple managed stands. However, most experience is with stands approximating

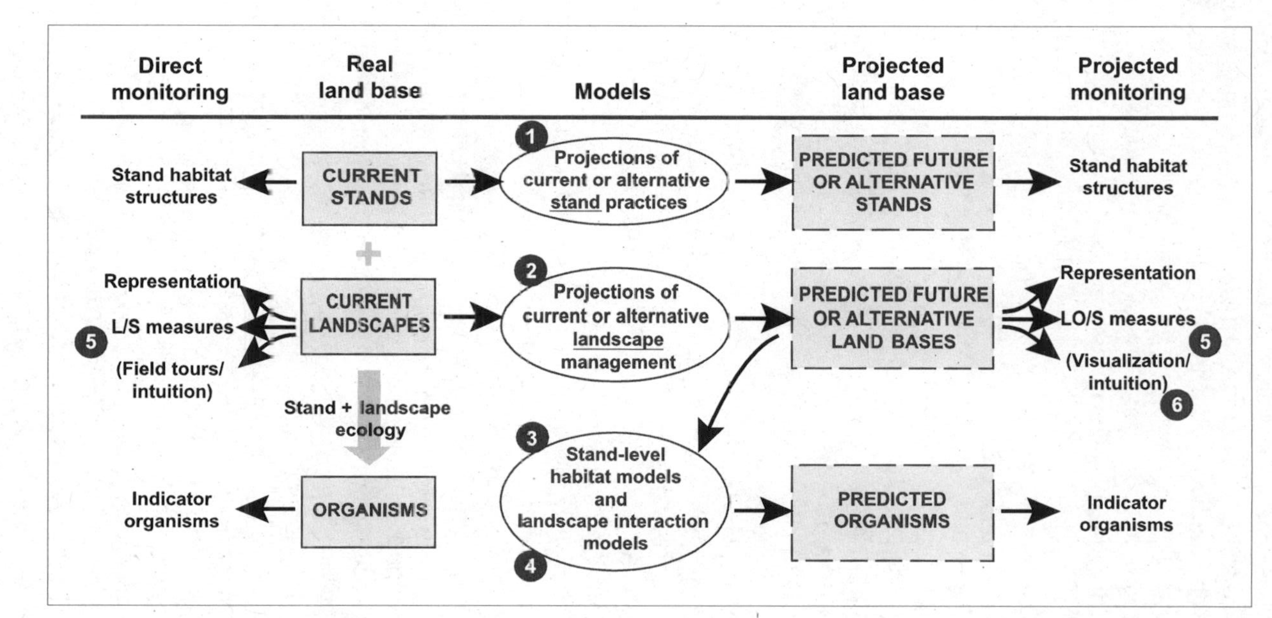

Figure 12.1 Schemata of how the modelling program allows the indicators used in the direct monitoring program (italics) to be projected to future or alternative conditions, for stands (S), landscapes (L/S), and indicator organisms. The kinds of models involved at each numbered point are discussed in the text. Ultimately, direct monitoring must be used to test the corresponding projections.

ideals for production forestry, so current models have limited ability to project habitat structures in VR stands. Interactions between retained large trees (in patches or dispersed) and regenerating trees are not as well known as the dynamics of even-aged plantations and require detailed spatial models of live trees. Stands with multiple species are often not modelled well for coastal conditions, particularly the ecologically important deciduous tree species. Furthermore, important heterogeneity from microsite variation is often ignored because silvicultural modelling assumes uniform site productivity. Probably most seriously, mortality from sources other than intertree competition is generally ignored because it is not expected to be important in productive stands under short rotations. That omission makes deadwood projections unreliable and probably affects modelled understorey growth and stand heterogeneity. Remeasurements of monitoring plots and use of older permanent sample plots in more complex silvicultural treatments can help to reduce these limitations, but more realistically complex live tree models incorporating other sources of information are needed. Realistic stand projection is a high priority because it is the basis of much of the adaptive management modelling.

At a broader scale, growth models based on managed stands may have little ability to predict habitat attributes in old, unmanaged stands, which comprise a substantial part of the land base. Assuming that habitat attributes are in dynamic equilibrium, in large patches of non-harvestable, older stands they could be based directly on monitoring measurements in different types of unmanaged forest, such as benchmark sites, riparian reserves, and scrub. However, modelling based on live trees is needed for smaller areas such as retention patches and dispersed retained trees, where the non-harvestable areas are affected by changing edge effects as the harvested area ages. Incorporating mortality sources for older trees in these non-harvestable patches is a priority for habitat modelling of retention stands.

Realistic live tree modelling in complex and older stands is required for stand-level habitat modelling. Other parameters, such as snag fall and decay rates and downed wood decay rates, can be estimated from the literature and validated or updated using remeasurements in the monitoring plots. Relationships between understorey vegetation and canopy trees are partially known through work on ungulate forage (Anderson 1984; Cowan 1945; Nelson and Leege 1982), but further information will be needed, particularly shade effects on post-harvest vegetation recovery. Providing such information could require modifying habitat structure plots. Model structure will change with feedback between the specific needs of organisms identified during monitoring of indicator organisms and measurement of parameters required to model these needs.

Landscape projection models: Early in the process, we developed a model to project stand types across the landscape over time (Dunsworth et al. 1997).

The model was used to explore planning of harvest entries and timber supply and to drive landscape visualization tools, but it had limitations. Blocking was done manually (since automated). The harvest scheduler permitted only one pass over the forest (since corrected). These issues were addressed by using the modelling platform Patchworks.[4] Other limitations were the projection of habitat elements over time, the features ascribed to the organisms projected, movement patterns, and the projection of natural disturbance regimes in the non-harvestable landscape. The first three of these were addressed to varying degrees by pilot studies and refined models (Huggard and Kremsater 2005). The approach to incorporating natural disturbance merits further refinement. Disturbances in non-harvestable areas can modify age-class distributions of different ecosystem types and can alter landscape patterns. Most importantly, naturally disturbed early seral areas are an important, and increasingly rare, habitat type used by many species and should be tracked and modelled to assess their contribution to meeting the goal of maintaining all species across the tenure.

The Zone Allocation Model (ZAM), a landscape-level optimization tool, was developed to evaluate optimal locations of stewardship zones, based on trade-offs among ecosystem representation, zone size and shape, timber supply, and other ecological attributes (Boyland, Nelson, and Bunnell 2004b). While not a projection model, ZAM can evaluate broad economic and ecological consequences of locations of zones under greater or lesser intensities of harvest.

Habitat suitability models (stand-level): Predicting the distribution and abundance of organisms across the landscape requires two stages, relating the organisms to the projected stand-level characteristics and then modelling organism use of these stands on the landscape. The landscape use model is discussed below. Stand-level models for organisms are based primarily on relationships with the projected habitat attributes. Habitat models for selected indicator organisms were based on the literature, supplemented with information collected during field monitoring. Predictions of habitat models often are not reliable when subtly different stand types are being compared. However, for broad-scale, long-term projections, the models need only to reliably predict associations of organisms across a broad range of stand types (e.g., young versus mid-rotation or clearcut versus retention blocks versus uncut stands). Incorporating stand type and age directly into habitat relationships may be necessary, particularly for organisms that are not clearly related to particular stand-level habitat elements (e.g., relating organism abundance to one-, ten-, or twenty-year-old retention stands rather than to density of large snags, shrubs, or another variable). Such an approach limits the ability to model effects of different stand-level management options, such as retaining 10 percent versus 20 percent of large trees in VR blocks.

Effects of within-stand features, such as edge effects or anchor type, will modify habitat models. Some edge effects within stands are already captured through their effects on habitat elements. For example, if snags are less abundant within 10 metres of the edge of a retention patch, that is reflected in the predicted use of organisms needing snags. However, other edge effects, or patch size effects, on organisms may not be directly related to modelled habitat elements. Some birds, for example, may simply avoid areas within 10 metres of an edge, regardless of habitat suitability, or epiphytic lichens may be damaged by wind near the edge, even if their preferred substrate is available. These non-habitat edge effects (based on information in the literature, supplemented by direct monitoring measurements) can be used to modify the predicted suitability based on habitat elements. Such edge effects are likely strongest across the highest-contrast edges and can be modelled as decreasing in proportion to the decreasing difference in live tree heights across the edge as the stand regenerates. Modelling based on habitat elements (e.g., large snags), modified by spatial factors within retention stands (e.g., patch size), extends the ability to project different management options within complex retention stands.

To project organisms across the landscape, suitability of older non-harvestable stands and naturally disturbed stands must be considered. As with stand-level habitat features, it is more effective to measure them by field monitoring than to model their suitability based on habitat elements. However, both habitat elements and organisms eventually must be monitored in non-harvestable stands to permit tests of the generality of habitat models used in the managed stands. Second-growth stands managed to restore old-growth conditions are another special case (e.g., variable density thinning). There we have little direct experience with organisms other than black-tailed deer (Bunnell 1985; Nyberg et al. 1986; Nyberg et al. 1989), and habitat attributes generally will be outside the range modelled in managed stands. Testing habitat models extrapolated to these conditions, using surveys in existing restored older stands, will be important for predicting the effects of restoration.

Neighbourhood models can extend the scale from within-stand correlation to correlate organisms with characteristics of surrounding stands. They have been used to evaluate both songbird and northern goshawk relations with larger areas of habitat (Vernier and Bunnell 2002; Vernier, Schmiegelow, and Cumming 2002) and provide one way of supplementing stand-level relationships to include local context. If neighbourhood models are used, then it is important not to confound local landscape effects when stand-level suitability is used with the landscape projections.

Landscape-level models of organisms: The stand-level models noted above project the suitability of each stand on the landscape for a range of indicator organisms. Whether the organisms actually use suitable stands depends on their ability to access the stands. That access is a function of the surrounding landscape and the movement ability and propensity of the organisms. Dispersal distances and barriers are generally poorly known, in part because they are so variable and affected in unknown ways by the habitat types through which the organism is dispersing. Fairly simple movement rules are used to model the overall suitability of landscapes for the variety of organisms (Huggard and Kremsater 2005). The landscape scale reveals the importance of considering all stand types, including non-harvestable areas of various types.

Hypothetical organisms as a landscape metric: The idea of hypothetical or idealized species as a tool to measure complex landscapes is discussed in §8.2.5. The approach is still under development and is needed for both direct and projected monitoring of landscape structure. Modelling idealized species parallels the modelling of actual organisms. However, it does not require the autecological detail or testing of predictions that modelling of real species requires because it is only used as an index of landscape structure. It is not intended to provide a direct means of accounting for where real biological diversity will be maintained on the tenure in the future.

We piloted landscape visualization tools (Dovetail Consulting 2001, 4). They appear to be useful for enabling a wide range of people to bring their experience to evaluating potential future conditions over large areas. They are less useful at predicting distribution of accessible habitat because movement patterns of most organisms are too poorly known. However, stand-level visualization allows experience-based evaluation of the predicted long-term consequences of stand management decisions. These intuitive, subjective evaluations of visualized projection should be seen as a complement to, not a substitute for, projections of quantitative indicators of biological diversity. The converse is equally true. Projections of indicators should be subject to some form of evaluation based on collective experience.

The above list of forms of modelling is incomplete and will grow as different questions are addressed.[5]

12.8.3 Implications of Incorporating Mechanisms

All models or projections invoke some mechanistic relationship or "explanation," even if it is primarily statistical. The apparent underlying mechanism or cause-effect relationship differs for each biological element studied. Incorporating explanations has implications for monitoring design, including

- providing the best current understanding, which can then be used to generalize or expose gaps to be filled by direct monitoring;

- sampling predicted features, measuring the variables invoked in the mechanisms, where possible;
- finding or creating sites to determine the relative influence of the different explanatory variables;
- creating models that have generality and wide applicability; and
- testing those models by measuring actual responses in the field in situations where the model makes widely different predictions.

Models that are explicitly explanatory may include variables such as habitat requirements of species, seral associations, dispersal and recolonization abilities, microclimate and substrate needs, or edge effects. The explanatory variables themselves need to be measured in a wide array of situations if the models are to have general applicability. The need for a wide array of conditions emphasizes the need to co-ordinate explanatory variables used by the different monitoring projects (recognizing that unique variables may also be suggested by a species' autecology).

Generally, studies to create or test explanatory models require more intensive work at fewer sites. Conversely, monitoring for direct comparisons or overall operational performance favours less intensive sampling across many sites. The proportion of effort spent on developing and testing explanations depends on the importance of being able to project findings or generalize results compared to making precise, specific comparisons. Individual monitoring studies should consider how they can produce results that generalize to all important situations, including future conditions. That includes adding explanatory variables to the measurements, seeking out anomalous situations to clarify relationships between the explanatory variables and predicted features, and sampling over a range of conditions to test the explanatory model.

The mechanistic models most needed to generalize results of habitat element sampling are those describing processes that change stands over time. These processes include growth and mortality of retained and regenerating live trees, decay and fall of snags, decay of down wood, changes in canopy structure, and growth and succession of understorey layers. Development of these projection models needs to be conducted in conjunction with the field monitoring, which is used to provide or refine needed parameter estimates and to test predictions of the model. The interdependency of models and sampling emphasizes the need to develop the framework of explanatory models early in the program.

12.9 Summary

The monitoring portion of an adaptive management program is about asking and answering questions. While attempting to answer our guiding questions (§3.4.2), we found that

- both making comparisons and evaluating mechanisms were necessary and helpful;
- context is needed for comparisons; we used natural benchmarks and, to a lesser extent, requirements of organisms;
- a cross-design or "what and how" matrix (Table 12.2) helps to confront the trade-off between number of sampled sites and number of sampled variables and establishes the basic monitoring program;
- effective use of blocking factors, experimental sites, and pre-treatment measures facilitates monitoring and makes it more cost effective (§12.6);
- design options for monitoring larger areas were limited, and few meaningful direct comparisons could be made at larger scales; we emphasized stand-level comparisons; and
- a modelling framework (Figure 12.1) that could project landscapes under different planning scenarios and measure landscape attributes is required to address larger scales and future conditions.

13
Summary: Progress and Lessons Learned

Fred L. Bunnell, David J. Huggard, and Laurie L. Kremsater

13.1 Context

We noted at the outset that attempts to assess success of efforts to design forest practices to sustain biodiversity fell into the class of "wicked" problems. Features of wicked problems (§1.1) impede neatly defined progress and impel some of the lessons we learned during the process. We assess progress through reference to the management questions intended to guide the process of designing adaptive management. We summarize the broadly applicable lessons learned while designing and implementing adaptive management for a forest tenure.

13.2 Progress

The most direct assessment of progress is in terms of the six major management questions determined at the outset of the adaptive management program (§3.4.2).

What is variable retention providing as habitat? Although stated broadly, this was a very high corporate priority because we believed that variable retention would provide habitat for many more species than clear-cutting. Monitoring Indicator 2 focused on documenting habitat structure in retention blocks, with natural benchmarks (older forest) providing a "scale" for the results. The question was successfully addressed for habitat structures, and improvements were suggested by the monitoring (Chapter 9). Progress relating the resultant habitat to a wide range of organisms (Indicator 3) was more modest, but different responses among species to types and amounts of retention were documented (Chapter 11).

Are there major edge effects within aggregated variable retention? This question was considered very high priority because group retention increases the amount of edge and associated edge effects. The question is also directly relevant to cutblock layout and to the basic definition of the retention silvicultural system (through forest influence). Edge effects were addressed well in terms of the patterns of retained structure around edges (Chapter 9) but were less rigorously documented for responses of resident species (Chapter 11). Various patterns with distance from edge were documented for bryophytes, epiphytic lichens, ectomycorrhizal fungi, vascular plants, and carabid beetles (Chapter 11), but most of these studies were of short duration. This question is a subset of the next question.

What is the best way of implementing variable retention (e.g., types and amounts of retention)? The phrasing of this question reflects initial enthusiasm. We would now avoid the term "best" (nature is too diverse) and phrase it as "What is gained or lost under different ways of implementing variable retention?" The question is fundamental to operational decisions on how to implement VR across the tenure and was ranked high, recognizing that individual aspects of the question could be addressed first. For key elements of habitat, monitoring effectively evaluated comparisons addressing major aspects of implementing variable retention: type and amount of retention, edge effects, and patch anchor type (Chapter 9). Monitoring of organisms again lagged, but some relationships with amount of retention, type of retention, and responses to edge were documented (Chapter 11). Our experience indicates that with focused monitoring it is possible to address the question but that, given the richness of forest-dwelling organisms, the answer will always be incomplete and will certainly never reveal a single "best" way.

Is stand restoration effective at creating desired structures and ultimately restoring species' distributions or numbers where old growth is rare? This question was assigned moderate priority because it applied to only a small portion of the tenure, although it could be of very high priority in localized areas. There are three broad aspects to effective restoration. The first is where to do it. The second is the probable biological response. The third is the "business case" for making an investment in restoration. Monitoring Indicator 1 documented which ecosystem types were least well represented by unmanaged or older stands. Within such types, riparian areas appeared as useful candidates because of their biological richness and potential economic gains from localizing restoration along streams. The business case involves projecting not only the likely biological responses but also the costs and benefits of creating those responses.

We explored restoration of old-growth attributes in riparian areas (Krcmar 2002; Perry and Muller 2002). Evaluating a business case for restoration used outputs from existing, somewhat crude, models projecting the probable responses of stands. Evaluation also required conventional economic analysis of the costs of restoration and resultant timber values and an assessment of the values of reduced risk to species (greater abundance) or increased species richness. Potential economic benefits and costs of silvicultural treatments to hasten old-growth attributes were evaluated. The costs of acquiring "old-growth" characteristics (mean stand dbh>50 cm) were lower when active intervention through early thinning was employed. These costs can be lowered by selecting variable density thinning treatments that maximize returns (Krcmar 2002).

Several hundred kilometres of variable density thinning along rivers were undertaken on public land before the restoration priority focused on the most productive private lands. The altered focus diminished corporate appetite, and, although significant potential was documented within the program, effort on private land never proceeded beyond the pilot stage.

Are stewardship zones established in the most appropriate locations? Answering this question initially was assigned very high priority because improperly located Old-Growth zones will not meet their primary purpose of sustaining late-seral conditions across the range of ecosystem types. Moreover, almost any allocation of Old-Growth zones incurs costs by reducing the harvestable area. If zones are inappropriately located, then the cost is incurred without the benefit. Given the controversy surrounding reservation of old growth, it was important to locate Old-Growth zones effectively. In the necessarily tenure-wide treatment, external targets were a potentially relevant comparison, but conventional or regulatory targets were already exceeded before implementation of the zone, so ecological representation was more critical (Chapter 7).

Zoning has major implications for the sustainability of a wide range of organisms and for the amount and schedule of volume available to harvest. Balancing ecological and economic objectives or values is a spatial optimization problem. We developed such optimization models (§12.8.2) but found no solution sufficiently compelling to encourage change.

We were hindered by equivocal guidance from the literature and the Scientific Advisory Panel in terms of evaluating the effectiveness of zones. Both landscape statistics and landscape projections using visualization were used to inform the panel. They agreed that initial sizes of some Old-Growth zones could be usefully reduced and some of the area reallocated within Habitat and Timber zones to improve ecosystem representation, but they believed that the size of the mini-zones should not go below 250 to 500 hectares. They also agreed that minimizing roads and stream crossings (assessed under Indicator 2) would be an important influence on the effectiveness of Old-Growth zones but that there was limited scientific guidance on targets or thresholds. The panel likewise concurred with our conclusion on the question of appropriate proportions of zones – science provides only limited guidance, so the decision is largely a question of social choice.

We developed monitoring (Indicator 1) and created tools capable of addressing the question but produced answers that were either unpalatable (reducing production on private land) or laden with sufficient scientific uncertainty to be of little help.

Is biological richness maintained over the tenure, given the mix of zoning, variable retention, and operational constraints? This is the overarching question that addresses the criterion for sustaining biological diversity. Initially, it was assigned high, rather than very high, priority because an answer would take time to develop and was dependent on other questions being answered first. Because the question is tenure wide, it addresses both the managed and the unmanaged land bases and the different types of managed stands as they change over time. That means that comparisons having less importance in previous questions now become more important (e.g., the contributions of "scrub" and the comparisons describing natural trends, such as the changes in retention stands over time; Table 12.2).

The question suggests that an overall goal of the monitoring and modelling program is an accounting system that demonstrates how we expect components of species richness to be maintained over the tenure (§12.8). Although this accounting system includes species-habitat relationships for a few species, it does not explicitly address every species but relies on surrogates such as ecological representation or habitat structure. Thus, a major part of the monitoring program remains to establish the credibility of these surrogates. To date, songbirds are the only organism group monitored tenure wide. Progress on this question has been made by linking representation and habitat structure to a few species (Huggard and Kremsater 2005), but assessing if and how variable retention and zoning accommodate most species remains a long-term project.

13.3 Lessons Learned

The monitoring portion of an adaptive management program is about asking and answering questions. While attempting to answer our guiding questions, we learned much about designing an effectiveness monitoring program for adaptive management. We summarize the more broadly applicable learnings under three headings: organizational structure, design, and feedback.

13.3.1 Organizational Structure

We were well aware that attempts at adaptive management fail less often in the design of the work than in closing the loop so that management actions are taken in response to findings (Lee 1993, 1999; Lee and Lawrence 1986; Ludwig, Hilborn, and Walters 1993; Stankey and Shindler 1997). Three new structures were created to help avoid such failure – the Adaptive Management Working Group, the Variable Retention Working Group, and the International Scientific Advisory Panel (§2.4). Despite these new structures, there was less response to findings of the monitoring program than was originally expected (e.g., Chapters 7, 9, 11). There were two broad reasons – naïveté with regard to the world of management decisions (discussed as part of feedback, §13.3.3), and failure to implement a shared vision within the world of researchers. The latter is best revealed by inadequacies within the initial pilot phase for organisms and reveals failings within the organizational structure. We noted inadequacies of the pilot phase in §11.2.6 and elaborate some here as lessons learned.

Commit time and resources to communication.
An underlying problem was poor communication of the goals of the pilot phase among the scientists helping to design the monitoring program, the company scientists running the project, and the researchers conducting the actual monitoring. Specific kinds of information are required to design and justify a long-term monitoring program (§13.3.2), but they were inadequately communicated and adopted as a shared vision. The key point – that pilot studies need to be specifically designed to provide information on sampling design options, costs, and expected precision – was not adequately communicated, and the required analyses were not emphasized as a critical product from the field researchers. Within the program, we allocated insufficient time and other resources to communication and ensuring a shared vision. A recent review (Jacobson et al. 2006) indicates that logistical and communication barriers (including the lack of clear goals and objectives) were the two most common failings in implementing adaptive land management programs. We recommend that the character and need for a shared vision be made clear to all participants at the start of any monitoring program and that the adoption of that vision be assessed early and strongly encouraged within contracts to consultant researchers.

Diverse funding sources can influence objectives.
There were genuine corporate needs to minimize costs in accord with a corporate philosophy that costs for research on public land should be partially supported by public funds. The result was a monitoring program that was partially driven by interests of individual researchers and funding agencies rather than being designed and rationalized explicitly within the context of long-term monitoring. Diverse funding sources compete directly with a shared vision. Reliance on external funding also meant that a focus on formal piloting was thwarted by an expectation that each project would produce a "stand-alone" product every year. This expectation derived not only from reviewers of external funding programs (mostly government funding that promoted research perceived as immediately useful to industry) but also from the perceived need to present yearly results to reviewers from the international science and environmental communities, from the need to produce information that seemed to be immediately useful to maintain operational support within the company, and from traditional expectations of field researchers themselves. A statistical analysis of sampling design, costs, and precision does not fulfill these expectations as well as studies that appear to provide answers to questions of immediate interest. Many of the yearly changes in individual projects reviewed in Chapter 11 derived from

chasing immediate answers to particular questions while losing sight of the longer-term purpose of pilot work. In the long run, however, pilot studies that contribute definitive information on sampling design, costs, and precision to a well-designed monitoring program are more likely to provide effective answers to important questions than more typical short-term, scattered ecological studies. Diverse funding sources also greatly increase issues of intellectual property rights that can thwart the analyses and syntheses necessary to design an effective monitoring program.

We had intended a three- to five-year pilot phase supported by diverse sources, followed by a smaller, more focused monitoring program that was corporately funded. Changes in corporate ownership modified those plans. The conflict between short-term expectations or demands and longer-term requirements in research and monitoring is not new and will not go away. We recommend that it be acknowledged at the outset and that the overall design carefully consider where focus can produce significant short-term contributions within necessarily longer-term efforts. Funding sources and award systems can then be matched more appropriately with intended products.

Recognize different reward systems.

Many creative researchers are also somewhat maverick in nature and nearly always have a different sense of reward and of risk than do practitioners (Bunnell 1989). As well, for academic researchers, pilot study work will not typically produce publishable papers, while conservation-oriented researchers do not feel that pilot studies directly support their interests.

Pilot studies for a long-term monitoring program have a different purpose, and hence a different design, than the shorter-term research projects with which biologists are familiar. A major difficulty in implementing the long-term monitoring program for organisms has been the mismatch between the need to repeat the same monitoring over several years to obtain useful results and various demands for short-term output from the work. Part of this difficulty derives from funding sources with short-term objectives and part from the very diversity of questions that monitoring is intended to address. The latter diversity encourages the tendency of researchers to chase an interesting new question rather than to build up the results needed to answer an older question well.

Communicating pilot study goals and finding researchers willing to address those goals are critical for successfully initiating an efficient long-term monitoring program. The communication is particularly critical when government commitment to long-term monitoring is lacking.

Stability is good.

In their review of barriers to the implementation of adaptive management, Jacobson et al. (2006) found that a perceived lack of resources, time, and staff was the most common barrier. These perceptions can only be overcome with stability in the funding and work environment. During the first six years of this project, the tenure was owned by four different companies; four different government programs focused on quick results were the main source of funding. Neither of these features encouraged the stable implementation required by a long-term monitoring program; both emphasized the importance of government support of monitoring.

13.3.2 Design

By designing and implementing an effectiveness monitoring program, we acquired a wealth of detail about appropriate sampling methods for different organisms, statistical blocking factors, questions more accessible to experimental approaches rather than to operational comparisons, conditions that encourage pre-treatment measures, and more. Much of that is summarized in Chapter 12. Here we review the broader, more generic lessons learned about designing a monitoring program.

Begin with the major questions and stay with them.

Some of the wandering progress of the organism pilot studies was due to the many possible issues that need to be addressed to establish biodiversity monitoring in a complex environment, including which groups are sensitive and feasible to work on, which of many variables define informative strata, and which management factors to examine. The relative novelty of monitoring lesser-known groups was a distraction. These groups require basic work, such as simply identifying which species were present and how they could be measured. Working on such non-vertebrate groups certainly provides a better perspective on the response of biodiversity, and these projects had considerable appeal to scientific and environmental reviewers. However, to contribute to an effective adaptive management program, monitoring of these groups must still provide sound quantitative information from relevant comparisons. That information has to be assessed along with the more basic natural history work. It was sometimes lacking within the organism pilot phase.

Throughout this complexity of relations and organisms is a critical tradeoff in all monitoring programs: examine many sites and few factors or few sites and many factors or variables. We believe that the approach we took can be effective when applied rigorously. We began by winnowing down the major management questions and ranking them (§3.4.2) and determining which potential comparisons were most relevant to which questions and ranking them as well (Table 12.1). We then matched indicator variables with comparisons through a cross-design (Table 12.2). The design used a suite of indicators for the highest-priority comparisons while ensuring that each important comparison was assessed with at least one or two suitable indicator organisms. That formed the basic monitoring program. We were unable, however, to apply that framework rigorously, primarily due to the difficulties noted under organizational structure above. A major challenge was dependence on diverse, sometimes opportunistic, funding sources that distracted from a focus on the highest-priority questions and comparisons. Perhaps such a lack of stability should be expected in a complex world. It does emphasize the importance of clarity in and commitment to a shared vision and of efforts to match reward systems to activities. If such projects are undertaken by other companies, then we recommend that they secure corporate funding commitment, complete the cross-design, then implement the species work as a package relying primarily on corporate funding.

Begin modelling early; you know you'll have to stay late.

Projection models are needed for at least two compelling reasons. First, the overarching question of whether sustained habitat is provided for biodiversity must be addressed over large areas and long periods of time. Second, beyond the need for mechanistic models to scale local findings up to larger scales and longer periods is the need for mechanistic models to assess comparisons or potential improvements for which field data are not currently available. To address these questions, we developed a modelling framework (Figure 12.1) that could project landscapes under different planning scenarios and measure landscape attributes. The framework is based on combining landscape-level projection tools with stand-level projection of habitat elements to populate projected stands on the landscape. That combination allows predictions of the habitat and landscape structures of Indicator 2. Within the framework, models of habitat and landscape relate organisms to the modelled structures to project Indicator 3.

It is obvious that development and refinement of useful question-based models is a long-term exercise. We made only a beginning (Huggard and Kremsater 2005). The important point is that it must be initiated early. Testing modelling components has implications for the design of a monitoring program, particularly the relative role of comparisons and mechanisms, and the allocation of funding to them. That is particularly true of habitat models and models projecting habitat elements. We initiated a modelling framework early (Dunsworth et al. 1997) and did refine and extend it (Huggard and Kremsater 2005). Its potential guiding role, however, was lessened in the face of a concerted effort to acquire information on a wide range of species. One result was that we developed relatively few sound relationships between organisms and their habitat. An equally important point to beginning early is to sustain the effort – there are major advantages to be gained by question- or model-oriented sampling (e.g., Edwards et al. 2005, 2006; Särndal, Swensson, and Wretman 1992; Welch and MacMahon 2005).

One size does not fit all.
The kinds of data required to assess effectiveness of forest planning and practice at sustaining biodiversity are diverse, but most involve relating organisms to their habitats. Relating habitat elements to the needs of a range of organisms through empirical habitat models incorporates many habitat elements into fairly simple, easily understood indices. Developing such habitat relationships often requires organism monitoring to use study designs and measurements that differ from those needed to make direct comparisons. Similarly, models required to project habitat over time require information on processes, such as growth and mortality of retained and regenerating live trees, decay and fall of snags, and decay of down wood (§12.8.2). Development of such models must be conducted in conjunction with the field monitoring so as to provide or refine needed parameter estimates and to test predictions of the model. Table 12.2 is a useful summary of the core of the monitoring program as it is expressed in evaluating comparisons. It omits some of the additional data that are required to answer larger-scale, longer-term questions. These questions rely on models. The interdependency of models and sampling emphasizes the need to develop the framework of explanatory models early in the program and to continue to evaluate progress on the modelling framework. It also means that if researchers are not closely involved in model formulation they must, at the least, understand why and how their data can inform modelling.

Monitoring and research are not discrete.
There is a tendency to view monitoring and research as discrete and different activities. More specifically, monitoring is viewed as repeatedly (thus boringly) recording conditions, while research involves asking more immediate questions – typically about explanatory mechanisms. This distinction ignores the important complementarity between the two activities, particularly within an adaptive management program (see Noss and Cooperrider 1994). Monitoring is necessarily a longer-term activity, while a well-designed research question may be answered relatively quickly. We found that we had to employ both major means of asking questions within the monitoring program – making comparisons and evaluating mechanisms. We emphasized the former because comparisons of operational sites proved to be fruitful in addressing the near-term management questions. Explanatory mechanisms were necessary to project long-term consequences of present management decisions, particularly for habitat elements. Such explanatory mechanisms were implemented into the simulation framework but neglected in the face of designing a cost-effective system of comparisons (Table 12.2). That neglect has impeded our ability to answer key questions about larger scales and the long term.

We also identified some questions that were better asked in experimental than in operational sites (§12.6.1). Doing that, we exposed difficulties in inserting experimental sites or active adaptive management into operational forest activities. It is likely that no company is prepared to abruptly alter its priorities, "drop tools," and implement a broad array of experimental treatments simultaneously across a large area. In our case, experimental sites were initiated over time and at the discretion of operating divisions within the company. Not all experimental sites were being laid out and harvested during the organism pilot phase, which had several consequences. Organisms that might benefit from pre-harvest measurements in experimental sites needed to be monitored on these blocks, thereby reducing the effort available for stated pilot study objectives. That in turn meant that anticipated contributions to operational comparisons were not made for some organism groups. The experimental sites have strong appeal to scientific reviewers, they help to involve operational staff directly in adaptive management, and they may produce more definitive information when the scattered replicates are eventually completed. However, the design of the overall monitoring program needs to recognize that there is a substantial cost to monitoring experiments. With dispersed replicates and implementation over many years, there is a long delay before results from experimental sites will be available and considerable uncertainty about how meaningful they will be.

There is likely little that can be done to avoid asynchronous initiation of experimental sites at operational scales within working forests. It is important, however, to recognize that (1) monitoring and research are inseparable, (2) explanatory models are required to both scale up local findings and generate better alternatives, (3) question-oriented modelling is efficient, and (4) only some questions benefit from experimental settings and pre-treatment measurements (§12.6.3).

Scale is essential.
We noted (§5.4) that even precise measurements are unhelpful without context or scale. To provide the context or scale for comparisons, we found regulatory targets to be in flux and inconsistently supported by data, so we emphasized natural benchmarks in older forest (Figure 5.1). Comparison of practices with each other and to benchmarks led to modifications in practice during the initial years of implementation, when the operational environment was conducive to improvement (Chapter 9). The implication for monitoring design is that resources must be allocated to such benchmarks, the "baseline" monitoring of Noss and Cooperrider (1994).

13.3.3 Feedback

The purpose of adaptive management is to improve management through information gained on actions taken. Feedback to management can take several forms. We emphasized planned comparisons designed to reveal areas of relative weakness where improvements to management would be most effective. For example, direct comparisons of the amounts and types of variable retention can inform decisions about appropriate mixes of practices. Feedback could also be based on comparisons to thresholds or targets established externally by government and certification groups or by requirements of organisms. A further mechanism linking monitoring to management is formal review of monitoring results and recommendations from scientific, operational, and public groups.

Simpler is not always better.
Most of our recommendations from monitoring to management involved selecting better practices and thus gradually increasing performance as assessed by the indicators. Specific targets based on ecological thresholds did not play a role in this feedback, although we used natural benchmarks to provide scale or context. Comparing retention results to benchmark sites to identify and focus on the weakest points in retention was a useful approach to simplifying the message for practitioners. Such simple comparisons are most effective when the goal is to compare responses of one or a few elements among a few alternative treatments. The approach does not evade the complexity inherent in relations among forest-dwelling organisms and their environment. By focusing on individual elements, we can lose sight of the fact that we want to maintain a diversity of habitat, not just a few focal elements. We suggest that simple comparisons be limited to simple questions and more complex issues be addressed by modelling approaches (§12.8) and by acknowledging the distinction between informed choice and general guidance (see below).

There are few clear thresholds.
Although there is considerable interest from government, environmentalists, and conservation biologists (Bűtler et al. 2004; Guénette and Villard 2005; Huggett 2005; Lindenmayer and Luck 2005) in identifying firm targets that should stimulate management action for each variable monitored, we believe that this approach is simplistic and does not reflect reality – either ecologically or in the way value-based decisions are made. Ecological thresholds (e.g., a hypothetical amount of habitat below which populations suddenly decrease) depend entirely on which organism is considered, what it considers habitat, how easily it recovers and disperses, and more.[1] With the myriad of creatures, habitats, and ecosystems present, there is no way to define one broadly applicable threshold. Moreover, it is likely that few species will show clear thresholds in their responses within complex, dynamic ecosystems. The "habitat alteration" threshold of Guénette and Villard (2005), for example, is comprised of at least four different variables.

Instead of searching for discrete thresholds, we suggest that describing response or marginal-value curves indicating the benefit of incremental gains

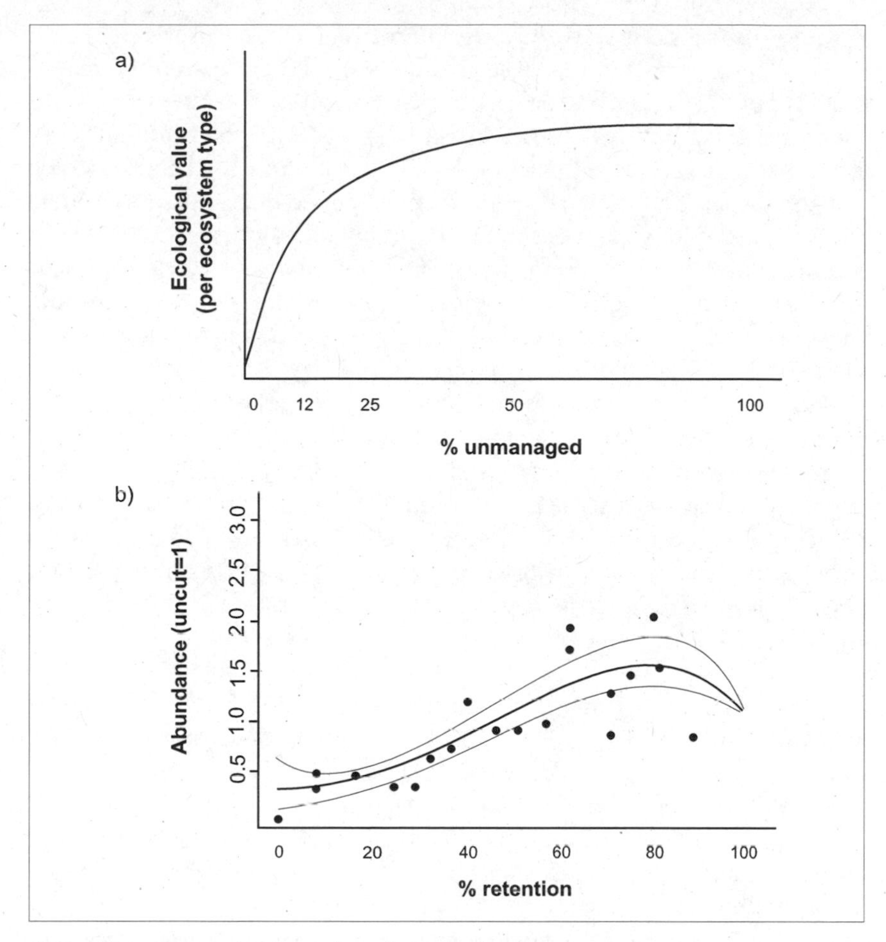

Figure 13.1 Illustrative examples of marginal value curves.
Note: (a) changes in ecological value with changes in the percent of area unmanaged. (b) response of mountain chickadee to amount of forest cover retained; outside lines are 5 percent and 95 percent quantiles. 100 percent represents the natural benchmark.

is our best option for presenting information (Figure 13.1). The converse, showing the marginal cost or risk to various organisms of potential options, is also informative. Note that in each case natural benchmarks provide the scale for the relationship. We have such curves implicitly in mind when we evaluate information. Making them explicit helps to inform decision makers.

Biological reasons that specific targets for amount of representation or size of area cannot be specified unequivocally were noted in §6.2. Most simply, assessing how much representation is adequate only makes sense when you ask *enough for what?* Setting a pass/fail target is therefore neither possible nor useful. However, more unmanaged area probably benefits most organisms and processes, but there are diminishing returns – the incremental value of adding more unmanaged area decreases as the amount increases. For example, going from 40 percent unmanaged in an ecosystem type to 50 percent is good, but it is not nearly as good as going from 1 percent to 11 percent (e.g., Figure 13.1a). The response of the mountain chickadee to the amount of forest cover retained is illustrated in Figure 13.1b. It is one example from a likelihood-based meta-analysis of sixty-nine bird species occurring in uncut forest using data from our own and fifty-one other North American studies.[2] As expected given the variety of bird species, the curves assume most conceivable shapes, including negative responses to increased forest cover. They thus illustrate why simpler is not always better.

The rationale for using marginal value curves applies to all indicators. Defining the shapes of such curves – for various indicators (Y-axis) against various management variables (X-axis) – is helpful in summarizing monitoring results for applied users. The absence of sharp boundaries applies equally to delineating classes of entities, such as age classes (Boyland et al. 2006). Evading artificial, clearly defined thresholds, unless they actually occur, provides less hindrance to people's abilities to consider multiple values when making decisions about complex issues.

Informed choice and general guidance are both useful.

In adaptive management, all monitoring has the goal of detecting responses and revealing appropriate management actions to alter undesirable trends. Despite this common goal, conflicting recommendations to management are inevitable when each monitoring project makes recommendations for its particular subject (Bunnell and Dunsworth 2004). The shapes of response curves for birds alone (Figure 13.1a) cover the entire range of response patterns. Conflicts can be exacerbated by economic and social objectives for the same land base. Numerous, conflicting recommendations from scientists can overwhelm managers and be ignored. Response curves derive from data. Decisions on best options or appropriate amounts are based on values, especially tolerance of risk and assumptions about the values of the managed land base. Stated differently, decisions on corporate or government direction depend on relative values placed on different consequences (e.g., sapsuckers versus timber loss versus worker safety) and do not rest with scientists. The insertion of values ensures that the problem remains "wicked" (§1.1). Valuation of results, however, is external to the data, and monitoring results can be presented independently of specific values.

Recognizing that data are not equivalent to a decision, we suggest that feedback of complex relationships to management should use two different mechanisms: informed choice, and general guidance. The first is most useful for evaluating choices among specific alternatives, the second for sustaining long-term improvements.

To separate scientific content from values inherent in decision making, we suggest "informed choice," a model for integrating conflicting feedback. Informed choice presents decision makers with predictions of the effects of different, clearly specified management options on a number of valued components based on the best available science. Selection of the Northwest Forest Plan ("President's Plan") for forest practices in the American Pacific Northwest is an example. The decision makers, who have responsibility for incorporating public values, decide which option or plan is best based on the information presented. The best available science includes general published literature and monitoring results. Results can be formally combined with literature results (e.g., Bayesian analysis) or presented as additional material, unincorporated into analysis. The options to evaluate can be suggested by managers, scientists, environmental groups, or other interested groups. Ideally, the values used by those making the decision are also evident. The important point is that the predicted effects of the management options on *each biodiversity component* are presented rather than having researchers' values implicitly incorporated by making recommendations directly or by using a summary score. The informed choice approach is useful when the number of clearly defined options available for a particular management decision is limited: for example, riparian restoration in dry ecosystems versus increased amounts of variable retention versus not cutting remaining old growth. The ecological values in a decision matrix should be represented by a wide range of subindicators or indices, along with the best-supported predictions of effects of different management alternatives.

At the simplest level, informed choice can be used to compare responses of single-habitat elements within different treatments relative to the scale provided by natural benchmarks. This approach is illustrated for variables measured for Indicator 2 (habitat) in Chapter 9 and can provide recommendations for improvement.

Although informed choice is helpful when specific management alternatives can be defined, it does not provide general guidance to management questions such as "What should we be doing more or less of or doing differently altogether?" or "What issues will we be facing in ten years?" Guidance on these kinds of questions happens best through a group of people who have ready access to monitoring results and other information and who can use their expertise and judgment to decide which are the most pressing problems or issues and what solutions or means of improvement are possible. The nature of this group or groups and the ways to "institutionalize" the approach should be considered (e.g., the International Scientific Panel and the Variable Retention Working Group; §2.4).

Both forms of overall feedback to management – informed choice and groups offering more general evaluation and general guidance – imply certain responsibilities for the monitoring program. The most obvious and common need is to clearly present relevant, scientifically valid monitoring results and other information to allow effective decisions, evaluations, and broader guidance. However, the types of information differ somewhat between the informed choice and the general guidance approaches.

An informed choice is selected from among clearly defined alternatives and requires focused monitoring on specific alternatives, generally existing management options. Because questions addressed by individual monitoring projects must be co-ordinated to provide a set of monitoring results focused coherently on the same set of options, foresight in directing the individual projects is required. The co-ordinated monitoring of many variables at the experimental sites is an example. For overall evaluation of efforts to sustain biological diversity, and to provide guidance for the future, the most basic information required is some form of species accounting system (§12.8). That requires a combination of monitoring results at different scales and other available information to document expectations about where species or their surrogates will be maintained on the tenure. The system should be sufficiently complete that it can identify weak aspects or causes for concern. Overall guidance should incorporate information *beyond* current practice. Examples include experiments that extend the range of variables, general mechanistic models, and efforts to examine emerging issues and viewpoints that might influence the monitoring program.

The differences between informed choice and more general evaluation and guidance reflect two underlying challenges to the monitoring program. The first relates to the length of view – short term versus long term. Specifically, is the monitoring program intended primarily to help with short-term management problems by choosing between currently available options, or is it intended to provide long-term guidance toward an ever-increasing ability to attain the overall goal for biodiversity conservation? The second distinction is whether the program reflects efforts to "mitigate" (find the least-bad current option) or to "do good" (work toward a better condition). Comparisons and the priorities assigned to them (e.g., Table 12.2) indicate that the program encompasses all four alternatives. Initial efforts are necessarily focused on the short term, but long-term capability is explicit in the development of projection models. The program attempts to identify potentially deleterious practices but is also designed to help improve existing practices. The major challenge to the program will be to maintain a commitment to long-term guidance and improvement while addressing short-term choices and attempting to detect deleterious effects quickly enough to mitigate them.

Expectations from monitoring may be naive.

It is probably naive to expect direct short-term response from monitoring results in the face of strong economic pressures on forest managers. Faced with the difficulties reviewed here and elsewhere in the book, we believe that careful design of a monitoring program can clarify thinking and provide some steering, but it is naive to think that we can fully implement a long-term monitoring program as designed. Instead, programs will evolve at a faster rate than they will provide answers. Even though approved by President Clinton, the Northwest Forest Plan has not been implemented as designed (Molina, Marcot, and Lesher 2006). As with this case study, the primary difficulty was monitoring organisms. At best, people overseeing a monitoring program can provide selective forces that help to keep the monitoring useful in the long term.

Our experience suggests that the appealing idea of a simple adaptive feedback loop – modifying management based on results from comparing options – may be unrealistic with management and response variables as complex as forest practices and organisms. The issue is too thoroughly an example of a wicked problem to permit simple solutions. Such loops may work for simple operational questions, especially when single, small-scale factors affect individual species of direct management concern. Changes in forest management are more likely to be based on information received less directly than specific recommendations and from people who integrate many information sources, such as the International Scientific Panel, market groups, or policy makers. Because such recommendations are partly informed by science, a monitoring program can make an important contribution by generating sound scientific results. They may ultimately have more impact on improving management – by whatever indirect route – than monitoring that attempts direct, immediate feedback to management decisions. For a company, conducting sound scientific projects (even if not directly applied) can be seen as "rent" for the use of the broader scientific knowledge that is the fundamental basis of informed management decisions. With an ever-changing organizational environment for monitoring, and the inherent interests of most researchers, good scientific contributions may be the main legacy of a long-term monitoring program.

Figure 13.2 A slug species new to science found on Haida Gwaii during the pilot phase of monitoring. It is one of twenty-seven species of slugs and snails recorded from the study tenure. The proposed scientific name for the new species is *Staala gwaii.* | *Photograph by K. Ovaska and L. Sopuck.*

Monitoring may often lead to improved practices not by direct feedback but by serving as a frequent reminder that particular forest practices are valued for more than their economic contributions. In that case, the main contribution of the organism-monitoring program – and one that already has occurred in this case study – is simply to serve as a reminder that maintaining species is a goal of the new forest practices and that this includes all manner of species, not just the well-known groups that have guided most previous thinking. A simple image of a slug (Figure 13.2) can remind forest managers of the breadth of their stewardship responsibilities once they have undertaken to sustain biodiversity.

13.4 Summary Thoughts

We addressed all six management questions posed at the outset of the program but less completely than anticipated. We note, as lessons learned, how greater success could be attained. Despite not fulfilling anticipations, it is clear that the adaptive management program did provide information that allowed the company to better meet its objectives (Chapters 7, 9, and 11).

Elements of an adaptive management program also provide value to companies or agencies in other ways. Participating in long-term trend monitoring is a worthwhile contribution, particularly as part of an established regional program such as the Breeding Bird Surveys. Even if the management feedback is not immediate or direct, these programs can warn of negative trends, with the regional context helping to show whether the effects are local or more widespread. Local problems should spur more creative thinking about management solutions. Contributing to regional monitoring also has clear public relations benefits for the company. Similarly, establishing well-designed, operational-scale experimental sites is a valuable contribution that companies can make to the accumulation of scientific knowledge. This contribution is likely to be most important if the sites are maintained for several decades. We delivered relatively quick and efficient feedback to management on ecosystem representation and habitat structure; feedback about organisms is necessarily longer term. To maintain funding and corporate interest, it is important to implement any experiments early and reasonably quickly. Ideally, replicates of the same treatments, focusing on one or two primary variables, would be established simultaneously and in similar ecosystem types to provide meaningful initial results within two or three years of harvesting. More elaborate designs, implemented over several years, are likely to exhaust funding for monitoring the treatments as well as the attention spans of company executives, researchers, and funding agencies.

An adaptive management program remains incomplete if management actions are not improved in response to the program's findings. Our review here and summaries of how indicators have contributed feedback (Chapters 7, 9, and 11) reveal four large points relevant to closing the loop back to management. First, the major indicators complement each other in answering the questions; few are answered effectively by a single indicator. Second, there are more future contributions than current contributions to feedback – monitoring is necessarily a long-term activity. Third, and following from the second, is the necessity of sustained focus on selected questions and comparisons so that the effort of monitoring does produce returns to management. Fourth, making recommendations is far easier than instigating management actions. Our review of lessons learned exposes the need for approaches that integrate the complexity of biodiversity and the utility of attempting to separate values from empirically documented relationships.

Given the diversity of forest-dwelling organisms and habitat requirements, there can be no simple pass/fail target for assessing the effectiveness of forest practice. Our criterion, to maintain all native species on the tenure, is a useful guide even with conflicting recommendations from different measures. We do not expect to sustain every species everywhere at all times, but we do expect every species to persist somewhere within the tenure at all times. The latter expectation emphasizes that the primary need in integrated feedback is a "species accounting system" to show where we expect each group of species now and in the future. Because most species are represented by their surrogates, habitat structures and non-harvestable areas, or by broader taxonomic groups, we need to account for where those surrogates are provided (e.g., how common and where those attributes are) and which groups' needs they encompass. Models projecting habitat attributes and linking them to species' needs are critical tools to link results to management. They are also critical for dealing with the longer term. In whatever specific manner it is implemented, a species accounting system emphasizes that all monitoring has the common goal of maintaining species across the tenure.

Ecological relationships are themselves sufficiently complex that values inherent in decision making should be separated from scientific information to the extent possible. We suggest "informed choice" to present decision makers with predictions of the effects of specific management options on a number of valued components based on the best available science. The predicted effects of the management options on each biodiversity component should be presented rather than implicitly incorporating researchers' values in recommendations for management actions. Informed choice does not answer more encompassing or general questions such as "Which issues will be facing us in ten years?" Guidance on these kinds of questions best happens through a group of informed people using their expertise and judgment to highlight pressing issues, solutions, or means of improvement.

Appendices

Appendix 1

Vascular Plants, Ferns, and Fern Allies Potentially Closely Associated with Forested Conditions in Coastal British Columbia

The following list is derived primarily from information in *The Vascular Plants of British Columbia* (Douglas, Straley, and Meidinger 1989-94; also Douglas, Meidinger, and Pojar 1999; Douglas, Straley, and Meidinger 1998) and *Plants of Coastal British Columbia* (Pojar and MacKinnon 1994).[1] Plants included are closely associated with or restricted to forest conditions. Species found in forests, but also in a variety of other habitats, are excluded. Because this list is intended to address focal species, species are grouped by their responses to potential impacts. Geographic restrictions are noted in square brackets, as are specific comments by reviewers. HG = Haida Gwaii; VI = Vancouver Island.

A. *Vascular plants, ferns, and fern allies frequently associated with canopy cover (preferring a degree of shade and moist conditions)*

Ferns and fern allies (need moist conditions for establishment)

Deer fern	*Blechnum spicant*
Giant chain fern	*Woodwardia fimbriata*
Green spleenwort	*Asplenium viride* [on shaded, rocky limestone][2]
Lady fern	*Athyrium filix-femina*
Maidenhair fern	*Adiantum pedatum* [also associated with humus-rich forests and spray zones]
Narrow beech fern	*Thelypteris phegopteris*
Oak fern	*Gymnocarpium dryopteris*
Oregon selaginella	*Selaginella oregana* [trunks and branches of maple trees; extreme west coast of VI only]
Spiny wood fern	*Dryopteris expansa*
Sword fern	*Polystichum munitum*

Vascular plants

Five-leaved bramble	*Rubus pedatus*
Beautiful bittercress	*Cardamine pulcherrima* var *tenella*

Bleeding heart	*Dicentra formosa* [not HG; common on edges]
Branching montia	*Montia diffusa* [not HG]
Brewer's mitrewort	*Mitella breweri* [not HG]
Bunchberry	*Cornus canadensis*
Clasping twistedstalk	*Streptopus amplexifolius*
Cut-leaved foam flower	*Tiarella trifoliata* var *laciniata*
Devil's club	*Oplopanax horridus*
Fairy slipper	*Calypso bulbosa* [dry to moist mossy forest]
False bugbane	*Trautvetteria caroliniensis*
False lily-of-the-valley	*Maianthemum dilatatum*
False Solomon's seal	*Smilacina racemosa* [not HG]
Fern-leaved gold thread	*Coptis asplenifolia*
Five-stemmed mitrewort	*Mitella pentandra*
Foam flower	*Tiarella trifoliata* (includes var *unifoliata, trifoliata,* and *laciniata*)
Fringecup	*Tellima grandiflora*
Heart-leaved twayblade	*Listera cordata*
Hooker's fairy bells	*Disporum hookeri* [not HG]
Long-bracted frog orchid	*Coeloglossum viride* [moist to wet forests]
Mitrewort	*Mitella* spp *caulescens, ovalis, trifida* [not HG]
Northwest twayblade	*Listera caurina*
One-sided wintergreeen	*Orthilia secunda*
Pathfinder	*Adenocaulon bicolor* [not HG]
Phantom orchid	*Cephalanthera austinae* [not HG; moist to mesic forest]
Pink fawn lily	*Erythronium revolutum* [not HG]
Pink wintergreen, white-veined wintergreen	*Pyrola* spp *picta* and *asarifolia* [not HG]
Queen's cup	*Clintonia uniflora*
Redwood sorrel	*Oxalis oregana*
Rosy twisted stalk	*Streptopus roseus*
Single delight	*Moneses uniflora*
Smith's fairybells	*Disporum smithii* [not HG]
Star-flowered false Solomon's seal	*Smilacina stellata* [not HG]
Stink current	*Ribes bracteosum*
Tooth-leaved monkey flower	*Mimulus dentatus* [not HG]
Twinflower	*Linnaea borealis* [in edges and openings as well]
Vaccinium spp (*membranaceum* and *ovatum*)	[not on HG; tolerant of disturbance]
Vanilla leaf	*Achlys triphylla* [not HG}
Vine maple	*Acer circinatum* [not HG or VI; adaptable/resilient]
Violets	*Viola* spp (*glabella, howellii, lancolata, sempervirens*) [*sempervirens* not HG]
Western trillium	*Trillium ovatum* [not HG]

White fawn lily	*Erythronium oregonum* [not HG; can be common in meadows]
Wild ginger	*Asarum caudatum* [not HG]
Yellow avens	*Geum aleppicum*
Youth on age	*Tolmiea menziesii*

B. *Vascular plants and ferns associated with down wood or rich humus*

Candystick	*Allotropa virgata* [not HG]
Fairy slipper	*Calypso bulbosa*
False azalea	*Menziesia ferruginea* [possibly limited to acid humus; similar to *Vaccinium* but in dry forests as well]
Gnome plant	*Hemitomes congestum* [not HG]
Huckleberry	*Vacinnium parvifolium*
Indianpipe	*Monotropa uniflora* [not HG]
Lady fern	*Athyrium filix-femina*
Licorice fern	*Polypodium glycyrrhiza* [also on maple trees and on rock in moist conditions]
Menzies' pipsissewa	*Chimaphila menziesii* [not HG]
Pinedrops	*Pterospora andromedea* [not HG]
Pinesap	*Hypopitys monotropa*
Prince's pine	*Chimaphila umbellata* [not HG]
Rattlesnake plantain	*Goodyera oblongifolia*
Salal	*Gaultheria shallon*
Small twisted stalk	*Streptopous streptopoides*
Striped coralroot	*Corallorhiza striata* [not HG]
Western coralroot	*Corallorhiza maculata* spp *mertensiana*

C. *Vascular plants that literature indicates may be associated with old forests*

Lax-flowered bluegrass	*Poa laxiflora* [uncommon; unclear if true in BC)

The saprophytes listed for down wood (B above), many ferns (A above), and most *Vaccinium* spp (not *V. parvifolium*) are often more abundant in older forests than in younger forests, but that may be related to canopy density more than stand age.

D. *Vascular plants that some literature suggests may be associated with forest interior*

Rattlesnake plantain	*Goodyera oblongifolia*
Trillium	*Trillium ovatum* [not HG]
Western yew	*Taxus brevifolia*

E. *Rare vascular plants and ferns on the study tenure that are associated with forests (from Conservation Data Centre list)*

Angled bittercress	*Cardamine angulata*

Dwarf bramble	*Rubus lasiococcus* [not HG, possibly restricted to eastern Fraser Valley and outside tenure]
Fringed pinesap	*Pleuricospora fimbriolata* [not HG; possibly not VI]
Howell's violet	*Viola howellii* [not HG]
Marginal wood fern	*Dryopteris marginalis*
Redwood sorrell	*Oxalis oregana* [locations provided suggest not limited to forests]
Scouler's corydali	*Corydalis scouleri* [not HG]
Smith's fairybells	*Disporum smithii* [not HG]
Snow dewberry	*Rubus nivalis* [not HG]
White wintergreen	*Pyrola elliptica* [not HG]

F. Forested streamsides or moist forests

American bittercress	*Barbarea orthoceras*
Angled bittercress	*Cardamine angulata*
Canadian butterweed	*Senecio pauperculus* [not HG]
False bugbane	*Trautvetteria caroliniensis* [not HG]
Palmate coltsfoot	*Petasites frigidus* var *palmatus*
Scoular's corydalis	*Corydalis scouleri* [not HG]

G. Specific to forest hosts

Mistletoe	*Arceuthobium campylopodum*

H. Exotic species that may be useful to track

Brass buttons	*Cotula coronopifolia*
Chickory	*Cichorium intybus*
Common burdock	*Arctium minus*
Common draba	*Draba verna*
Dandelions	*Taraxacum officinale*
Dune tansy	*Tanacetum bipinnatum*
English daisy	*Bellis perennis*
Field bindweed	*Convolvulus arvensis*
Field mustard	*Brassica campestris*
Field pennycress	*Thlaspi arvense*
Foxglove	*Digitalis purpurea* [restricted to dry openings and edges]
Gorse	*Ulex europaeus*
Hedge mustard	*Sisymbrium officinale*
Hemp nettle	*Galeopsis tetrahit*
Himalayan blackberry	*Rubus discolor*
Nipplewort	*Lapsana communis*
Pineapple weed	*Matricaria discoidea*
Prairie pepper-grass	*Lepidium densiflorum*
Purple loosestrife	*Lythrum salicaria*
Ribwort	*Plantago lanceolata*
Scotch broom	*Cytisus scoparius*

Smooth hawksbeard	*Crepis capillaris*
Tufted vetch	*Vicia cracca*
Wall lettuce	*Lactuca muralis*
Wormseed mustard	*Erysimum cheiranthoides*
Yellow salsify	*Tragopogon dubius*

Notes:

1 We reviewed the list developed by Thomas et al. (1993), referred to by FEMAT (1993) and the Interior Columbia Basin Project. Those projects used the Delphi approach to assess species to monitor.

2 Two other ferns largely restricted to calcareous substrates may occur on the tenure: *Polypodium hesperinum* and *Polystichum lonchitis* (Lewis and Inselberg 2001).

Appendix 2

Rare Bryophytes (Liverworts and Mosses) Potentially Occurring on the Study Tenure

Species	CWH	CDF	MH	Ext.[1]	S1[2]
A. Listed liverworts by BEC zone potentially occurring on the study tenure					
Blepharostoma arachmoideum	×			×	
Calycularia crispula	×		×		×
Cephalozia connivens	×				×
Cephaloziella phyllacantha	×		×		
Chandonanthus hirtellus	×				×
Cololejeunea macounii	×				×
Dendrobazzania griffithia	×			×	
Herbertus sendtneri	×				×
Jungermannia confertissima	×				×
Jungermannia hyalina	×				×
Jungermannia subelliptica	×				×
Kurzia trichoclados	×				×
Lophozia ascendens	×				×
Lophozia bantriensis	×				×
Lophozia collaris	×				×
Lophozia elongata	×				×
Lophozia gillmanii[3]	×				
Lophozia sudetica	×				×
Lunularia cruciata	×				×
Nardia insecta	×				×
Riccia frostii	×				×
Scapania gymnostomophila	×				×
Scapania hians subsp *salishensis*	×			×	
Scapania mucronata	×				×
Sphenolobopsis pearsonii	×		×		×
Tritomaria exsecta	×				×
Sphaerocarpos texanus		×			×
Phaeoceros hallii (a horn wort)		×			

Species	CWH	CDF	MH	Ext.[4]	S1[5]	G1[6]
B. Listed mosses by BEC zone potentially occurring on the study tenure						
Atrichum tenellum	×				×	
Barbula amplexifolia	×				×	
Barbula convoluta var *gallinula*	×				×	
Brachydontium olympicum	×				×	
Bryhnia hultenii	×				×	
Bryum amplyodon	×		×		×	
Bryum gemmiparum	×	×	×		×	
Callicladium haldanianum	×				×	
Campylopus japonicus	×				×	
Claopodium bolanderi[7]	×					
Claopodium pellucinerve	×				×	
Daltonia splachnoides	×				×	
Dichodontium pellucidum[7]	×					
Didymodon vinealis[7]	×					
Discelium nudum	×			×		
Encalyptera procera[7]	×					
Entodon concinnus	×				×	
Fabronia pusilla	×				×	
Fissidens fontanus	×				×	
Fissidens pauperculus	×				×	
Hypotergium fauriei[7]	×					
Isopterygiopsis muelleriana	×				×	
Micromitrium tenerum	×					
Mnium marginatum[7]	×					
Oedipodium griffithianum	×		×	×	×	
Orthotrichum diaphanum	×				×	
Orthotrichum hallii	×				×	
Orthotrichum pylaisii	×				×	
Orthotrichum rivulare	×				×	
Orthotrichum tenellum	×				×	
Physcomitrium immersum	×				×	
Pleuroziopsis ruthenica	×	×			×	
Pohlia sphagnicola	×				×	
Pseudephemerum nitidum	×			×		
Pseudoleskea julacea	×				×	
Rhizomnium punctatum	×				×	
Rhodobryum roseum	×				×	
Seligeria acutifolia	×				×	
Seligeria careyana	×					×
Schistidium apocarpum[7]	×					
Schistidium trichodon[7]	×					
Sphagnum junghuhnianum var *pseudomolle*	×			×		
Sphagnum schofieldii	×					×
Sphagnum subobesum	×				×	
Sphagnum wilfii	×				×	

Species	CWH	CDF	MH	Ext.[4]	S1[5]	G1[6]
Steerecleus serrulatus	×				×	
Syntrichia laevipila	×	×			×	
Tortella tortulosa[7]	×					
Tortula nevadensis	×				×	
Trichostomum recurvifolium	×		×		×	
Warnstorfia pseudostraminea	×				×	
Zygodon gracilis	×				×	
Andreaea schofieldiana			×			×
Andreaea sinuosa			×		×	
Gollania turgens			×		×	
Pohlia cardotii			×		×	
Trematodon boasii			×			×
Batramia stricta		×			×	
Bryum violaceum		×			×	
Tortella humilis		×			×	

C. Habitat or substrate of liverworts and mosses

Species	Habitat	Potentially influenced by forestry
Liverworts		
Blepharostoma arachmoideum	damp cliff	
Calycularia crispula	alpine/subalpine	
Cephalozia connivens	rotten wood	×
Cephaloziella phyllacantha	shaded rock	×
Chandonanthus hirtellus	humus	×
Cololejeunea macounii	rock/maple bark	×
Dendrobazzania griffithia	cliff	
Herbertus sendtneri	humus	×
Jungermannia confertissima	stone/soil	
Jungermannia hyalina	stone/soil	
Jungermannia subelliptica	stone/soil	
Kurzia trichoclados	humus bank	×
Lophozia ascendens	tree trunk/soil/rock	×
Lophozia bantriensis	rock	
Lophozia collaris	rock	
Lophozia elongata	sand/cliff	
Lophozia sudetica	rock	
Nardia insecta	soil	
Riccia frostii	mud/riparian	
Scapania gymnostomophila	soil/rock	
Scapania hians spp *salishensis*	soil	
Scapania mucronata	rock	
Sphenolobopsis pearsonii	soil	
Tritomaria exsecta	soil/rock/rotten wood	×
Sphaerocarpos texanus	soil	
Phaeoceros hallii (a horn wort)	soil	

Species	Habitat	Potentially influenced by forestry
Mosses		
Atrichum tenellum	soil	
Barbula amplexifolia	?	×
Barbula convoluta var *gallinula*	?	×
Brachydontium olympicum	rock	
Bryhnia hultenii	log	×
Bryum amplyodon	rock	
Bryum gemmiparum	soil/rock	
Callicladium haldanianum	rotten log	×
Campylopus japonicus	bog	
Claopodium pellucinerve	cliff	
Daltonia splachnoides	tree trunk	
Discelium nudum (extirpated)	soil	×
Entodon concinnus	soil	
Fabronia pusilla	soil/*Populus* tree trunk	×
Fissidens fontanus	concrete	
Fissidens pauperculus	soil	
Isopterygiopsis muelleriana	rock	
Micromitrium tenerum[1] (extirpated)	humus	
Oedipodium griffithianum	humus	×
Orthotrichum diaphanum	concrete wall	
Orthotrichum hallii	rock/alder bark	×
Orthotrichum pylaisii	limestone/base of trees	×
Orthotrichum rivulare	on shrubs/rocks/roots (riparian?)	×
Orthotrichum tenellum (maple, *Populus*)	deciduous trees	×
Phsycomitrium immersum	soil	
Pleuroziopsis ruthenica	soil	
Pohlia sphagnicola	sphagnum bog	
Pseudephemerum nitidum[2] (extirpated)	soil	
Pseudoleskea julacea	limestone	
Rhizomnium punctatum	rock/soil	
Rhodobryum roseum	?	×
Seligeria acutifolia	cliff	
Seligeria careyana	cliff	
Sphagnum junghuhnianum var *pseudomolle*	cliff	
Sphagnum schofieldii	seepy cliff	
Sphagnum subobesum	bog/cliff	
Sphagnum wilfii	soil	
Steerecleus serrulatus	soil	
Syntrichia laevipila	soil	

Species	Habitat	Potentially influenced by forestry
Tortula nevadensis	soil	
Trichostomum recurvifolium	soil/rock	
Warnstorfia pseudostraminea	fen	
Zygodon gracilis	limestone cliff	
Andreaea schofieldiana	soil/rock	
Andreaea sinuosa	rock	
Gollania turgens	cliff	
Pohlia cardotii	soil/riparian	
Trematodon boasii	soil	
Batramia stricta	rock	
Bryum violaceum	soil	
Tortella humilis	dry log/rock	×

Notes:
1 Extirpated.
2 Critically imperilled provincially.
3 Rare because largely restricted to calcareous substrate.
4 Extirpated.
5 Critically imperilled provincially.
6 Critically imperilled globally.
7 Rare because largely restricted to calcareous substrate.
Source: Lewis and Inselberg (2001).

Notes

Chapter 1: The Problem

1 See the various proceedings of conferences hosted by the International Union of Forest Research Organizations (IUFRO) and the International Federation of Operational Research Societies (IFORS).
2 The notion of a "wicked problem" was first formulated by Kunz and Rittel (1970) and Rittel (1972). See also Rittel and Webber (1984). A rich literature has followed.
3 We intend a tradition of joint management of forests and related resources; Native Americans burned forests to encourage a variety of plants and animals but not the forests themselves (see Kay 1995).
4 The three documents from the United Nations Conference on Environment and Development (UNCED '92) were the *Convention on Biodiversity*, *Agenda 21*, and *Guiding Principles on Forests* (the term "guiding principles" reflects the document's complete title: *Non-Legally Binding Authoritative Principles for Global Consensus on the Management, Conservation, and Sustainable Development of All Types of Forests*).
5 These values are from the late 1980s (Pearson 1989). The disparity between wood products and other materials has since grown larger. It is also cheaper to build with wood (Stener 2001).
6 Forestry may not have been practised long enough in North America to effectively assess long-term trends in site productivity. Recent reviews suggest that appropriate management does not reduce site productivity (Evans 1999; Johnson and Curtis 2001, Weetman 1998).
7 The seminal works on adaptive management are Holling (1978) and Walters (1986), but the literature has since burgeoned. Recent treatments specific to forests and biodiversity include Bormann et al. (1993), Halbert (1993), Walters (1997).
8 For example, Breeding Bird Surveys for songbirds and Frogwatch for amphibians.
9 For example, the Montreal Process, Canadian Standards Associate, Forest Stewardship Council, and Sustainable Forest Initiative (which uses "objectives" rather than "criteria").
10 On Vancouver Island, mixtures of western hemlock and amabilis fir are self-perpetuating through the interaction of their rooting habit and wind (see §2.1.2).

Chapter 2: The Example

1 MacMillan Bloedel was acquired by Weyerhaeuser in 1999, which in turn was acquired by Brascan Corporation specifically for the tenure's private lands in 2005. Brascan (now Brookfield Asset Management) formed a new company, Cascadia Forest Products, to manage the crown lands that were subsequently sold to Western Forest Products.
2 As part of the softwood lumber dispute with the United States, the public forest tenures of major companies in British Columbia were reduced in size by about 20 percent on average in 2004.

3 Officially known as the Scientific Panel for Sustainable Forest Practices in Clayoquot Sound, it was created by the provincial government in the fall of 1993. CSP (1995a) provided an overview of First Nations' perspectives; CSP (1995b) recommended active engagement of the Nuu-Chah-Nulth and other local people in all phases of planning and managing the resources of Clayoquot Sound.
4 See Meidinger and Pojar (1991) and Pojar, Klinka, and Meidinger (1987) for comprehensive treatments of the BEC system, including attributes of the zones.
5 The coastal temperate rainforest as defined by Alaback (1991) and Ecotrust, Pacific GIS, and Conservation International (1995).
6 The spirit or ghost bear is an almost whitish colour phase of the black bear that differs from other colour phases of the black bear by a single recessive nucleotide in one receptor gene of unknown adaptive significance (Ritland, Newton, and Marshall 2001) – much like the difference between black and golden Labrador retrievers. This particular colour phase is limited to one area of British Columbia's coast.
7 This was the Scientific Panel for Sustainable Forest Practices in Clayoquot Sound. The independent chair of the panel was Fred Bunnell. Once it became clear that scientists and First Nations could work respectfully together, the panel was co-chaired by Bunnell and Richard Atleo of the Nuu-Chah-Nulth.
8 Recommendations regarding variable retention were developed by the Scientific Panel for Sustainable Forest Practices in Clayoquot Sound. Additional reviews were done by Beese and Zielke (1998) and Bunnell, Kremsater, and Boyland (1998).
9 Given the changes in ownership, it is unclear to what degree the Coast Forest Strategy will be sustained.
10 See Franklin et al. (1997) and Mitchell and Beese (2002). The company made a distinction between the VR approach of the Clayoquot panel and the retention silvicultural system (see the Glossary).
11 This portion of the rationale is discussed most completely in Bunnell, Kremsater, and Boyland (1998).
12 International scientists who have served on the panel include Per Angelstam, Jerry Franklin, David Lindenmayer, Bruce Marcot, Reed Noss, David Perry, and M.A. Sanjayan.

Chapter 3: The Approach

1 The Convention on Biological Diversity (UN 1992) defines biological diversity as "variability among living organisms" (Article 2) and notes that "each contracting party shall identify components of biological diversity important for its conservation and sustainable use"(Article 7a).
2 The United Nations Conference on Environment and Development, June 1992, Rio de Janeiro, also known as the Earth Summit.
3 See Bunnell (1998b) for a brief review of concepts around the question "are all species equal?" For woodpeckers, see Aitken, Wiebe, and Martin (2002); Aubrey and Raley (2002); and Martin, Aitken, and Wiebe (2004); for beaver, see Jones, Lawton, and Shachak (1994) and Naiman, Melillo, and Hobbie (1986).
4 But Boyland, Nelson, and Bunnell (2004a) and Bunnell (1998b) note that plans are rarely followed closely even over five- to twenty-year periods.
5 They include interactions within and among genes, species, and ecosystems and vexing issues of scale (e.g., over long time periods, genetic isolation begets new species; over short periods, it can lead to local extinctions). See Bunnell (1998b) and Delong (1996).
6 The coastal operations of Weyerhaeuser in British Columbia received certification under several systems, including ISO 14001, CSA, SFI, and Kerhout (based in the Netherlands).
7 Priorities were assigned in a workshop in November 2001. For each question, the role of modelling in understanding short-term responses and responses over the longer term or for larger areas was noted and can be found in Appendix 5.2 of *Learning to Sustain Biological Diversity* at www.forestbiodiversityinbc.ca. Note that, although Question 6 is the overarching question, it was assigned high, not very high, priority in 2001 because other questions contributing to the answer could be addressed more quickly.

Chapter 4: Implementing the Approach

1 For the first time in about a century, legislation recognized a new silvicultural system – the retention silvicultural system. See Mitchell and Beese (2002).

2 The Province of British Columbia has implemented subregional or local land use plans that are collectively referred to as higher-level plans.

3 The cross-sectional area of a tree trunk measured in centimetres squared or inches squared; see the Glossary.

4 The removal of the most commercially valuable trees (high-grade trees), often leaving a residual stand composed of trees of poor condition or species composition (Helms 1998). See also Beese et al. (2003).

5 A height (often 3 metres) at which trees within a cutblock are assumed to have become satisfactorily established. See the Glossary.

Chapter 5: Effectiveness Monitoring

1 The most commonly used typology of monitoring (Noss and Cooperrider 1994) distinguishes between effectiveness and validation monitoring; we consider validation monitoring one aspect of refinement monitoring, which itself is often indistinguishable from research.

2 All comparisons that were considered within the project, including those rejected, and their rationales are addressed in Chapter 12 and summarized in Table 12.1 for easy reference.

3 Site series is a finely discriminated unit within the ecological classification system; see §2.1.2, §7.3.3, or the Glossary.

Chapter 6: Ecosystem Representation

1 We use "non-harvestable" rather than "unmanaged" to recognize that many areas not available for timber harvest are being "managed" for values other than timber (e.g., slope stability, stream protection, visual quality, biological diversity, fire suppression). "Inoperable" areas, often of low volume, are one part of non-harvestable forest, but some non-harvestable forest is physically operable, of high value, yet constrained from harvesting for other reasons. Note also that we are not referring to "unharvested" forest, which includes areas that will be harvested but have not yet been harvested.

2 Even though favoured species, such as black-tailed deer, are a component of biological diversity, such species-specific approaches can work against sustaining all biological diversity. It is important to assess how areas set aside for a single species contribute to the broader representation goals.

3 D.J. Huggard (2000) found similar relations in original BEC plot data for dry interior subzones of British Columbia: if a species occurred in more than two plots, then it also occurred in two or more site series.

4 Simberloff and Abele (1976, 1982) review the initial debates; Meffe and Carroll (1994) and many others have subsequently reviewed the debate.

5 Other effects may extend much farther, such as introduced species or effects of human access (Laurance 2000). Global climate change, or the unknown causes of global amphibian declines, could be viewed as extended edge effects from developed areas. These changes are not discussed as edge effects because the total area of even the largest non-harvestable areas is potentially affected. Such regional effects reinforce the need for some wide-scale monitoring of species, even where non-harvestable area is abundant.

6 Mapping of soil types is often so limited that defining rarity is a challenge; nonetheless, some soil types, such as calcareous soils, host unique assemblages of organisms.

7 For example, isolated patches in agricultural areas versus larger, more connected ones in forested areas (Saab 1999).

8 Since 1995, British Columbia has mandated the designation of these management areas on public forest land to help retain late-seral forest conditions.

Chapter 7: Learning from Ecosystem Representation

1 Private lands have since been reconstituted as a separate company with new owners.

2 On the coast, subzone names are derived from classes of relative precipitation and continentality. The first part of the subzone name describes the relative precipitation (x = very dry, m = moist, w = wet, v = very wet), and the second part describes relative continentality (h = hypermaritime, m = maritime). Within a subzone, variants ending in 1 are at lower elevations than variants ending in 2. Nomenclature of BEC units is discussed in §2.1.2; see also Figure 6.1.

3 The numbers (e.g., 01, 07) represent the soil moisture regime. Mesic (01) is considered the representative or "modal" condition; 04, 03, 02 are increasingly drier; 05 and higher are increasingly wetter. See Figures 6.1 and 7.4.

4 This situation is unlikely to improve now that private lands have been reconstituted into a separate company.

Chapter 8: Sustaining Forested Habitat

1 We use the common-language, operational meaning of stand – an area of forest perceived by people as relatively homogeneous and discrete, typically between 1 hectare and 100 hectares. Practically, "stands" are a basic unit of decision making for forest managers. We also recognize that the term has no purely objective ecological meaning because bacteria, beetles, and bears undoubtedly use the word differently.

2 Bunnell, Kremsater, and Wind (1999) reviewed specific attributes of some of these elements (e.g., size and decay state of snags) that have been shown to influence vertebrate use of habitat in the Pacific Northwest.

3 When stands have different disturbance histories (structures), expert opinion and empirical evidence often differ when examining species' responses to stand age (see Bunnell 1999 for review).

4 For example, breeding blue grouse numbers often burgeon when expansive, early-seral stages are created, as by fire.

5 Some attributes of these habitat elements that are important to vertebrates in this region were reviewed by Bunnell, Kremsater, and Boyland (1998); Bunnell, Kremsater, and Wind (1999); Thomas (1979). Primary sources are not repeated here.

6 Dbh is diameter at breast height.

7 Living on wood surfaces.

8 See the review of 179 studies in Trombulak and Frissell (2000).

9 This would be similar to the "hypothetical species" of Richards, Wallin, and Schumaker (2002).

Chapter 9: Learning from Habitat Elements

1 Details of methods and results can be found in Huggard (2004).

2 Akaike's Information Criterion; see Burnham and Anderson (1998).

3 Ecosystem representation by variant within subzones is shown in Figure 7.1; a brief description of the position of subzones within the ecological classification is presented in §2.1.2 and §6.2.2.

4 This finding may reflect the fact that drier subzones were predominantly second growth, so trees retained were more likely to be smaller when compared with older benchmark sites.

5 Adoption of large-patch variable retention was primarily motivated by operational and economic concerns about trying to implement typical retention in all sites.

Chapter 10: Sustaining Forest-Dwelling Species

1 The BEC system of classification was introduced in §2.1.2, and the approach to using site series was discussed in §7.3.3; see also the Glossary.

2 Kremsater and Bunnell (1999) note relatively short distances, but Matlack (1994) reports strong effects of forest continuity on the presence of species that disperse poorly, such as ants.

3 Species listed by the BC Conservation Data Centre.

4 The riparian program is not treated here. Within this program, some experimental sites would lack riparian buffers.

5 Monitoring rare or geographically restricted plants ideally would track changes in distribution. That is possible only if distributions are already known and sites can be revisited after disturbance and in the absence of disturbance. Because these plants are uncommon, records from the Conservation Data Centre (CDC) show scattered locations that prohibit effective sampling. The few geographically restricted plants living in forests are listed in Appendix 1, as are the rare, forest-dwelling plants. Known locations of these plants have been acquired from the CDC but are not summarized here. We do not recommend specialized inventories searching for rare plants.

6 Hornworts are represented by only one order and family. Local representatives appear to have cosmopolitan distributions and are not useful for monitoring.

7 Schofield (1976) noted almost two dozen from British Columbia.

8 See Schofield (1988) for a treatment of the role of mosses in characterizing BEC zones in British Columbia.

9 The data of Andersson and Hytteborn (1991), Rambo and Muir (1998), and Söderström (1988) illustrate the problem; we found only one study that attempted to separate effects of stand age from amounts of favourable substrate (Crites and Dale 1998).

10 Nitrogen-fixing or cyanolichens comprise less than 15 percent of all lichen species. Their non-fungal components "fix" nitrogen in the air, making it available for other uses. Some researchers argue that the nitrogen contribution on nutrient-poor sites is significant (Pike 1978; Rhoades 1995); others disagree (Nash 1996). While lichen litterfall in old-growth forests is often higher in percent nitrogen than conifer needles, it typically makes up only 5 percent of total litterfall, whereas needles comprise 30-40 percent (e.g., Abee and Lavender 1972; Stevenson and Rochelle 1984). Nonetheless, lichen contributions represent an additional input of nitrogen.

11 Areas that are reasonably well surveyed include Haida Gwaii (Queen Charlotte Islands; Brodo 1995; Brodo and Ahti 1996; Brodo and Santesson 1997; Brodo and Wirth 1998), southeast Vancouver Island (Bird and Bird 1973; Noble 1982; Ryan 1991), and Wells Gray Provincial Park (Goward and Ahti 1992).

12 Several cyanolichen genera in British Columbia's Interior Cedar Hemlock zone are essentially restricted to old growth: *Collema, Fuscopannaria, Lichinodium, Lobaria, Nephroma, Parmeliella, Polychidium, Pseudocyphellaria, Sticta.*

13 Studies in British Columbia are limited to Atwood (1998) and Williston (2000), who worked in the Okanagan Valley and Kamloops area.

14 Mycorrhizal fungi can be ectomycorrhizal, growing on the outsides of roots, or endomycorrhizal, growing on the insides of roots. Most fungi on conifers are ectomycorrhizal; western redcedar and big-leaf maple have endomycorrhizal fungi.

15 Earlier work is summarized in FEMAT (1993, IV-79).

16 They studied only two stands of each age class. See also Smith et al. (2002).

17 Red-backed voles, flying squirrels, and wood rats are among the best-known dispersers of these fungi. None of these occurs on Vancouver Island or Haida Gwaii, where we believe their role is assumed by other rodents.

18 Our impression is that researchers surveying fruiting bodies are more likely to report dramatic declines following logging, while those sampling root tips report low-level persistence and gradual recovery. The latter is more consistent with evidence of sustained plant growth.

19 Bunnell et al. (2008) reported that legacies of down wood were equally important in sustaining crustose lichens.

20 September was the peak time of production over three years near Hazelton, British Columbia (Kranabetter and Kroeger 2001).

21 An average well-decayed stump contained 48,800 individual collembola from thirty to forty species.

22 Results of pilot studies on Vancouver Island indicate that some gastropods show affinity for deciduous cover (Ovaska and Sopuck 2005).

23 Okland (1996) did not consider these forests "old growth"; rather, they represented areas with "continuity" of forest cover.

24 For some questions, however, identification to the species level may not be necessary (see Derrail et al. 2002).

25 Many authors, however, have reported decreases to be temporary and, despite being relatively slow dispersers, recovery to be rapid (e.g., Niemelä, Haila, and Punttila 1996; Niemelä, Langor, and Spence 1993; Prezio et al. 1999; Spence et al. 1996; Strayer et al. 1986).

26 With few exceptions, primarily hunted species, all forest-dwelling vertebrates designated "at risk" in British Columbia could be associated with specific structural features of forests, as could the large majority of all forest-dwelling vertebrates (Bunnell, Kremsater, and Wind 1999; see also Table 10.5). That association permits the design of specific forest practices, predictions about outcomes, and monitoring success in attaining desired outcomes.

27 Berg et al. (1994) reviewed 1,434 endangered species of cryptogams, macrofungi, and invertebrates in Swedish forests. A full 96 percent of them were associated with the structural features or habitat elements of forests originally derived from vertebrate studies and employed by the effectiveness-monitoring program (see Chapter 8).

28 A list of all species believed to be present is found in Appendix 4.4 of Bunnell et al. (2003) http://www.forestbiodiversityinbc.ca.

Chapter 12: Designing a Monitoring Program

1 At the time of writing, all provisions and standards of the new Forest and Range Practices Act are unknown.

2 The sampling protocol used by the summer bird surveys is applied continent wide in the Breeding Bird Surveys; see http://www.mbr-pwrc.usgs.gov/bbs/bbs.html.

3 Several different forms of modelling have already been used within the adaptive management program, including those of Boyland (2002); Boyland et al. (2006); Dunsworth et al. (1997); Huggard and Kremsater (2005); and Vernier, Schmiegelow, and Cumming (2002).

4 That is, the Patchworks Landscape Modelling Tool, developed by Spatial Planning Systems; see http://www.swanvalleyforest.ca/scenarioplanning/PatchWorks%20Overview.pdf.

5 Dynamic programming was used to evaluate economic and ecological costs and benefits of old-forest restoration (Krcmar 2002). The model Saltus was developed to explore the amount of land available to timber harvest after the reservation and recruitment of older forest patches under an age-dependent disturbance regime (Boyland 2002).

Chapter 13: Summary

1 There is little evidence that the generalization of Andrén (1994) derived from oceanic islands – 30 percent removal of habitat represents a significant threshold – holds in forested "mainland" habitats.

2 Huggard (2006, 2007) illustrates response curves synthesized from 123 treatments reported from fifty-one studies.

Glossary

anchor type: Or anchor point. Ecological features such as wetlands, rock outcrops, or specific wildlife habitat features around which retention patches are anchored.
basal area: The cross-sectional area of a tree trunk measured in square centimetres or inches; basal area is normally measured at 137 cm (4.5 feet) above ground level.
Biogeoclimatic Ecosystem Classification (BEC): A hierarchical land classification system that groups together ecosystems on the basis of similar climate, soil, and vegetation. Biogeoclimatic zones, subzones, and variants describe geographic units with similar climate. Local site units define areas capable of producing similar plant communities in a mature or climax stage (see site series).
bryophytes: Non-vascular plants typically found in moist or wet habitats: includes mosses, hornworts, and liverworts.
chronosequence: As used here, a sequence of forest stands that differ from one another primarily in terms of age.
clearcut: The removal of all trees in a single harvest from a cutblock with the intention of creating an even-aged stand; the minimum size is an area larger than four tree heights across.
climax: As used here, the end point in ecological development of plant communities that the climate of a region will permit. The entire sequence of plant communities successively occupying an area from initial disturbance to climax is termed a sere.
coarse woody debris (or down wood): Tree trunks or larger limbs left on the ground after natural disturbance or harvest.
cutblock: Or simply block (compartment in Europe). A specific contiguous area of land within which timber removal has occurred or will occur; a harvest unit or compartment. A retention block refers to an area harvested using variable retention.
cyanolichens: Lichens containing cyanobacteria, which are capable of fixing nitrogen from the air.
ectomycorrhizal fungi: Fungi that live symbiotically on the roots of vascular plants. Mycorrhizal fungi can be ectomycorrhizal (growing between the outer cells of roots and forming a mantle) or endomycorrhizal (also called vesicular-arbuscular, growing both within and between root cells). Most fungi on live conifers are ectomycorrhizal.
epigeous: Fruiting above the ground, referring to fungi.
epiphyte: A plant that grows on the above-ground organs of other plant species but is not parasitic.
epixylic: Growing and living on wood surfaces.
folivorous: Eating foliage.
forb: A broad-leaved herb, excluding grasses and grass-like plants.
Forest Ecosystem Network: A contiguous network of old-growth and mature forests (some of which provide forest interior habitat conditions) delineated in a managed landscape.

forest influence: The biophysical effects of forests or individual trees on the environment of the surrounding land, such as microclimate and soil. Forest influence is greatest in closed canopies but extends variable distances into non-treed areas. For defining the retention system, forest influence is considered to extend one tree height into non-treed areas.

gastropods: A class of mollusks typically having a one-piece, coiled shell (sometimes much reduced) and flattened muscular foot with a head bearing stalked eyes. Includes slugs and snails.

green-up: A height (often 3 metres) at which trees within a cutblock are assumed to have become sufficiently established to grow without intervention. May also refer to the period during which trees become established.

group selection: A silvicultural system that removes mature timber in small groups at relatively short intervals, repeated indefinitely, where the continual establishment of regeneration is encouraged and an uneven-aged stand is maintained. Harvest openings are generally less than twice the height of mature trees in the stand.

hardwoods: Broad-leaved trees; in the Pacific Northwest, most are also deciduous.

heli-logging: Helicopter logging. Harvested trees are removed by helicopter, eliminating road construction within the harvested area.

hygric: A moisture class or descriptor for sites with permanent seepage, where water is removed slowly enough to keep the soil wet for most of the growing season (see site series).

hypogeous: Fruiting below the ground, referring to fungi such as truffles.

individual tree selection: See single tree selection.

late seral: Or late successional. The stage of forest succession through a sere in which most of the trees are mature or older (see climax).

matrix: The vegetation type that is most continuous over a landscape. Also used to refer to the harvested parts of variable retention cutblocks.

mesic: A moisture class or descriptor for sites with intermediate moisture regime, determined mainly by regional climate rather than overriding soil or topographic conditions. A moist or fresh soil moisture regime (see site series).

multi-pass systems: Harvest systems that remove trees in time intervals termed passes. Intervals may be short or long.

mycorrhizal fungi: Fungi that live symbiotically on the roots of vascular plants. The fungi form mutually beneficial relationships with plant roots by acting as an extension of the root system.

retention silvicultural system: A silvicultural system designed to retain individual trees or groups to maintain structural diversity over the cutblock for at least one rotation and to leave more than half the total area of the cutblock within one tree height from the base of a tree or group of trees.

rotation: The planned number of years between the initiation of a tree crop or stand and its final cutting.

silviculture: The art and science of guiding the establishment, growth, composition, health, and quality of forests to meet diverse needs and values on a sustainable basis. Silvicultural practices include any of the activities to meet these ends, including the form of harvest (e.g., variable retention).

single tree selection: A silvicultural system that removes mature timber as single scattered individuals at relatively short intervals, repeated indefinitely. The continual establishment of regeneration is encouraged, and an uneven-aged stand is maintained.

site index: A measure of the height that trees are expected to reach at a reference age, in this case fifty years; a measure of site productivity.

site series: The most commonly used designation for ecosystem site types within the BEC system; areas with site moisture and nutrient conditions capable of producing similar plant communities (site associations) at a mature or climax seral stage within the same biogeoclimatic variant. The moisture regime varies from hydric (at the wettest) to very xeric (at the driest).

snag: Standing dead tree; includes whole or broken dead trees in all stages of decay.

species richness: The number of species in an area or ecosystem.
sporocarp: Fruiting body of a fungus – mushroom.
stocking: A measure of the area occupied by trees, usually measured in terms of well-spaced trees per hectare, or basal area per hectare, relative to an optimum or desired level.
subhydric: A moisture class or descriptor for sites where water is removed slowly enough to keep the water table at or near the surface for most of the year.
subxeric: A moisture class or descriptor for sites where water is removed rapidly in relation to supply; soil is moist for short periods following precipitation (see site series).
succession: The process by which a series of different plant communities and associated animals and microbes successively occupy and replace each other over time following disturbance of an ecosystem or landscape location. Recognizably distinct communities are termed "successional stages."
thalli: The main growth structure of fungi and lichens – it is a cellular expansion taking the place of stem and foliage.
uneven-aged management: A silvicultural system designed to create or maintain and regenerate an uneven-aged stand structure, a stand of trees containing three or more age classes. Single tree and group selection are uneven-aged silvicultural systems.
variable retention: An approach to silvicultural systems and forest harvesting in which structural elements of the existing stand are retained throughout a harvested area for at least the next rotation to achieve specific management objectives.
vascular plant: A major taxonomic division of plants containing green plants with tubular food and water transport systems; includes trees, shrubs, herbs, graminoides, and ferns.
xeric: A moisture class or descriptor for sites with a dry, moisture-shedding physical environment. Soil is moist for very short periods following precipitation.
yarding: In logging, the hauling of felled timber to the landing or temporary storage site from where trucks (usually) transport it to the mill site. Yarding methods include ground skidding, hoe-forwarding, various cable techniques, and aerial methods such as helicopter and balloon yarding.

Literature Cited

Abee, A., and D. Lavender. 1972. Nutrient cycling in throughfall and litterfall in 450-year-old Douglas fir stands. Pp. 133-43 *in* J.F. Franklin, L.J. Dempster, and R.H. Waring, eds., Proceedings – Research on Coniferous Forest Ecosystems – a Symposium. USDA Forest Service, Portland, OR.

Abensperg-Traun, M., G.T. Smith, G.W. Arnold, and D.E. Steven. 1996. The effects of habitat fragmentation and livestock grazing on animal communities in remnants of gimlet *Eucalyptus salubris* woodland in the Western Australian wheatbelt. I. Arthropods. Journal of Applied Ecology 33: 1281-1301.

ABMP (Alberta Biodiversity Monitoring Program Management Board). 2005. The Alberta biodiversity monitoring program: Program overview and consultation backgrounder. http://www.abmp.arc.ab.ca/Documents/ABMPSummary.pdf.

Ackoff, R.L. 1974. Redesigning the Future: A Systems Approach to Societal Problems. John Wiley and Sons, New York, NY.

Agee, J.K. 1993. Fire Ecology of Pacific Northwest Forests. Island Press, Washington, DC.

Aitken, K.E.H., K.L. Wiebe, and K. Martin. 2002. Nest site reuse patterns for a cavity-nesting bird community in interior British Columbia. Auk 119: 391-402.

Alaback, P.B. 1991. Comparative ecology of temperate rainforests of the Americas along analogous climatic gradients. Revista Chilena de Historia Natural 64: 399-412.

Allen, G.M., and E.M. Gould Jr. 1986. Complexity, wickedness, and public forests. Journal of Forestry 84(4): 20-23.

Allen, T.R., S.M. Berch, and M.L. Berbee. 2000. Diversity of ericoid mycorrhizal fungi on *Gaultheria shallon*. P. 785 *in* L.M. Darling, ed., At Risk: Proceedings of a Conference on Biology and Management of Species and Habitats at Risk, 15-19 February 1999, Kamloops, BC. Ministry of Environment, Lands, and Parks, Victoria, BC.

Anderson, A.E. 1984. *Odocoileus hemionus*. Mammalian Species 219: 1-9.

Anderson, J.M., S.A. Huish, P. Ineson, M.A. Leonard, and P.R. Splatt. 1985. Interactions of invertebrates, microorganisms, and tree rots in nitrogen and mineral element fluxes in deciduous woodland soils. Pp. 377-92 *in* A.H. Fitter, ed., Ecological Interactions in the Soil. Blackwell Scientific, Oxford, UK.

Andersson, L.I., and H. Hytteborn. 1991. Bryophytes and decaying wood: A comparison between managed and natural forest. Holarctic Ecology 14: 121-30.

Andrén, H. 1994. Effects of habitat fragmentation on birds and mammals in landscapes with different proportions of suitable habitat: A review. Oikos 71: 355-66.

Angelstam, P. 1997. Landscape analysis as a tool for the scientific management of biodiversity. Ecological Bulletin (Copenhagen) 46: 140-70.

Arcese, P., and A.R.E. Sinclair. 1997. The role of protected areas as ecological baselines. Journal of Wildlife Management 61: 587-602.

Arsenault, A., and T. Goward. 2000. Ecological characteristics of inland rainforests. Pp. 437-39 *in* L.M. Darling, ed., At Risk: Proceedings of a Conference on Biology and Management of Species and Habitats at Risk, 15-19 February 1999, Kamloops, BC. Ministry of Environment, Lands, and Parks, Victoria, BC.

Arseneau, M.J., J.P. Ouellet, and L. Sirois. 1998. Fruticose arboreal lichen biomass accumulation in an old growth balsam fir forest. Canadian Journal of Botany 76: 1669-76.

Ås, S. 1993. Are habitat islands islands? Woodliving beetles (Coleoptera) in deciduous forest fragments in boreal forest. Ecography 16: 219-28.

Askins, R.A. 1993. Population trends in grassland, shrubland, and forest birds in eastern North America. Pp. 1-34 *in* D.M. Power, ed., Current Ornithology. Volume 11. Plenum Publishing Corporation, New York, NY.

Atwood, L.B. 1998. Ecology of the microbiotic crust of the antelope-brush (*Purshia tridentata*) shrub steppe of the south Okanagan, British Columbia. MSc thesis, University of British Columbia, Vancouver, BC.

Aubrey, K., and C. Raley. 2002. The pileated woodpecker as a keystone species. Pp. 257-74 *in* W.F. Laudenslayer Jr., P.J. Shea, B.E. Valentine, C.P. Weatherspoon, and T.E. Lisle, tech. co-ords., Proceedings of the Symposium on the Ecology and Management of Dead Wood in Western Forests, 2-4 November 1999, Reno, Nevada. USDA Forest Service, General Technical Report PSW-GTR-181, Albany, CA.

Aubry, K.B., M.P. Amaranthus, C.B. Halpern, J.D. White, B.L. Woodard, C.E. Peterson, C.A. Lagoudakis, and A.J. Horton. 1999. Evaluating the effects of varying levels and patterns of green-tree retention: Experimental design of the DEMO study. Northwest Science 73 (special issue): 12-26.

Aubry, K.B., C.B. Halpern, and D.A. Maquire. 2004. Ecological effects of variable-retention harvest in the northwestern United States: The DEMO study. Forest, Snow, and Landscape Research 78: 119-34.

Austin, M.P., and C.R. Margules. 1986. Assessing representativeness. Pp. 45-67 *in* M.B. Usher, ed., Wildlife Conservation Evaluation. Chapman and Hall, London, UK.

Awimbo, J.A., D.A. Norton, and F.B. Overmars. 1996. An evaluation of representativeness for nature conservation, Hokitika Ecological District, New Zealand. Biological Conservation 75: 177-86.

Bader, P., S. Jansson, and B.G. Jonsson. 1995. Wood-inhabiting fungi and substrate decline in selectively logged boreal spruce forests. Biological Conservation 72: 355-62.

Bailey, S.-A., C. Horner-Devine, G. Luck, L.A. Moore, K.M. Carney, S. Anderson, C. Betrus, and E. Fleishman. 2004. Primary productivity and species richness: Relationships among functional guilds, residency groups, and vagility classes at multiple spatial scales. Ecography 27: 207-17.

Baldwin, L. 2000. The Effect of Remnant Size, Forest Age, and Landscape Position on Bryophyte Forest Communities in Isolated Forest Remnants. Final report submitted to Weyerhaeuser, Nanaimo, BC.

Baldwin, L.K., and G.E. Bradfield. 2005. Bryophyte community differences between edge and interior environments in temperate rain-forest fragments in coastal British Columbia. Canadian Journal of Forest Research 35: 580-92.

Barkman, J.J. 1958. Phytosociology and Ecology of Cryptogamic Epiphytes. Van Gorcum, Assen, Netherlands.

Barrie, W.T., S.F. McCool, and G.H. Stankey. 1998. Protected area planning principles and strategies. Pp. 133-54 *in* K. Lindberg, M. Epler Wood, and D. Engeldrum, eds., Ecotourism: A Guide for Planners and Managers. Volume 2. The Ecotourism Society, North Bennington, VT.

Bayne, E.M., and K.A. Hobson. 1997. Comparing the effects of landscape fragmentation by forestry and agriculture on predation of artificial nests. Conservation Biology 11: 1418-29.

Beese, W.J. 2003. Rare Plant Awareness Training Manual – South Island Timberlands. Weyerhaeuser BC Coastal Group technical manual, Nanaimo, BC.

Beese, W.J., B.G. Dunsworth, K. Zielke, and B. Bancroft. 2003. Maintaining attributes of old-growth forests in coastal B.C. through variable retention. Forestry Chronicle 79: 570-78.
Beese, W., and K. Zielke. 1998. A Forest Management Strategy for the 21st Century: Silviculture – a New Balance. MacMillan Bloedel, Nanaimo, BC.
Beier, P., and R. Noss. 1998. Do habitat corridors provide connectivity? Conservation Biology 12: 1241-52.
Belbin, L. 1993. Environmental representativeness: Regional partitioning and reserve selection. Biological Conservation 66: 223-30.
Belnap, J. 1994. Potential role of cryptobiotic crusts in semiarid rangelands. Pp. 179-85 *in* S.B. Monsen and S.G. Kitchen, comps., Proceedings: Ecology and Management of Annual Rangelands. USDA Forest Service, General Technical Report INT-GTR-313, Logan, UT.
Berg, Å., B. Ehnström, L. Gustafsson, T. Hallingbäck, M. Jonsell, and J. Weslien. 1994. Threatened plant, animal, and fungus species in Swedish forests: Distribution and habitat associations. Conservation Biology 8: 718-31.
–. 1995. Threat levels and threats to red-listed species in Swedish forests. Conservation Biology 9: 1629-33.
Bergeron, Y., S. Gauthier, V. Kafka, P. Lefort, and D. Lesieur. 2001. Natural fire frequency for the eastern Canadian boreal forest: Consequences for sustainable forestry. Canadian Journal of Forest Research 31: 384-91.
Bergeron, Y., and B. Harvey. 1997. Basing silviculture on natural ecosystem dynamics: An approach applied to the southern boreal mixedwood forest of Quebec. Forest Ecology and Management 92: 235-42.
Bergeron, Y., A. Leduc, B.D. Harvey, and S. Gauthier. 2002. Natural fire regime: A guide for sustainable management of the Canadian boreal forest. Silva Fennica 36: 1-15.
Bernes, C., ed. 1994. Biological Diversity in Sweden: A Country Study. Monitor 14. Swedish Environmental Protection Agency, Stockholm, Sweden.
Berryman, S., and B. McCune. 2006. Estimating epiphytic macrolichen biomass from topography, stand structure, and lichen community data. Journal of Vegetation Science 17: 157-70.
Bieringer, G., and K.P. Zulka. 2003. Shading out species richness: Edge effect of a pine plantation on the Orthoptera (Tettigoniidae and Acrididae) assemblage of an adjacent dry grassland. Biodiversity and Conservation 12: 1481-95.
Binkley, C.S. 1992. Forestry after the end of nature. Journal of Forestry 90: 33-37.
Bird, C.D., and R.D. Bird. 1973. Lichens of Saltspring Island, British Columbia. Syesis 6: 57-80.
Bonham, K.J., R. Mesibov, and R. Bashford. 2002. Diversity and abundance of some ground-dwelling invertebrates in plantation vs native forests in Tasmania, Australia. Forest Ecology and Management 158: 237-47.
Borcard, D., P. Legendre, and P. Drapeau. 1992. Partialling out the spatial component of ecological variation. Ecology 73: 1045-55.
Bormann, B.T., P.G. Cunningham, M.H. Brookes, V.W. Manning, and M.W. Collopy. 1993. Adaptive Ecosystem Management in the Pacific Northwest. USDA Forest Service, General Technical Report PNW-GTR-341, Portland, OR.
Boudreault, C., Y. Bergeron, S. Gaulthier, and P. Drapeau. 2002. Bryophyte and lichen communities in mature to old-growth stands in eastern boreal forests of Canada. Canadian Journal of Forest Research 32: 1080-93.
Boutin, S., and D. Hebert. 2002. Landscape ecology and forest management: Developing an effective partnership. Ecological Applications 12: 390-97.
Bowan, G.D. 1980. Disturbance effects on fungi. Pp. 165-90 *in* P. Mikola, ed., Tropical Mycorrhizal Research. Oxford University Press, New York, NY.
Bowman, C.M.J.S. 1993. Biodiversity: Much more than biological inventory. Biodiversity Letters 1: 163.
Boychuk, D., A. Perera, M. Ter-Mikaelian, D.L. Martell, and C. Li. 1997. Modelling the effect of spatial scale and correlated fire disturbances on forest age distribution. Ecological Modelling 95: 145-64.

Boyland, M. 2002. Management of seral lands under harvesting regimes. PhD diss., University of British Columbia, Vancouver, Canada.
Boyland, M., J. Nelson, and F. Bunnell. 2004a. A test for robustness in harvest scheduling models. Forest Ecology and Management 207(1-2): 121-32.
–. 2004b. Creating zones for forest management: A simulated annealing approach. Canadian Journal of Forest Research 34: 1669-82.
Boyland, M., J. Nelson, F. Bunnell, and R. D'Eon. 2006. An application of fuzzy set theory for seral-class constraints in forest planning models. Forest Ecology and Management 223: 395-402.
Bratton, S.P. 1994. Logging and fragmentation of broadleaf deciduous forests: Are we asking the right ecological questions? Conservation Biology 8: 295-97.
Breininger, D.R., M.J. Barkaszi, R.B. Smith, D.M. Oddy, and J.A. Provancha. 1998. Prioritizing wildlife taxa for biological diversity conservation at the local scale. Environmental Management 22: 315-21.
Brett, R.B., and K. Klinka. 1998. A transition from gap to tree-island regeneration patterns in the subalpine forest of south-coastal British Columbia. Canadian Journal of Forest Research 28: 1825-31.
Brink, V.C. 1959. A directional change in the subalpine forest-heath ecotone in Garibaldi Park, British Columbia. Ecology 40: 10-16.
British Columbia Forest Service. 1990. Old Growth Forests: Problem Analysis. Ministry of Forests, Victoria, BC.
Broberg, L.E. 1999. Will management of vulnerable species protect biodiversity? Journal of Forestry 97(7): 12-18.
Brodo, I.M. 1995. Lichens and lichenicolous fungi of the Queen Charlotte Islands, British Columbia, Canada. 1. Introduction and new records for B.C., Canada, and North America. Mycotaxon 56: 135-73.
Brodo, I.M., and T. Ahti. 1996. Lichens and lichenicolous fungi of the Queen Charlotte Islands, British Columbia, Canada. 2. The Cladoniaceae. Canadian Journal of Botany 74: 1147-80.
Brodo, I.M., and R. Santesson. 1997. Lichens of the Queen Charlotte Islands, British Columbia, Canada. 3. Marine species of *Verrucaria* (Verrucariaceae, Ascomycotina). Journal of the Hattori Botanical Laboratory 82: 27-37.
Brodo, I.M., and V. Wirth. 1998. Lichens and lichenicolous fungi of the Queen Charlotte Islands, British Columbia, Canada. 4. The genus *Fuscidea* (Fuscideaceae). Pp. 149-62 *in* M.G. Glennn, R.C. Harris, R. Dirig, and M.S. Cole, eds., *Lichenographia Thomsoniana:* North American Lichenology in Honor of John W. Thomson. Mycotaxon, Ithaca, NY.
Brooke, R.C., E.B. Peterson, and V.J. Krajina. 1970. The subalpine Mountain Hemlock zone. Ecology of Western North America 2: 148-349.
Brouat, C., S. Meusnier, and J-Y. Rasplus. 2004. Impacts of forest management practices on carabids in European fir forests. Forestry 77(2): 85-97.
Brown, D.H., and J.W. Bates. 1990. Bryophytes and nutrient cycling. Botanical Journal of the Linnean Society 104: 129-47.
Brown, G.W., and J.T. Krygier. 1970. Effects of clear-cutting on stream temperature. Water Resources Research 6: 1133-39.
Brumwell, L.J., K.G. Craig, and G.E. Scudder. 1998. Litter spiders and carabid beetles in a successional Douglas-fir forest in British Columbia. Northwest Science 72: 94-95.
Brundtland, G.H. 1987. Our Common Future: World Commission on Environment and Development. Oxford University Press, Oxford, UK.
Brunet, J. 1993. Environmental and historical factors limiting the distribution of rare forest grasses in south Sweden. Forest Ecology and Management 61: 263-75.
Buchanan, R. 1992. Wicked problems in design thinking. Design Issues 8(2): 5-21.
Budd, W.W., P.L. Cohen, P.R. Saunders, and F.R. Steiner. 1987. Stream corridor management in the Pacific Northwest: I. Determination of stream-corridor widths. Environmental Management 11: 587-97.
Bunnell, F.L. 1985. Forestry and black-tailed deer: Conflicts, crises, or cooperation. Forestry Chronicle 61: 180-84.

–. 1989. Alchemy and Uncertainty: What Good Are Models? USDA Forest Service, General Technical Report PNW-GTR-232, Portland, OR.
–. 1995. Forest-dwelling vertebrate faunas and natural fire regimes in British Columbia: Implications to conservation. Conservation Biology 9: 636-44.
–. 1997. Operational criteria for sustainable forestry: Focusing on the essence. Forestry Chronicle 73: 679-84.
–. 1998a. Evading paralysis by complexity when establishing operational goals for biodiversity. Journal of Sustainable Forestry 7(3-4): 145-64.
–. 1998b. Managing forests to sustain biodiversity: Substituting accomplishment for motion. Forestry Chronicle 74: 822-27.
–. 1998c. Setting goals for biodiversity in managed forests. Pp. 117-53 *in* F.L. Bunnell and J.F. Johnson, eds., The Living Dance: Policy and Practices for Biodiversity in Managed Forests. UBC Press, Vancouver, BC.
–. 1999. What habitat is an island? Pp. 1-31 *in* J.A. Rochelle, L.A. Lehmann, and J. Wisniewski, eds., Forest Fragmentation: Wildlife and Management Implications. Brill, Leiden, Netherlands.
–. 2005. Adaptive management for biodiversity in managed forests: It can be done. Pp. 3-11 *in* C.E. Peterson, and D.A. Maguire, eds. Balancing Ecosystem Values: Innovative Experiments for Sustainable Forestry: Proceedings of a Conference. USDA Forest Service, Gen. Tech. Rep. PNW-GTR-635. Portland, OR.
Bunnell, F.L., and M. Boyland. 2002. Decision-support systems: It's the question not the model. Journal for Nature Conservation 10: 269-79.
Bunnell, F.L., and A.C. Chan-McLeod. 1997. Terrestrial vertebrates. Pp. 103-30 *in* P.K. Schoonmaker, B. von Hagen, and E.C. Wolf, eds., The Rain Forests of Home: Profile of a North American Bioregion. Island Press, Washington, DC.
–. 1998. Forestry and biological diversity: Elements of the problem. Pp. 3-18 *in* F.L. Bunnell and J.F. Johnson, eds., The Living Dance: Policy and Practices for Biodiversity in Managed Forests. UBC Press, Vancouver, BC.
Bunnell, F.L., and B.G. Dunsworth. 2004. Making adaptive management for biodiversity work: The example of Weyerhaeuser in coastal British Columbia. Forestry Chronicle 80: 37-43.
Bunnell, F., G. Dunsworth, D. Huggard, and L. Kremsater. 2003. Learning to Sustain Biological Diversity on Weyerhaeuser's Coastal Tenure. Weyerhaeuser, Nanaimo, BC. http://www.forestbiodiversityinbc.ca.
Bunnell, F.L., and D.S. Eastman. 1976. Effects of forest management practices on the wildlife in the forests of British Columbia. Pp. 631-89 *in* Proceedings of Division I, 15th IUFRO (International Union of Forest Research Organizations) World Congress, Oslo, Norway.
Bunnell, F.L., R. Ellis, S. Stevenson, and D.S. Eastman. 1978. Evaluating ungulate populations and range in British Columbia. Transactions of the North American Wildlife and Natural Resources Conference 43: 311-22.
Bunnell, F.L., T. Goward, I. Houde, and C. Björk. 2007. Larch seed trees sustain arboreal lichens and encourage re-colonization of regenerating stands. Western Journal of Applied Forestry 22(2): 94-98.
Bunnell, F.L., I. Houde, B. Johnston, and E. Wind. 2002. How dead trees sustain live organisms in western forests. Pp. 291-318 *in* P. Shea, W. Laudenslayer, P. Weatherspoon, and B. Valentine, tech. co-ords., Proceedings of Symposium on the Ecology and Management of Dead Wood in Western Forests, 1-4 November 1999, Reno, Nevada. USDA Forest Service, General Technical Report PSW-GTR-181, Albany, CA.
Bunnell, F.L., and D.J. Huggard. 1999. Biodiversity across spatial and temporal scales: Problems and opportunities. Forest Ecology and Management 115: 113-26.
Bunnell, F.L., and L.L. Kremsater. 1990. Sustaining wildlife in managed forests. Northwest Environmental Journal 6: 243-69.
Bunnell, F.L., L.L. Kremsater, and M. Boyland. 1998. An Ecological Rationale for Changing Forest Management on MacMillan Bloedel's Forest Tenure. Publication R-22. Centre for Applied Conservation Biology, University of British Columbia, Vancouver, BC.

Bunnell, F.L., L.L. Kremsater, and E. Wind. 1999. Managing to sustain vertebrate diversity in forests of the Pacific Northwest: Relationships within stands. Environmental Reviews 7: 97-146.

Bunnell, F.L., T. Spribille, I. Houde, T. Goward, and C. Björk. 2008. Lichens on downed wood in young and old stands. Canadian Journal of Forest Research 3(8): 1033-41.

Bunnell, F.L., and K.A. Squires. 2005. Forest-Dwelling Endemics of British Columbia. Extension pamphlet, Centre for Applied Conservation Research, University of British Columbia, Vancouver, BC.

Burnham, K.P., and D.R. Anderson. 1998. Model Selection and Inference: A Practical Information-Theoretic Approach. Springer-Verlag, New York, NY.

Bűtler, R., P. Angelstam, P. Ekelund, and R. Schlaepfer. 2004. Dead wood threshold values for the three-toed woodpecker presence in boreal and sub-alpine forest. Biological Conservation 119: 305-18.

Cadenasso, M.L., and S.T.A. Pickett. 2001. Effects of edge structure on the flux of species into forest interiors. Conservation Biology 15: 91-97.

Caicco, S.L., J.M. Scott, B. Butterfield, and B. Csuti. 1995. A gap analysis of the management status of vegetation of Idaho (U.S.A.). Conservation Biology 9: 498-511.

Canadian Council of Forest Ministers. 1997. Criteria and Indicators of Sustainable Forest Management in Canada. Technical report, Ottawa, ON.

Carey, A.B., D.R. Thysell, and A.W. Brodie. 1999. The Forest Ecosystem Study: Background, Rationale, Implementation, Baseline Conditions, and Silvicultural Assessment. USDA Forest Service, General Technical Report PNW-GTR-457, Portland, OR.

Caro, T.M., and G. O'Doherty. 1999. On the use of surrogate species in conservation biology. Conservation Biology 13: 805-14.

Castelle, A.J., A.W. Johnson, and C. Conolly. 1994. Wetland and stream buffer size requirements: A review. Journal of Environmental Quality 23: 878-81.

Chan-McLeod, A., and P. Vernier. 2005. Effects of Variable Retention on Songbirds 2000-2004. Data consolidation report prepared for TimberWest and Weyerhaeuser Coastal Group, Nanaimo, BC.

Chapman, B., J. Battigelli, L. Paul, and G. Xiao. 2000. Disappearing large woody substrate: A habitat change of biblical proportions? Pp. 775-78 *in* L.M. Darling, ed., At Risk: Proceedings of a Conference on Biology and Management of Species and Habitats at Risk, 15-19 February 1999, Kamloops, BC. Ministry of Environment, Lands, and Parks, Victoria, BC.

Chen, J., and J.F. Franklin. 1992. Vegetation responses to edge environments in old-growth Douglas-fir forests. Ecological Applications 3: 387-96.

Chen, J., J.F. Franklin, and J.S. Lowe. 1996. Comparison of abiotic and structurally defined patch patterns in a hypothetical forest landscape. Conservation Biology 10: 854-62.

Chiarucci, A., D. D'Auria, V. de Dominicis, A. Lagana, C. Perini, and E. Salerni. 2005. Using vascular plants as a surrogate taxon to maximize fungal species richness in reserve design. Conservation Biology 19: 1644-52.

Christie, G.C., and H.A. Regier. 1988. Measures of optimal thermal habitat and their relationship to yields for four commercial species. Canadian Journal of Fisheries and Aquatic Sciences 45: 301-14.

Chust, G., J.L. Pretus, D. Ducrot, A. Bedòs, and L. Deharveng. 2003. Response of soil fauna to landscape heterogeneity: Determining optimal spatial scales for biodiversity modeling. Conservation Biology 17: 1712-23.

Countess, R.E., B. Kendrick, and J.A. Trofymow. 1998. Macrofungal diversity in successional Douglas-fir forests. Northwest Science 72: 110-12.

Cowan, I.M. 1945. The ecological relationships of the food of the Columbian black-tailed deer, *Odocoileus hemionus columbianus* (Richardson), in the coastal forest region of southern Vancouver Island, British Columbia. Ecological Monographs 15(2): 110-39.

Coxson, D., S. Stevenson, and J. Campbell. 2003. Short-term impacts of partial cutting on lichen retention and canopy microclimate in an Engelmann spruce-subalpine fir forest in north-central British Columbia. Canadian Journal of Forest Research 33: 830-84.

Crisp, P.N., J.M. Dickinson, and G.W. Gibbs. 1998. Does native invertebrate diversity reflect native plant diversity? A case study from New Zealand and implications for conservation. Biological Conservation 83: 209-20.
Crites, S., and M.R.T. Dale. 1998. Diversity and abundance of bryophytes, lichens, and fungi in relation to woody substrate and successional stage in aspen mixedwood boreal forests. Canadian Journal of Botany 76: 641-51.
Crum, H.A. 1972. The geographic origins of the mosses of North America's eastern deciduous forest. Journal of the Hattori Botanical Laboratory 35: 269-98.
CSP (Scientific Panel for Sustainable Forest Practices in Clayoquot Sound). 1995a. Report 3. First Nations' Perspectives Relating to Forest Practices Standards in Clayoquot Sound. Victoria, BC.
–. 1995b. Report 5. Sustainable Ecosystem Management in Clayoquot Sound: Planning and Practices. Victoria, BC.
Cumming, S.G., P.J. Burton, and B. Klinkenberg. 1996. Boreal mixedwood forests may have no "representative" areas: Some implications for reserve design. Ecography 19: 162-80.
Cushman, S.A., and K. McGarigal. 2004. Hierarchical analysis of forest bird species-environment relationships in the Oregon Coast Range. Ecological Applications 14: 1090-1105.
Daniels, L.D., J. Dobry, K. Klinka, and M.C. Feller. 1997. Determining year of death of logs and snags of *Thuja plicata* in southwestern coastal British Columbia. Canadian Journal of Forest Research 27: 1132-41.
Davies, K. 1995. The Assessment of Cumulative Environmental Effects: Selected Bibliography. Prepared for the Environmental Assessment Branch, Environment Canada, Hull, QC.
Davies, K.F., and C.R. Margules. 1998. Effects of habitat fragmentation on carabid beetles: Experimental evidence. Journal of Animal Ecology 67: 460-71.
Davis, L.S., K.N. Johnson, P.S. Bettinger, and T.E. Howard. 2001. Forest Management to Sustain Ecological, Economic, and Social Values. 4th ed. McGraw-Hill, New York, NY.
De Vrie, H.H., and P.J. Den Boer. 1990. Survival of populations of *Agonum ericetipanz* (*Col: Carabidae*) in relation to fragmentation of habitats. Netherlands Journal of Zoology 40: 484-98.
Delong, D.C. Jr. 1996. Defining biodiversity. Wildlife Society Bulletin 24: 738-49.
Derrail, J.G.B., G.P. Closs, K.J.M. Dickinson, P. Sirivd, B.I.P. Barratt, and B.N. Patrick. 2002. Arthropod morphospecies versus taxonomic species: A case study with Araneae, Coleoptera, and Lepidotera. Conservation Biology 16: 1015-23.
Dettki, H., M. Edman, P-A. Esseen, H. Hedenås, B.G. Jonnson, N. Kruys, J. Moen, and K-E. Renhorn. 1998. Screening for species potentially sensitive to habitat fragmentation. Ecography 21: 649-52.
DeVelice, R.L., and J.R. Martin. 2001. Assessing the extent to which roadless areas complement the conservation of biological diversity. Ecological Applications 11: 1008-18.
Diamond, J.M. 1975. The island dilemma: Lessons of modern biogeographic studies for the design of natural reserves. Biological Conservation 7: 129-46.
–. 1984. "Normal" extinctions of isolated populations. Pp. 191-246 *in* M.H. Nitecki, ed., Extinctions. University of Chicago Press, Chicago, IL.
Dobson, A.P., J.P. Rodriguez, W.M. Roberts, and D.S. Wilcove. 1997. Geographic distribution of endangered species in the United States. Science 275: 550-53.
Douglas, G.W., D. Meidinger, and J. Pojar. 1999. Illustrated Flora of British Columbia. Volumes 1 to 3. Ministry of Environment; Ministry of Forests, Victoria, BC.
Douglas, G.W., G.B. Straley, and D.V. Meidinger. 1989-94. The Vascular Plants of British Columbia: Part 1-4. Ministry of Environment; Ministry of Forests, Victoria, BC.
–. 1998. Rare Native Vascular Plants of British Columbia. Ministry of Environment; Ministry of Forests, Victoria, BC.
Dovetail Consulting. 2001. Summary of the Third Year Critique Workshop on the Weyerhaeuser BC Coastal Forest Project, 16-18 July 2001, Vancouver, BC.
Duffy, D.C., K. Boggs, R.H. Hagenstein, R. Lipkin, and J.A. Michaelson. 1999. Landscape assessment of the degree of protection of Alaska's terrestrial biodiversity. Conservation Biology 13: 1332-43.

Dunn, E.H., D.J.T. Hussell, and D.A. Welsh. 1999. Priority-setting tool applied to Canada's landbirds based on concern and responsibility for species. Conservation Biology 13: 1404-15.

Dunsworth, B.G., S.M. Northway, N.J. Smith, and G. Sutherland. 1997. Modeling the Impact of Harvest Schedules on Biodiversity: 1996 Final Report. FRBC FR-96/97-002.

Durall, D.M., M.D. Jones, E.F. Wright, P. Kroeger, and K.D. Coates. 1999. Species richness of ectomycorrhizal fungi in cutblocks of different sizes in the interior cedar-hemlock forests of northwestern British Columbia: Sporocarps and ectomycorrhizae. Canadian Journal of Forest Research 29: 1322-32.

Ecotrust, Pacific GIS, and Conservation International. 1995. The Rain Forests of Home: An Atlas of People and Place. Part 1: Natural Forests and Native Languages of the Coastal Temperate Rain Forest. Portland, OR.

Edwards, M.E. 1986. Disturbance histories of four Snowdonian woodlots and relation to Atlantic bryophyte distributions. Biological Conservation 37: 301-20.

Edwards Jr., T.C., D.R. Cutler, N.E. Zimmermann, L. Geiser, and J. Alegria. 2005. Model-based stratifications for enhancing the detection of rare ecological events: Lichens as a case study. Ecology 86: 1081-90.

Edwards Jr., T.C., D.R. Cutler, N.E. Zimmermann, L. Geiser, and G.G. Moisen. 2006. Effects of sample survey design on the accuracy of classification tree models in species distribution models. Ecological Modelling 199: 132-41.

Eldridge, D.J. 1996. Distribution and floristics of terricolous lichens in soil crusts in arid and semi-arid New South Wales, Australia. Australian Journal of Botany 44: 581-99.

Esseen, P.A., and K.E. Renhorn. 1998. Edge effects on an epiphytic lichen in fragmented forests. Conservation Biology 12: 1307-17.

Esseen P.A., K.E. Renhorn, and R.B. Petersson. 1996. Epiphytic lichen biomass in managed and old-growth boreal forests: Effect of branch quality. Ecological Applications 6: 228-38.

Evans, J. 1999. Sustainability of forest plantations: A review of evidence and future prospects. International Forestry Review 1(3): 153-62.

Fahrig, L. 1999. Forest loss and fragmentation: Which has the greater effect on persistence of forest-dwelling animals? Pp. 87-95 *in* J. Wisniewski and J.A. Rochelle, eds., Forest Fragmentation: Wildlife and Management Implications. Brill, Leiden, Netherlands.

–. 2002. Effect of habitat fragmentation on the extinction threshold: A synthesis. Ecological Applications 12: 346-53.

Fearnside, P.M., and J. Ferraz. 1995. A conservation gap analysis of Brazil's Amazonian vegetation. Conservation Biology 9: 1134-47.

FEMAT (Forest Ecosystem Management Assessment Team). 1993. Forest Ecosystem Management: An Ecological, Economic, and Social Assessment. US Government Printing Office, Washington, DC.

Ferris, R., and J.W. Humphrey. 1999. A review of potential biodiversity indicators for application in British forests. Forestry 72: 313-28.

Frankel, O.H., and M.E. Soulé. 1981. Conservation and Evolution. Cambridge University Press, Cambridge, UK.

Franklin, J.F. 1993. Preserving biodiversity: Species, ecosystems, or landscapes? Ecological Applications 3: 202-5.

Franklin, J.F., D.R. Berg, D.A. Thornburgh, and J.C. Tappeiner. 1997. Alternative silvicultural approaches to timber harvesting: Variable retention harvest systems. Pp. 111-39 *in* K.A. Kohn and J.F. Franklin, eds., Creating a Forestry for the 21st Century: The Science of Ecosystem Management. Island Press, Washington, DC.

Franklin, J.F., M.E. Harmon, and F.J. Swanson. 1999. Complementary roles of research and monitoring: Lessons from the U.S. LTER Program and Tierra Del Fuego. Pp. 284-91 *in* C. Aguirre-Bravo and C.R. Franco, comps., North American Science Symposium: Toward a Unified Framework for Inventorying and Monitoring Forest Ecosystem Resources, Guadalajara, Mexico, 1-6 November 1998. USDA Forest Service, Proceedings RMRS-P-12, Fort Collins, CO.

Franklin, J.F., T. Spies, D. Perry, M. Harmon, and A. McKee. 1986. Modifying Douglas-fir management for nontimber objectives. Pp. 373-79 *in* C.D. Oliver, D.P. Hanley, and J.A. Johnson, eds., Douglas-Fir: Stand Management for the Future. College of Forest Resources, University of Washington, Seattle, WA.

Frego, K.A., and T.J. Carleton. 1995. Microsite tolerance of four bryophytes in a mature black spruce stand: Reciprocal transplants. The Bryologist 98: 452-58.

Galindo-Leal, C., and F.L. Bunnell. 1995. Ecosystem management: Implications and opportunities for a new paradigm. Forestry Chronicle 71: 601-6.

Gaston, K.J. 1991. The magnitude of global insect species richness. Conservation Biology 5: 283-96.

Gavin, D.G., L.B. Brubaker, and K.P Lertzman. 2003. Holocene fire history of a coastal temperate rain forest based on soil charcoal radiocarbon dates. Ecology 84: 186-210.

Gibb, H.J., J.P. Ball, O. Atlegrim, R.B. Pettersson, J. Hilszczaòski, T. Johansson, and K. Danell. 2006. Effects of landscape composition and substrate availability on saproxylic beetles in boreal forests: A study using experimental logs for monitoring assemblages. Ecography: 191-204.

Gibert, J., D.L. Danielopol, and J.A. Stanford, eds. 1998. Groundwater Ecology. Academic Press, San Diego, CA.

Gignac, L.D., and M.R.T. Dale. 2005. Effects of fragment size and habitat heterogeneity on cryptogam diversity in the low-boreal forest of western Canada. Bryologist 108: 50-66.

Gilfedder, L., and J.B. Kirkpatrick. 1998. Distribution, disturbance tolerance, and conservation of *Stackhousia gunnii* in Tasmania. Australian Journal of Botany 46: 1-13.

Gillis, M.D., A.Y. Omule, and T. Brierley. 2005. Monitoring Canada's forests: The National Forest Inventory. Forestry Chronicle 81: 214-20.

Gilpin, M.E., and M.E. Soulé. 1986. Minimum viable populations: The processes of species extinctions. Pp. 13-34 *in* M.E. Soulé, ed., Conservation Biology: The Science of Scarcity and Diversity. Sinauer Associates, Sunderland, MA.

Glacken, C.J. 1967. Traces on the Rhodian Shore. University of California Press, Berkeley, CA.

Glime, J.M., and D.H. Vitt. 1987. A comparison of bryophyte species diversity and niche structure of montane streams and stream banks. Canadian Journal of Botany 65: 1824-37.

Goodman, D.M., and J.A. Trofymow. 1998. Comparison of communities of ectomycorrhizal fungi in old-growth and mature stands of Douglas-fir on southern Vancouver Island. Northwest Science 72: 91-93.

Goward, T. 1994. Notes on old growth-dependent epiphytic macrolichens in inland British Columbia, Canada. Acta Botanica Fennica 150: 31-38.

–. 1995. Lichens of British Columbia: rare species and priorities for inventory. B.C. Ministry of Forests/Ministry of Environment, Victoria, BC.

Goward, T., and T. Ahti. 1992. Macrolichens and their zonal distribution in Wells Gray Provincial Park and its vicinity, British Columbia, Canada. Acta Botanica Fennica 147: 1-60.

–. 1997. Notes on the distributional ecology of the Cladoniaceae (lichenized Ascomycetes) in temperate and boreal western North America. Journal of the Hattori Botanical Laboratory 82: 143-55.

Goward, T., and A. Arsenault. 1997. Notes on the assessment of lichen diversity in old-growth Engelmann spruce-subalpine fir forests. Pp. 67-78 *in* C. Hollstedt and A. Vyse, eds., Sicamous Creek Silvicultural Systems Project: Workshop Proceedings, 24-25 April 1996, Kamloops, BC. Ministry of Forests, Victoria, BC.

Goward, T., O. Breuss, B. Ryan, B. McCune, H. Sipman, and C. Scheidegger. 1996. Notes on the lichens and allied fungi of British Columbia, III. Bryologist 99: 439-49.

Grant, G.E., and F. Swanson. 1991. Cumulative effects of forest practices. Forest Perspectives 1(4): 9-11.

Green, R.N., and K. Klinka. 1994. A Field Guide for Site Identification and Interpretation for the Vancouver Forest Region. BC Ministry of Forests Land Management Handbook 28. Victoria, BC.

Greenberg, C.H., and T.G. Forrest. 2003. Seasonal abundance of ground-occurring microarthropods in forest and canopy gaps in the southern Appalachians. Southeastern Naturalist 2: 591-608.

Greene, D.F., D.D Kneeshaw, C. Messier, V. Lieffers, D. Cormier, R. Doucet, K.D. Coates, A. Groot, G. Grover, and C. Calogeropoulos. 2002. Modelling silvicultural alternatives to the conventional clearcut/plantation prescription in boreal mixedwood stands (aspen/white spruce/balsam fir). Forestry Chronicle 78: 281-95.

Grove, S. 2002a. The influence of forest management history on the integrity of the saproxylic beetle fauna in an Australian lowland tropical rainforest. Biological Conservation 104: 149-71.

–. 2002b. Saproxylic insect ecology and the sustainable management of forests. Annual Review of Ecology and Systematics 33: 1-23.

Grumbine, R.E. 1994. What is ecosystem management? Conservation Biology 8: 27-39.

Guénette, J-S., and N-A. Villard. 2005. Thresholds in forest bird response to habitat alteration as quantitative targets for conservation. Conservation Biology 19: 1168-80.

Gustafsson, L. 1994. A comparison of biological characteristics and distribution between Swedish threatened and non-threatened forest vascular plants. Ecography 17: 39-49.

Gustafsson, L., A. Fiskesjo, T. Hallinbäck, and T. Ingelog. 1992a. Semi-natural deciduous broadleaved woods in southern Sweden: Habitat factors of importance to some bryophyte species. Biological Conservation 59: 175-81.

Gustafsson, L., A. Fiskesjo, T. Ingelog, B. Pettersson, and G. Thor. 1992b. Factors of importance to some lichen species of deciduous broad-leaved woods in southern Sweden. Lichenologist 24: 255-66.

Haeussler, S., and D. Kneeshaw. 2003. Comparing forest management to natural processes. Pp. 307-68 *in* P.J. Burton, C. Messier, D.W. Smith, and W.L. Adamovich, eds., Towards Sustainable Management of the Boreal Forest. NRC Press, Ottawa, ON.

Halbert, C.L. 1993. How adaptive is adaptive management? Implementing adaptive management in Washington State and British Columbia. Reviews in Fisheries Science 1: 261-83.

Halme, E., and J. Niemelä. 1993. Carabid beetles in fragments of coniferous forest. Annales Zoologici Fennici 30: 17-30.

Hamer, K.C., J.K. Hill, L.A. Lace, and A.M. Langdon. 1997. Ecological and biogeographical effects of forest disturbance on tropical butterflies of Samba, Indonesia. Biogeography 24: 67-75.

Hammon, P.C., and J.C. Miller. 1998. Comparison of the biodiversity of Lepidoptera within three forested ecosystems. Annales of the Entomological Society of America 91: 323-28.

Hammond, H.E.J., D.W. Langor, and J.R. Spence. 2004. Saproxylic beetles (Coleoptera) using *Populus* inboreal aspen stands of western Canada: Spatiotemporal variation and conservation of assemblages. Canadian Journal of Forest Research 34: 1-19.

Hamrick, J.L., J.B. Mitton, and Y.B. Linhart. 1979. Relationships between life history characteristics and electrophoretically detectable genetic variation in plants. Annual Review of Ecology and Systematics 10: 173-200.

Hansen, A.J., J.J. Rotella, M.P.V. Kraska, and D. Brown. 1999. Dynamic habitat and population analysis. An approach to resolve the biodiversity manager's dilemma. Ecological Applications 9: 1459-76.

Hansen, A.J., T.A. Spies, F.J. Swanson, and J.L. Ohmann. 1991. Conserving biodiversity in managed forests. BioScience 41: 382-92.

Harmon, M.E., J.F. Franklin, F.J. Swanson, P. Sollins, S.V. Gregory, J.D. Lattin, N.H. Anderson, S.P. Cline, N.G. Aumen, J.R. Sedell, G.W. Lienkaemper, K. Cromack Jr., and K.W. Cummins. 1986. Ecology of coarse woody debris in temperate ecosystems. Advances in Ecological Research 15: 133-302.

Harper, K.A., D. Lesiieur, Y. Bergeron, and P. Drapeau. 2004. Forest structure and composition at young fire and cut edges in black spruce boreal forest. Canadian Journal of Forest Research 34: 289-302.

Harvey, A.E., M.J. Larsen, and M.F. Jurgensen. 1979. Comparative distribution of ectomycorrhizae in soils of three western Montana forest habitat types. Forest Science 25: 350-58.

Hawkins, C.P., M.L. Murphy, and N.H. Anderson. 1982. Effects of canopy, substrate composition, and gradient on the structure of macroinvertebrate communities in Cascade Range streams of Oregon. Ecology 63: 1840-56.

Hawkins, C.P., M.L. Murphy, N.H. Anderson, and M.A. Wilzbach. 1983. Density of fish and salamanders in relation to riparian canopy and physical habitat in streams of the northwestern United States. Journal of Fisheries and Aquatic Sciences 40: 1173-85.

Hayes, J.F. 1991. How mammals become endangered. Wildlife Society Bulletin 19: 210-15.

Hayes, J.P., S.P. Cross, and P.W. McIntire. 1986. Seasonal variation in mycophagy by the western red-backed vole, *Clethrionomys californicus,* in southwestern Oregon. Northwest Science 60: 250-57.

Hayward, G., and R. Rosentreter. 1994. Lichens as nesting material for northern flying squirrels in the northern Rocky Mountains. Journal of Mammalogy 75: 663-73.

Hazell, P., and L. Gustafsson. 1999. Retention of trees at final harvest: Evaluation of a conservation technique using epiphytic bryophyte and lichen transplants. Biological Conservation 90: 133-42.

Hazell, P., O. Kellner, H. Rydin, and L. Gustaffson. 1998. Presence and abundance of four epiphytic bryophytes in relation to density of aspen (*Populus tremula*) and other stand characteristics. Forest Ecology and Management 107: 147-58.

Heilmann-Clausen, J., and M. Christensen. 2003. Fungal diversity on decaying beech logs: Implications for sustainable forestry. Biodiversity and Conservation 12: 953-73.

Helms, J.A., ed. 1998. Dictionary of Forestry. Society of American Foresters, Bethesda, MD.

Hemstrom, M.A., and J.F. Franklin. 1982. Fire and other disturbances of the forests in Mount Rainier National Park. Quaternary Research 18: 32-51.

Hessburg, P.F., B.G. Smith, and R.B. Salter. 1999. Using Estimates of Natural Variation to Detect Ecologically Important Change in Forest Spatial Patterns: A Case Study, Cascade Range, Eastern Washington. USDA Forest Service, Research Paper PNW-RP-514, Portland, OR.

Heyer, W.R., M.A. Donnelly, R.W. McDiarmid, L.C. Hayek, and M.S. Foster. 1994. Measuring and Monitoring Biological Diversity: Standard Methods for Amphibians. Smithsonian Institution Press, Washington, DC.

Higgins, J.V., T.H. Ricketts, J.D. Parrish, E. Dinerstein, G. Powell, S. Palminteri, J.M. Hoekstra, J. Morrison, A. Tomasek, and J. Adams. 2004. Beyond Noah: Saving species is not enough. Conservation Biology 18: 1672-73.

Hilmo, O., and H. Holien. 2002. Epiphytic lichen response to the edge environment in a boreal *Picea abies* forest in central Norway. Bryologist 105: 48-56.

Hilmo, O., H. Holien, and H. Hytteborn. 2005. Logging strategy influences colonization of common chlorolichens on branches of *Picea abies*. Ecological Applications 15: 983-96.

Hilmo, O., H. Hytteborn, and H. Holien. 2005. Do different logging strategies influence the abundance of epiphytic chlorolichens? Lichenologist 37: 543-53.

Hodgman, T.P., and R.T. Bowyer. 1985. Winter use of arboreal lichens, Ascomycetes, by white-tailed deer, *Odocoileus virginianus,* in Maine. Canadian Field-Naturalist 99: 313-16.

Holien, H. 1996. Influence of site and stand factors on the distribution of crustose lichens of the Caliciales in a suboceanic spruce forest area in central Norway. Lichenologist 28: 315-30.

Holling, C.S., ed. 1978. Adaptive Environmental Assessment and Management. John Wiley and Sons, London, UK.

Hollstedt, C., and A. Vyse, eds. 1997. Sicamous Creek Silvicultural Systems Project: Workshop Proceedings, 24-25 April 1996, Kamloops, BC. Ministry of Forests, Victoria, BC.

Honnay, O., M. Hermy, and P. Coppin. 1999. Effects of area, age, and diversity of forest patches in Belgium on plant species richness, and implications for conservation and reforestation. Biological Conservation 87: 73-84.

Hopkins, P.J., and N.R. Webb. 1984. The composition of the beetle and spider faunas on fragmented heathlands. Journal of Applied Ecology 21: 935-46.

Huggard, D. 2001. Habitat Attributes in the Arrow IFPA Non-Harvestable Landbase. Report to Arrow Innovative Forest Practices Agreement.
–. 2004. Habitat Monitoring 1999 to 2004: Summary and Data Report. Report to Weyerhaeuser, Nanaimo, BC.
–. 2006. Synthesis of Studies of Forest Bird Responses to Partial-Retention Forest Harvesting. Centre for Applied Conservation Research, University of British Columbia, Vancouver, BC. http://www.forestbiodiversityinbc.ca/downloads/asp.
–. 2007. Stand-Level Retention and Forest Birds: A Synthesis of Studies. Centre for Applied Conservation Research, University of British Columbia, Vancouver, BC.
Huggard, D., and J. Herbers. 2000. Squirrel Monitoring Pilot Study, 2000. Final report submitted to Weyerhaeuser, Nanaimo, BC.
Huggard, D., and L. Kremsater. 2005. Linking Multiple Indicators: Summary with Technical Reports Appended (Progress during 2004/05). Report to Forest Sciences Program. http://www.forestbiodiversityinbc.ca.
Huggett, A.J. 2005. The concept and utility of "ecological thresholds" in biodiversity conversation. Biological Conservation 124: 301-10.
Humble, L.M., N.H. Winchester, and R.A. Ring. 2001. The potentially rare and endangered terrestrial arthropods in British Columbia: Revisiting British Columbia's biodiversity. Pp. 101-8 *in* L.M. Darling, ed., At Risk: Proceedings of a Conference on Biology and Management of Species and Habitats at Risk, 15-19 February 1999, Kamloops, BC. Ministry of Environment, Lands, and Parks, Victoria, BC.
Hunt, G.A., and J.M. Trappe. 1987. Seasonal hypogeous sporocarp production in a western Oregon Douglas-fir stand. Canadian Journal of Botany 65: 438-45.
Hunter Jr., M.L. 1993. Natural fire regimes as spatial models for managing boreal forests. Biological Conservation 65: 115-20.
Hylander, K. 2005. Aspect modifies the magnitude of edge effects on bryophyte growth in boreal forests. Journal of Applied Ecology 42: 518-25.
Hyvarinen, M., P. Halonen, and M. Kauppi. 1992. Influence of stand age and structure on the epiphytic lichen vegetation in the middle-boreal forests of Finland. Lichenologist 24: 165-80.
IUCN (World Conservation Union), UNEP (United Nations Environment Programme), and WWF (World Wildlife Fund). 1990. Caring for the World: A Strategy for Sustainability. World Conservation Union, Gland, Switzerland.
Jacobson, S.K., J.K. Morris, J.S. Sanders, E.N. Wiley, M. Brooks, R.E. Bennetts, H.F. Percival, and S. Marynowski. 2006. Understanding barriers to implementation of an adaptive land management program. Conservation Biology 20: 1516-27.
Johnson, D.W., and P.S. Curtis. 2001. Effects of forest management on soil carbon and nitrogen storage: Meta analysis. Forest Ecology and Management 140: 227-38.
Johnson, R.K. 1999. Regional representativeness of Swedish reference lakes. Environmental Management 23: 115 24.
Jones, C.G., J.H. Lawton, and M. Shachak. 1994. Organisms as ecosystem engineers. Oikos 69: 373-86.
Jones, M., and D. Durall. 1997. Effects of several silvicultural systems on ectomycorrhizal diversity and hypogeous sporocarp biomass in the Engelmann Spruce – Subalpine Fir zone: Preliminary results. Pp. 101-9 *in* C. Hollstedt and A. Vyse, eds., Sicamous Creek Silvicultural Systems Project: Workshop Proceedings, 24-25 April 1996, Kamloops, BC. Research Branch, Ministry of Forests, Victoria, BC.
Jonsson, B.G. 1996. Riparian bryophytes of the H.J. Andrews Experimental Forest in the West Cascades, Oregon. Bryologist 99: 226-35.
Jonsson, B.G., N. Kruys, and T. Ranius. 2005. Ecology of species living on dead wood: Lessons for dead wood management. Silva Fennica 39: 289-309.
Jules, E.S. 1998. Habitat fragmentation and demographic changes for a common plant: *Trillium* in old-growth forest. Ecology 79: 1645-56.
Jules, E.S., and B.J. Rathcke. 1999. Mechanisms of reduced *Trillium* recruitment along edges of old-growth forest fragments. Conservation Biology 13: 784-93.

Kay, C.E. 1995. Technical commentary: Aboriginal overkill and native burning: Implications for modern ecosystem management. Western Journal of Applied Forestry 10(4): 121-26.

Keenan, R.J., and J.P. Kimmins. 1993. Ecological effects of clearcutting. Environmental Reviews 1: 121-44.

Kenkel, N.C., and G.E. Bradfield. 1986. Epiphytic vegetation on *Acer macrophyllum*: A multivariate study of species-habitat relationships. Vegetatio 68: 43-53.

Khanna, K.R. 1964. Differential evolutionary activity in bryophytes. Evolution 18: 652-70.

Kimmins, J.P. 1997. Balancing Act: Environmental Issues in Forestry. 2nd ed. UBC Press, Vancouver, BC.

Kimmins, J.P., R.S. Remple, C.V.J. Welham, B. Seely, and K.C.J. Van Rees. 2007. Biophysical sustainability, process-based monitoring and forest ecosystem management decision support systems. Forestry Chronicle 83: 502-14.

Kindvall, O., and I. Ahlen. 1992. Geometrical factors and metapopulation dynamics of the bush cricket, *Metrioptera bicolor* Philippi (*Orthoptera: Tettigoniidae*). Conservation Biology 6: 520-29.

King, J. 1993. Learning to solve the right problems: The case of nuclear power in America. Journal of Business Ethics 13: 105-16.

Kivisto, L., and M. Kuusinen. 2000. Edge effects on the epiphytic lichen flora of *Picea abies* in middle boreal Finland. Lichenologist 32: 387-98.

Klein, G.C. 1989. Effects of forest fragmentation on dung and carrion beetle communities in central Amazonia. Ecology 70: 1715-25.

Kneeshaw, D.D., A. Leduc, C. Messier, P. Drapeau, D. Paré, S. Gauthier, R. Carignan, R. Doucet, and L. Bouthillier. 2000. Developing biophysical indicators of sustainable forest management at an operational scale. Forestry Chronicle 76: 482-93.

Knuchel, H. 1953. Planning and Control in the Managed Forest. Trans. M.L. Anderson. Oliver and Boyd, Edinburgh, Scotland.

Koen, J.H., and T.M. Crowe. 1987. Animal-habitat relationships in the Knysna Forest, South Africa: Discrimination between forest types by birds and invertebrates. Oecologia 72: 414-22.

Koivula, M. 2002. Boreal carabid-beetle (Coleoptera, Carabidae) assemblages in thinned uneven-aged and clear-cut spruce stands. Annales Zoologici Fennici 39: 131-49.

Kondla, N.G., C.S. Crispin, and J.H. Shepard. 2001. Butterflies of conservation interest in Alberta, British Columbia, and the Yukon. Pp. 95-100 *in* L.M. Darling, ed., At Risk: Proceedings of a Conference on Biology and Management of Species and Habitats at Risk, 15-19 February 1999, Kamloops, BC. Ministry of Environment, Lands, and Parks, Victoria, BC.

Kranabetter, J.M., and P. Kroeger. 2001. Ectomycorrhizal mushroom response to partial cutting in a western hemlock-western redcedar forest. Canadian Journal of Forest Research 31: 978-87.

Krcmar, E. 2002. Framework for Economic Analysis of Old Growth Restoration. Report to Weyerhaeuser Adaptive Management Working Group, Nanaimo, BC.

Krebs, C.J. 1989. Ecological Methodology. Harper and Row, New York, NY.

Kremen, C. 1992. Assessing the indicator properties of species assemblages for natural areas monitoring. Ecological Applications 2: 203-17.

Kremen, C., R.K. Colwell, T.L. Erwin, D.D. Murphy, R.F. Noss, and M.A. Sanjayan. 1993. Terrestrial arthropod assemblages: Their use in conservation planning. Conservation Biology 7: 798-808.

Kremsater, L. 2000. The Role of Habitat Representation in Design of Conservation Strategies with Specific Application to the Southern Okanagan of British Columbia. Report to Canadian Wildlife Service, Delta, BC.

Kremsater, L.L., and F.L. Bunnell. 1999. Edges: Theory, evidence, and implications for management of western forests. Pp. 117-53 *in* J.A. Rochelle, L.A. Lehmenn, and J. Wisniewski, eds., Forest Fragmentation: Wildlife and Management Implications. Brill, Leiden, Netherlands.

Kunz, W., and H. Rittel. 1970. Issues as Elements of Information Systems. Technical Report S-78-2, Insitut für Gundlagen Der Planung I.A., Universität Stuttgart, Germany. Also available as Working Paper 131, Institute of Urban and Regional Development, University of California at Berkeley, CA.

Kuusinen, M. 1994a. Epiphytic lichen diversity on *Salix caprea* in old growth southern and middle boreal forests of Finland. Annales Botanica Fennici 31: 77-92.

–. 1994b. Epiphytic lichen flora and diversity on *Populus tremula* in old growth and managed forests of southern and middle boreal Finland. Annales Botanica Fennici 31: 245-60.

Kuusinen, M., and J. Siitonen. 1998. Epiphytic lichen diversity in old-growth and managed *Picea abies* stands in southern Finland. Journal of Vegetation Science 9: 283-92.

La Roi, H.H., and M.H.L. Stringer. 1976. Ecological studies in the boreal spruce-fir forests of the North American taiga. II. Analysis of the bryophyte flora. Canadian Journal of Botany 54: 619-43.

Lambeck, R.J. 1997. Focal species: A multi-species umbrella for nature conservation. Conservation Biology 11: 849-57.

Landres, P.B., J. Verner, and J.W. Thomas. 1988. Ecological uses of vertebrate indicator species: A critique. Conservation Biology 2: 316-28.

Larsson, T-B., ed. 2001. Biodiversity evaluation tools for European forests. Ecological Bulletins 50: 1-236.

Laudenslayer Jr., W.F., P.J. Shea, B.E. Valentine, C.P. Weatherspoon, and T.E. Lisle, tech. co-ords. 2002. Proceedings of the Symposium on the Ecology and Management of Dead Wood in Western Forests, 2-4 November 1999, Reno, NV. USDA Forest Service, General Technical Report PSW-GTR-181, Albany, CA.

Laurance, W.F. 2000. Do edge effects occur over large spatial scales? Trends in Ecology and Evolution 15: 134-35.

Laurance, W.F., and E. Yensen. 1991. Predicting the impacts of edge effects in fragmented habitats. Biological Conservation 55: 77-92.

Leake, R.J. 1994. The biology of myco-heterotrophic ("saprophytic") plants. New Phytologist 127: 171-216.

Lee, K.N. 1993. Compass and Gyroscope: Integrating Science and Politics for the Environment. Island Press, Washington, DC.

–. 1999. Appraising adaptive management. Conservation Ecology 3(2): 3. http//:www.consecol.org/vol3/iss2/art3.

Lee, K.N., and J. Lawrence. 1986. Adaptive management: Learning from the Columbia River basin fish and wildlife program. Environmental Law 16: 431-60.

Lehmkuhl, J.F., L.E. Gould, E. Cazares, and D.R. Hosford. 2004. Truffle abundance and mycophagy by northern flying squirrels in eastern Washington forests. Forest Ecology and Management 200: 49-65.

Lertzman, K.P. 1992. Patterns of gap-phase replacement in a subalpine, old-growth forest. Ecology 73: 657-69.

Lertzman, K.P., J. Fall, and B. Dorner. 1998. Three kinds of heterogeneity in fire regimes: At the crossroads of fire history and landscape ecology. Northwest Science 72: 4-23.

Lertzman, K.P., and C.J. Krebs. 1991. Gap-phase structure of a subalpine old-growth forest. Canadian Journal of Forest Research 21: 1730-41.

Lertzman, K.P., G. Sutherland, A. Inselberg, and S. Saunders. 1996. Canopy gaps and landscape mosaic in a temperate rain forest. Ecology 77: 1254-70.

Lesica, P., B. McCune, S.V. Cooper, and W.S. Hong. 1991. Differences in lichen and bryophyte communities between old-growth and managed second-growth forests in the Swan Valley, Montana. Canadian Journal of Botany 69: 1745-55.

Lewis, T., and A. Inselberg. 2001. Survey of Limestone Species in the Holberg Operation. Report prepared for Western Forest Products, Campbell River Office, Campbell River, BC.

Lindenmayer, D.B. 1999. Future directions for biodiversity conservation in managed forests: Indicator species, impact studies, and monitoring programs. Forest Ecology and Management 115: 277-87.

Lindenmayer, D.B., R.B. Cunningham, C.F. Donnelly, and R. Leslie. 2002. On the use of landscape surrogates as ecological indicators in fragmented forests. Forest Ecology and Management 159: 203-16.

Lindenmayer, D.B., and J. Fischer. 2003. Sound science or social hook: A response to Brooker's application of the focal species approach. Landscape and Urban Planning 62: 149-58.
Lindenmayer, D.B., and J.F. Franklin. 2002. Conserving Forest Biodiversity. Island Press, Washington, DC.
Lindenmayer, D.B., and G. Luck. 2005. Synthesis: Thresholds in conservation and management. Biological Conservation 124: 311-16.
Lindenmayer, D.B., C.R. Margules, and D. Botkin. 2000. Indicators of biodiversity for ecologically sustainable forest management. Conservation Biology 14: 941-50.
Loertscher, M., A. Erhardt, and J. Zettl. 1995. Microdistribution of butterflies in a mosaic-like habitat: The role of nectar sources. Ecography 18: 15-26.
Lofroth, E.C. 1993. Scale dependent analyses of habitat selection by marten in the sub-boreal spruce biogeoclimatic zone, British Columbia. MSc thesis, Simon Fraser University, Burnaby, BC.
Lomolino, M.V. 1994. An evaluation of alternative strategies for building networks of nature reserves. Biological Conservation 69: 243-49.
Lonsdale, W.M. 1999. Concepts and synthesis: Global patterns of plant invasions and the concept of invisibility. Ecology 80: 1522-36.
Lovejoy, T.E. 1985. Forest fragmentation in the Amazon: A case study. Pp. 243-51 *in* H. Messel, ed., The Study of Populations. Pergamon Press, New York, NY.
Ludwig, D., R. Hilborn, and C. Walters. 1993. Uncertainty, resource exploitation, and conservation: Lessons from history. Science 260: 17, 36.
Luoma, D.L. 1989. Biomass and community structure of sporocarps formed by hypogeous ectomycorrhizal fungi within selected forest habitats of the H.J. Andrews Experimental Forest. PhD diss., Oregon State University, Corvallis, OR.
Luoma, D.L., J.L. Eberhart, R. Molina, and M.P. Amaranthus. 2004. Response of ectomycorrhizal fungus sporocarp production to varying levels and patterns of green-tree retention. Forest Ecology and Management 202: 337-54.
Luoma, D.L., R.E. Frankel, and J.M. Trappe. 1991. Fruiting of hypogeous sporocarps in Oregon Douglas-fir forests: Seasonal and habitat variation. Mycologia 83: 335-53.
MacArthur, R.H., and J.W. MacArthur. 1961. On bird species diversity. Ecology 42: 594-98.
Mack, R.N., D. Simberloff, W.M. Lonsdale, H. Evans, M. Clout, and F.A. Bazzaz. 2000. Biotic invasions, causes, epidemiology, global consequences, and control. Ecological Applications 10: 689-710.
Mackey, B.G., H.A. Nix, J.A. Stein, and S.E. Cork. 1989. Assessing the representativeness of the wet tropics of Queensland World Heritage property. Biological Conservation 50: 279-303.
MacKinnon, A., and T. Trofymow. 1998. Structure, processes, and diversity in successional forests of British Columbia. Northwest Science 72(2): 1-3.
Magura, T., B. Tóthmérész, and Z. Elek. 2003. Diversity and composition of carabids during a forestry cycle. Biodiversity and Conservation 12: 73-85.
Marcot, B.G. 1998. Selecting appropriate statistical procedures and asking the right questions: A synthesis. Pp. 129-42 *in* V. Sit and B. Taylor, eds., Statistical Methods for Adaptive Management Studies. Research Branch, Ministry of Forests, Victoria, BC. http://www.for.gov.bc.ca/hfd/pubs/docs/lmh/lmh42.htm.
Margules, C.R. 1993. The Wog Wog habitat fragmentation experiment. Environmental Conservation 19: 316-25.
Margules, C.R., A.O. Nicholls, and R.L. Pressey. 1988. Selecting networks of reserves to maximize biological diversity. Biological Conservation 43: 63-76.
Margules, C.R., and R.L. Pressey. 2000. Systematic conservation planning. Nature 405: 243-53.
Margules, C.R., and J.L. Stein. 1989. Patterns in the distributions of species and the selection of nature reserves: An example from *Eucalyptus* forests in south-eastern New South Wales. Biological Conservation 50: 219-38.
Margules, C.[R.], and M.B. Usher. 1981. Criteria used in assessing wildlife conservation potential: A review. Biological Conservation 21: 79-109.

Marshall, V.G. 2001. Sustainable forestry and soil fauna diversity. Ecoforestry 16(2): 29-34.

Marshall, V.G., H. Setala, and J.A. Trofymow. 1998. Collembolan succession and stump decomposition in Douglas-fir. Northwest Science 72: 84-85.

Martikainen, P., J. Siitonen, P. Punttila, L. Kaila, and J. Rauh. 2000. Species richness of Coleoptera in mature managed and old-growth boreal forests in southern Finland. Biological Conservation 94: 199-209.

Martin, K., K.E.H. Aitken, and K.L. Wiebe. 2004. Nest sites and nest webs for cavity-nesting communities in interior British Columbia, Canada: Nest characteristics and niche partitioning. Condor 106: 5-19.

Marzluff, J.M., M.G. Raphael, and R. Sallabanks. 2000. Understanding the effects of forest management on avian species. Wildlife Society Bulletin 28: 1132-43.

Marzluff, J.M., and M. Restani. 1999. The effects of forest fragmentation on avian nest predation. Pp. 155-69 *in* J.A. Rochelle, L.A. Lehmann, and J. Wisniewski, eds., Forest Fragmentation: Wildlife and Management Implications. Brill, Leiden, Netherlands.

Maser, C., J.M. Trappe, and R.A. Nussbaum. 1978. Fungal-small mammal relationships with emphasis on Oregon coniferous forests. Ecology 59: 799-809.

Massicotte, H.B., R. Molina, L.B. Tackaberry, J.E. Smith, and M.P. Amaranthus. 1999. Diversity and host specificity of ectomycorrhizal fungi retrieved from three adjacent forest sites by five host species. Canadian Journal of Botany 77: 1053-76.

Matlack, G.R. 1994. Plant species migration in a mixed-history forest landscape in eastern North America. Ecology 75: 1491-1502.

Mato, K., and S. Sato. 2004. Impacts of forestry on ant species richness and composition in warm-temperate forests of Japan. Forest Ecology and Management 187: 213-23.

Mayr, E. 1997. This Is Biology: The Science of the Living World. Belknap Press of Harvard University Press, Cambridge, MA.

Mazurek, M.J., and W.J. Zielinksi. 2004. Individual legacy trees influence vertebrate wildlife diversity in commercial forests. Forest Ecology and Management 193: 321-34.

McCracken, J.D. 2005. Where bobolinks roam: The plight of North America's grassland birds. Biodiversity 6(3): 20-29.

McCune, B. 1993. Gradients in epiphytic biomass in three *Pseudotsuga-Tsuga* forests of different ages in Oregon and Washington. Bryologist 96: 405-11.

McCune, B., and L. Geiser. 1997. Macrolichens of the Pacific Northwest. Oregon State University Press, Corvallis, OR.

McGarigal, K., and S.A. Cushman. 2002. Comparative evaluation of experimental approaches to the study of habitat fragmentation effects. Ecological Applications 12: 335-45.

McGarigal, K., and B.J. Marks. 1995. FRAGSTATS. Spatial Pattern Analysis Program for Quantifying Landscape Structure. USDA Forest Service, General Technical Report GTR-PNW-285, Portland, OR.

McKenzie, N.L., L. Belbin, C.R. Margules, and G.J. Keighery. 1989. Selecting representative reserve systems in remote areas: A case study in the Nullarbor Region, Australia. Biological Conservation 50: 239-61.

McLean, I.F.G., and M.C.D. Speight. 1993. Saproxylic invertebrates: The European context. Pp. 21-32 *in* K.J. Kirby and C.M. Drake, eds., Dead Wood Matters. English Nature Science 7. British Nature, Peterborough, UK.

McNeeley, J.A., K.R. Miller, M.V. Reid, R.A. Mittermeier, and T.B. Werner. 1990. Conserving the World's Biological Diversity. International Union for Conservation of Nature and Natural Resources, World Resources Institute, Conservation International, World Wildlife Fund (US), and World Bank.

McRae, D.J., L.C. Duchesne, B. Freedman, T.J. Lynham, and S. Woodley. 2001. Comparisons between wildfire and forest harvesting and their implications in forest management. Environmental Reviews 9: 223-60.

Meffe, G.K., and C.R. Carroll, eds. 1994. Principles of Conservation Biology. Sinauer Associates, Sunderland, MA.

Meidinger, D., and J. Pojar. 1991. Ecosystems of British Columbia. Special Report Series No. 6. Ministry of Forests, Victoria, BC.

Meier, A.J., S.P. Bratton, and D.C. Duffy. 1995. Possible ecological mechanisms for loss of vernal-herb diversity in logged eastern deciduous forests. Ecological Applications 5: 935-46.

Meijer, W. 1970. Temperature effects on tropical ectomycorrhizae. Malayan Forester 33: 204-29. [Original not seen; cited from Molina et al. 1993.]

Mellen, K., and A. Ager. 2002. A coarse wood dynamics model for the western Cascades. Pp. 503-16 *in* W.F. Laudenslayer Jr., P.J. Shea, B.E. Valentine, C.P. Weatherspoon, and T.E. Lisle, tech. co-ords., Proceedings of the Symposium on the Ecology and Management of Dead Wood in Western Forests, 2-4 November 1999, Reno, NV. USDA Forest Service, General Technical Report PSW-GTR-181, Albany, CA.

Mellen, K., B. Marcot, J. Ohmann, K.L. Waddell, E.A. Willhite, B.B. Hostetler, S.A. Livingston, and C. Ogden. 2002. DecAID: A decaying wood advisory model for Oregon and Washington. Pp. 527-33 *in* W.F. Laudenslayer Jr., P.J. Shea, B.E. Valentine, C.P. Weatherspoon, and T.E. Lisle, tech. Co-ords., Proceedings of the Symposium on the Ecology and Management of Dead Wood in Western Forests, 2-4 November 1999, Reno, NV. USDA Forest Service, General Technical Report PSW-GTR-181, Albany, CA.

Mendel, L.C., and J.B. Kirkpatrick. 2002. Historical progress of biodiversity conservation in the protected-area system of Tasmania, Australia. Conservation Biology 16: 1520-29.

Meretsky, V.J., D.L. Wegner, and L.E. Stevens. 2000. Balancing endangered species and ecosystems: A case study of adaptive management in the Grand Canyon. Environmental Management 25: 579-86.

Merriam, G. 1998. Biodiversity at the population level: A vital paradox. Pp. 45-65 *in* F.L. Bunnell and J.F. Johnson, eds., The Living Dance: Policy and Practices for Biodiversity in Managed Forests. UBC Press, Vancouver, BC.

Mills, S.L., M.E. Soulé, and D.F. Doak. 1993. The keystone-species concept in ecology and conservation. Bioscience 43: 219-24.

Mitchell, S.J., and W.J. Beese. 2002. The variable retention system: Reconciling variable retention with the principles of silvicultural systems. Forestry Chronicle 78: 397-403.

Mitchell, S.J., T. Hailemariam, and Y. Kulis. 2001. Empirical modeling of cutblock edge windthrow risk on Vancouver Island, Canada, using stand level information. Forest Ecology and Management 154: 117-30.

Mittelbach, C.G., C.F. Steiner, S.M. Scheiner, K.L. Gross, H.L. Reynolds, R.B. Waide, M.R. Willig, S.I. Dodson, and L. Gough. 2001. What is the observed relationship between species richness and productivity? Ecology 82: 2381-96.

Moen, J., and B.G. Jonsson. 2003. Edge effects on liverworts and lichens in forest patches in a mosaic of boreal forest and wetlands. Conservation Biology 17: 380-88.

Moldenke, A.R. 1989. Creatures of the Forest Soil [educational video]. Oregon State University Media Center, Corvallis, OR. [Original not seen; cited from Moldenke 1999.]

–. 1999. Soil-dwelling arthropods: Their diversity and functional roles. Pp. 33-44 *in* R.T. Meurisse, W.G. Ypsilantis, and C. Seybold, tech. eds., Proceedings: Pacific Northwest Forest and Rangeland Soil Organism Symposium. USDA Forest Service, General Technical Report PNW-GTR-461, Portland, OR.

Moldenke, A., M. Pajutee, and E. Ingham. 1996. The functional roles of forest soil arthropods: The soil is a lively place. Pp. 7-22 *in* R.F. Powers, D.L. Hauxwell, and G.M. Nakamura, tech. eds., Proceedings of the California Forest Soils Council Conference on Forest Soils Biology and Forest Management. USDA Forest Service, General Technical Report PSW-GTR-178, Albany, CA.

Molina, R., B.G. Marcot, and R. Lesher. 2006. Protecting rare, old-growth, forest-associated species under the survey and manage program guidelines of the Northwest Forest Plan. Conservation Biology 20: 306-18.

Molina, R., T. O'Dell, D. Luoma, M. Amaranthus, M. Castellano, and K. Russell. 1993. Biology, Ecology, and Social Aspects of Wild Edible Mushrooms in the Forests of the Pacific Northwest: A Preface to Managing Commercial Harvest. USDA Forest Service, General Technical Report PNW-GTR-309, Portland, OR.

Molnar, J., M. Marvier, and P. Kareiva. 2004. The sum is greater than the parts. Conservation Biology 18: 1670-71.

Moola, F.M., and L. Vasseur. 2004. Recovery of late-seral vascular plants in a chronosequence of post-clearcut forest stands in coastal Nova Scotia, Canada. Plant Ecology 172: 183-97.

Morrison, M.L., and M.G. Raphael. 1993. Modeling the dynamics of snags. Ecological Applications 3: 322-30.

Muhle, H., and F. LeBlanc. 1975. Bryophyte and lichen succession on decaying logs. I. Analysis along an evapotranspirational gradient in eastern Canada. Journal of the Hattori Botanical Laboratory 39: 1-33.

Mulder, B.S., B.R. Noon, T.A. Spies, M.G. Raphael, C.J. Palmer, A.R. Olsen, G.H. Reeves, and H.H. Welsh. 1999. The Strategy and Design of the Effectiveness Monitoring Program for the Northwest Forest Plan. USDA Forest Service, General Technical Report PNW-GTR-437, Portland, OR.

Murcia, C. 1995. Edge effects in fragmented forests: Implications for conservation. Trends in Ecology and Evolution 10: 58-62.

Murphy, M.L., C.P. Hawkins, and N.H. Anderson. 1981. Effects of canopy modification and accumulated sediment on stream communities. Transactions of the American Fisheries Society 110: 469-78.

Nadkarni, N.M. 1984a. Biomass and mineral capital of epiphytes in an *Acer macrophyllum* community of a temperate moist coniferous forest, Olympic Peninsula, Washington State. Canadian Journal of Botany 62: 2223-28.

–. 1984b. Epiphyte biomass and nutrient capital of a neotropical elfin forest. Biotropica 16: 249-56.

Naeem, S., L.J. Thompson, S.P. Lawler, J.H. Lawton, and R.M. Woodfin. 1994. Declining biodiversity can alter the performance of ecosystems. Nature 368: 735-37.

Naiman, R.J., J.M. Melillo, and J.M. Hobbie. 1986. Ecosystem alteration of boreal forest streams by beaver (*Castor canadensis*). Ecology 67: 1254-69.

Namkoong, G. 1993. Integrating science and management at the University of British Columbia. Journal of Forestry 91(10): 24-27.

–. 1998. Genetic diversity for forest policy and management. Pp. 30-44 *in* F.L. Bunnell and J.F. Johnson, eds., The Living Dance: Policy and Practices for Biodiversity in Managed Forests. UBC Press, Vancouver, BC.

Nash III, T.H. 1996. Nitrogen, its metabolism, and potential contribution to ecosystems. Pp. 121-35 *in* T.H. Nash III, ed., Lichen Biology. Cambridge University Press, Cambridge, UK.

Nelson, J.R., and T.A. Leege. 1982. Nutritional requirements and food habits. Pp. 323-68 *in* J.W. Thomas and D.E. Toweill, eds., Elk of North America: Ecology and Management. Stackpole Books, Harrisburg, PA.

Newmaster, S.G., R. Belland, A. Arsenault, and D.H. Vitt. 2003. Patterns of bryophyte diversity in humid coastal and inland cedar-hemlock forests of British Columbia. Environmental Reviews 11: S159-89.

Niemelä, J. 1997. Invertebrates and boreal forest management. Conservation Biology 11: 601-10.

Niemelä, J., Y. Haila, E. Halme, T. Pajunen, and P. Punttila. 1992. Small-scale heterogeneity in the spatial distribution of carabid beetles in the southern Finnish taiga. Journal of Biogeography 19: 173-81.

Niemelä, J., Y. Haila, and P. Punttila. 1996. The importance of small-scale heterogeneity in boreal forests: Variation in diversity in forest floor invertebrates across the succession gradient. Ecography 19: 352-68.

Niemelä, J., D. Langor, and J.R. Spence. 1993. Effects of clear-cut harvesting on boreal ground beetle assemblages *(Coleoptera: Carabidae)* in western Canada. Conservation Biology 7: 551-61.

Nilsson, S.G., U. Arup, R. Baranowski, and S. Ekman. 1994. Tree-dependent lichens and beetles as indicators in conservation forests. Conservation Biology 9: 1208-15.

Nilsson, S.G., and R. Baranowski. 1997. Habitat predictability and the occurrence of wood beetles in old growth beech forests. Ecography 20: 491-98.

Noble, W.J. 1982. The lichens of the coastal Douglas-fir dry subzone of British Columbia. PhD diss., University of British Columbia, Vancouver, BC.

Norse, E.W. 1990. Ancient Forests of the Pacific Northwest. Island Press, Washington, DC.

Norton, B.G. 1986. On the inherent danger in undervaluing species. Pp. 110-37 *in* B.G. Norton, ed., The Preservation of Species: The Value of Biological Diversity. Princeton University Press, Princeton, NJ.

Norton, D.A. 2002. Edge Effects in a Lowland Temperate New Zealand Rainforest. DOC Science Internal Series 27. Department of Conservation, Wellington, New Zealand.

Noss, R.F. 1990. Indicators for monitoring biodiversity: A hierarchical approach. Conservation Biology 4: 356-64.

–. 1993. A conservation plan for the Oregon Coast Range: Some preliminary suggestions. Natural Areas Journal 13: 276-90.

–. 1996. Protected areas: How much is enough? Pp. 91-120 *in* R.G. Wright, ed., National Parks and Protected Areas: Their Role in Environmental Protection. Blackwell Science, Cambridge, MA.

–. 1999. Assessing and monitoring forest biodiversity: A suggested framework and indicators. Forest Ecology and Management 115: 135-46.

Noss, R.F., and A.Y. Cooperrider. 1994. Saving Nature's Legacy: Protecting and Restoring Biodiversity. Island Press, Washington, DC.

Noss, R.F., J.R. Strittholt, L. Vance-Borland, K.C. Carroll, and P.A. Frost. 1999. A conservation plan for the Klamath-Siskiyou ecoregion. Natural Areas Journal 19: 392-411.

Nyberg, J.B., F.L. Bunnell, D.W. Janz, and R.M. Ellis. 1986. Managing Young Forests as Black-Tailed Deer Winter Ranges. Land Management Report 37. Ministry of Forests, Victoria, BC.

Nyberg, J.B., R.S. McNay, M.D. Kirchoff, R.D. Forbes, F.L. Bunnell, and E.L. Richardson. 1989. Integrated Management of Timber and Deer: Coastal Forests of British Columbia and Alaska. USDA Forest Service, General Technical Report PNW-GTR-226, Portland, OR.

Ødergaard, F., O.L. Diserud, S. Engen, and K. Aagaard. 2002. The magnitude of local host specificity for phytophagous insects and its implications for estimates of global species richness. Conservation Biology 14: 1182-86.

O'Hara, K.L., R.S. Seymour, S.D. Tesch, and J.M. Guldin. 1994. Silviculture and our changing profession: Leadership for shifting paradigms. Journal of Forestry 92: 8-13.

Okland, B. 1996. Unlogged forests: Important sites for preserving the diversity of mycetophilids (*Diptera: Sciaroidea*). Biological Conservation 76: 297-310.

Okland, B., A. Bakke, S. Hagvar, and T. Kvamme. 1996. What factors influence the diversity of saproxylic beetles? A multiscale study from a spruce forest in southern Norway. Biodiversity and Conservation 5: 75-100.

Oliver, C.D. 1992. A landscape approach: Achieving and maintaining biodiversity and economic productivity. Journal of Forestry 90(9): 20-25.

Oliver, I., A.J. Beattie, and A. York. 1998. Spatial fidelity of plant, vertebrate, and invertebrate assemblages in multiple use forest in eastern Australia. Conservation Biology 12: 822-35.

O'Neill, R.V., J.R. Kummel, R.H. Gardner, G. Sugihara, B. Jackson, D.L. DeAngelis, B.T. Milne, M.G. Turner, B. Zygmunt, S.W. Christensen, V.H. Dale, and R.L. Graham. 1988. Indices of landscape pattern. Landscape Ecology 1: 153-62.

OTA (Office of Technological Assessment). 1987. Technologies to Maintain Biological Diversity. Report OTA-F-330. Government Printing Office, Washington, DC.

Outerbridge, R., J.A. Trofymow, and H. Kope. 2001. Diversity of Ectomycorrhizae on Planted Seedlings in Variable Retention Forestry Sites: Results of Pilot Study and Description of Design of Future Experiment. Report to Weyerhaeuser Adaptive Management Working Group, Weyerhaeuser BC Coastal Group, Nanaimo, BC.

Outerbridge, R.A., and J.A. Trofymow. 2004. Diversity of ectomycorrhizae on experimentally planted Douglas-fir seedlings in variable retention forestry sites in southern Vancouver Island. Canadian Journal of Botany 82: 1671-1681.

Ovaska, K., and L. Sopuck. 2005. Terrestrial Gastropods as Indicators for Monitoring Ecological Effects of Variable-Retention Logging Practices: Synthesis of Field Data, Fall 1999-

2003. Report to Weyerhaeuser Adaptive Management Working Group, Weyerhaeuser BC Coastal Group, Nanaimo, BC.

Pajunen, T., Y. Haila, E. Halme, J. Niemelä, and P. Punttila. 1995. Ground-dwelling spiders (*Arachnida, Araneae*) in fragmented old forests and surrounding managed forests in southern Finland. Ecography 18: 62-72.

Parminter, J. 2004. Natural fire regimes in British Columbia and the summer of 2003. Botanical Electronic News 329. http://www.ou.edu/cas/botany-micro/ben/ben329.html.

Parsons, G.L., G. Cassis, A.R. Moldenke, J.D. Lattin, N.H. Anderson, J.C. Miller, P. Hammond, and T.D. Schowalter. 1991. Invertebrates of the H.J. Andrews Forest, Cascade Range, Western Oregon. Part V: An Annotated List of Insects and Other Arthropods. USDA Forest Service, General Technical Report PNW-GTR-290, Portland, OR.

Parsons, W.F.J., S.L. Miller, and D.H. Knight. 1994. Root-gap dynamics in lodgepole pine forest: Ectomycorrhizal and nonmycorrhizal fine root activity after experimental gap formation. Canadian Journal of Forest Research 24: 1451-1538.

Pauchard, A., and P.B. Alaback. 2004. Influence of elevation, land use, and landscape context on patterns of alien plant invasions along roadsides in protected areas of south-central Chile. Conservation Biology 18: 238-48.

Pearsall, I.A. 2005. Study to Assess Ground Beetles (*Coleoptera: Carabidae*) as Ecological Indicators in Weyerhaeuser's West Vancouver Island Operational Sites. Year 4 final report to Weyerhaeuser Coastal Group, Nanaimo, BC.

Pearson, D. 1989. The Natural House Book: Creating a Healthy, Harmonious, and Ecologically Sound Home Environment. Simon and Schuster, New York, NY.

Pellikka, J., H. Rita, and H. Lindén. 2005. Monitoring wildlife richness: Finnish applications based on wildlife triangle censuses. Annales Zoologici Fennici 42: 123-34.

Perera, A.H., L.J. Buse, and M.G. Weber, eds. 2004. Emulating Natural Forest Landscape Disturbances: Concepts and Applications. Columbia University Press, New York, NY.

Perry, D.A., R. Molina, and M.P. Amaranthus. 1987. Mycorrhizae, mycorrhizospheres, and reforestation: Current knowledge and research needs. Canadian Journal of Forest Research 17: 929-40.

Perry, J., and R. Muller. 2002. Riparian Restoration Program. Weyerhaeuser Forest Project Technical Project Summary, Report 3. Nanaimo, BC.

Pharo, E.J., A.J. Beattie, and D. Binns. 1998. Vascular plant diversity as a surrogate for bryophyte and lichen diversity. Conservation Biology 13: 282-92.

Pike, L.H. 1978. The importance of epiphytic lichens in mineral cycling. Bryologist 81: 247-57.

Pike, L.H., W.C. Denison, D.M. Tracey, M.A. Sherwood, and R.F. Rhodes. 1975. Floristic survey of epiphytic lichens and bryophytes growing on old-growth conifers in western Oregon. Bryologist 78: 389-402.

Pilz, D., N.S. Weber, M.C. Carter, C.G. Parks, and R. Molina. 2004. Productivity and diversity of morel mushrooms in healthy, burned, and insect damaged forests of northeastern Oregon. Forest Ecology and Management 198: 367-86.

Plochmann, R. 1989. The forests of central Europe: A changing view. Pp. 1-9 *in* Oregon's Forestry Outlook: An Uncertain Future. College of Forestry, Oregon State University, Corvallis, OR.

Pojar, J., K. Klinka, and D.V. Meidinger. 1987. Biogeoclimatic ecosystem classification in British Columbia. Forest Ecology and Management 22: 119-54.

Pojar, J., and A. MacKinnon, eds. 1994. *Plants of the Pacific Northwest coast.* Lone Pine, Vancouver, BC.

Poole, K.G., A.D. Porter, A. de Vries, C. Maundrell, S.D. Grindal, and C.C. St. Clair. 2004. Suitability of a young deciduous-dominated forest for American marten and the effects of forest removal. Canadian Journal of Zoology 82: 423-35.

Powell, G.V.N., J. Barborak, and M. Rodrigues. 2000. Assessing representativeness of protected natural areas in Costa Rica for conserving biodiversity: A preliminary gap analysis. Biological Conservation 93: 35-41.

Pressey, R.L. 2004. Conservation planning and biodiversity: Assembling the best data for the job. Conservation Biology 18: 1677-81.

Pressey, R.L., T.C. Hager, K.M. Ryan, J. Schwarz, S. Wall, S. Ferrier, and P.M. Creaser. 2000. Using abiotic data for conservation assessments over extensive regions: Quantitative methods applied across New South Wales, Australia. Biological Conservation 96: 55-82.

Pressey, R.L., and A.O. Nicholls. 1989. Application of a numerical algorithm to the selection of reserves in semi-arid New South Wales. Biological Conservation 50: 263-78.

Preston, M.I., and R.W. Campbell. 2001. Forest Owls as Indicators of Retention of Biodiversity in Coastal British Columbia. Progress report for Weyerhaeuser, Nanaimo, BC.

–. 2005. A 5-Year Summary of Summer Bird Surveys on Vancouver Island, the Sunshine Coast, and the Queen Charlotte Islands. Progress report for Weyerhaeuser, Nanaimo, BC.

Preston, M.I., and A.S. Harestad. 2007. Community and species responses by birds to group retention in a coastal temperate forest on Vancouver Island, British Columbia. Forest Ecology and Management 243: 156-67.

Prezio, J.R., M.W. Lankester, R.A. Lautenschlager, and F.W. Bell. 1999. Effects of alternative conifer release treatments on terrestrial gastropods in regenerating spruce plantations. Canadian Journal of Forest Research 29: 1141-48.

Province of British Columbia. 1994. Cave/karst Management Handbook for the Vancouver Forest Region. Ministry of Forests, Victoria, BC.

–. 2000. Vancouver Island Summary Land Use Plan. Government of British Columbia, Victoria, BC.

Ralph, C.J., J.R. Sauer, and S. Droege. 1995. Monitoring Birds by Point Counts. USDA Forest Service, General Technical Report PSW-GTR-149, Albany, CA.

Rambo, T.R., and P.S. Muir. 1998. Forest floor bryophytes of *Pseudostuga menziesii-Tsuga heterophylla* stands in Oregon: Influences of substrate and overstory. Bryologist 10: 116-30.

Rebelo, A.G., and W.R. Siegfried. 1992. Where should nature reserves be located in the Cape floristic region, South Africa? Models for the spatial configuration of a reserve network aimed at maximizing the protection of floral diversity. Conservation Biology 6: 243-52.

Rempel, R.S., D.W. Andison, and S.J. Hannon. 2004. Guiding principles for developing an indicator and monitoring framework. Forestry Chronicle 80: 82-90.

Rhoades, F.R. 1995. Non-vascular epiphytes in forest canopies: World wide distribution, abundance, and ecological roles. Pp. 353-408 *in* M.D. Lowman and N.M. Nadkarni, eds., Forest Canopies. Academic Press, Toronto, ON.

Richards, W.H., D.O. Wallin, and A.H. Schumaker. 2002. An analysis of late-seral forest connectivity in western Oregon, U.S.A. Conservation Biology 16: 1409-21.

Richardson, D.H.S. 1975. The Vanishing Lichens: Their History, Biology, and Importance. David and Charles, Newton Abott, UK.

Richardson, D.H.S., and C.M. Young. 1977. Lichens and vertebrates. Pp. 121-44 *in* M.R.D. Seaward, ed., Lichen Ecology. Academic Press, London, UK.

Rikkinen, J. 1995. What's behind the pretty colours? A study on the photobiology of lichens. Bryobrothera 4: 1-239.

–. 2003a. Calicioid lichens and fungi in the forests and woodlands of western Oregon. Acta Botanica Fennica 175: 1-41.

–. 2003b. *Chaenothecopsis nigripunctata,* a remarkable new species of resinicolous Mycocaliciaceae from western North America. Mycologia 95: 98-103.

–. 2003c. New resinicolous ascomycetes from beaver scars in western North America. Annales Botanici Fennici 40: 443-50.

Ritland, K., L.A. Dupuis, F.L. Bunnell, W.L.Y. Hung, and J.E. Carlson. 2000. Phylogeography of the tailed frog (*Ascaphus truei*) in British Columbia. Canadian Journal of Zoology 78: 1749-58.

Ritland, K., C. Newton, and H.D. Marshall. 2001. Inheritance and population structure of the white-phased "Kermode" black bear. Current Biology 11: 1468-72.

Rittel, H. 1972. On the planning crisis: Systems analysis of the "first and second generations." Bedrifts Okonomen 8: 390-96.

Rittel, H.W.J., and M.M. Webber. 1973. Dilemmas in a general theory of planning. Policy Sciences 4: 155-69.
–. 1984. Planning problems are wicked problems. Pp. 135-44 *in* N. Cross, ed., Developments in Design Methodology. John Wiley and Sons, New York, NY.
Roberge, J.M., and P. Angelstam. 2004. Usefulness of the umbrella species concept as a conservation tool. Conservation Biology 18: 76-85.
Rominger, E.M., L. Allen-Johnson, and J.L. Oldemeyer. 1994. Arboreal lichen in uncut and partially cut subalpine fir stands in woodland caribou habitat, northern Idaho and southwestern British Columbia. Forest Ecology and Management 70: 195-202.
Rominger, E.M., and J.L. Oldemeyer. 1989. Early-winter habitat of woodland caribou, Selkirk Mountains, British Columbia. Journal of Wildlife Management 53: 238-43.
Rose, F. 1976. Lichenological indicators of age and environmental continuity in woodlands. Pp. 279-307 *in* D.H. Brown, D.L. Hawksworth, and R.H. Bailey, eds., Lichenology: Progress and Problems. Systematics Association Special Volume 8. Academic Press, New York, NY.
–. 1992. Temperate forest management: Its effects on bryophyte and lichen floras and habitats. Pp. 211-33 *in* J.W. Bates and A.M. Farmer, eds., Bryophytes and Lichens in a Changing Environment. Clarendon Press, Oxford, UK.
Rudnicky, T.C., and M.L. Hunter. 1993. Avian nest predation in clearcuts, forest, and edges in a forest-dominated landscape. Journal of Wildlife Management 57: 358-64.
Ryan, M.W. 1991. Distribution of bryophytes and lichens in Garry oak. MSc thesis, University of Victoria, Victoria, BC.
–. 1996. Bryopyhtes of British Columbia: Rare Species and Priorities for Inventory. Ministry of Forests, Victoria, BC.
Rydin, H., M. Diekmann, and T. Hallinbäck. 1997. Biological characteristics, habitat associations, and distribution of macrofungi in Sweden. Conservation Biology 11: 628-40.
Ryti, R.T. 1992. Effects of the focal taxon on selection of nature reserves. Ecological Applications 2: 404-10.
Saab, V. 1999. Importance of spatial scale to habitat use by breeding birds in riparian forests: A hierarchical analysis. Ecological Applications 9: 135-51.
Sadler, K. 2004. Vegetation Monitoring in Coastal Douglas-Fir Zone Forests of Vancouver Island: Influence of Age Class and Edge Proximity on Vascular Plant and Bryophyte Distributions. Prepared for Weyerhaeuser, Nanaimo, BC.
Sætersdal, M., and H.J.B. Birks. 1993. Assessing the representativeness of nature reserves using multivariate analysis: Vascular plants and breeding birds in deciduous forests, western Norway. Biological Conservation 65: 121-32.
Sætersdal, M., J.M. Line, and H.J.B. Birks. 1993. How to maximize biological diversity in nature reserve selection: Vascular plants and breeding birds in deciduous woodlands, western Norway. Biological Conservation 66: 131-38.
Salafsky, N., R. Margoluis, K.H. Redford, and J.G. Robinson. 2002. Improving the practice of conservation: A conceptual framework and research agenda for conservation science. Conservation Biology 16: 1469-79.
Särndal, C.E., B. Swensson, and J. Wretman. 1992. Model Assisted Survey Sampling. Springer-Verlag, New York, NY.
Schmidt, R.L. 1960. Factors controlling the distribution of Douglas-fir in coastal British Columbia. Quarterly Journal of Forestry 54: 155-60.
–. 1970. A history of pre-settlement fires on Vancouver Island as determined from Douglas-fir ages. Pp. 107-8 *in* J.H.G. Smith and J. Worrall, eds., Tree-Ring Analysis with Special Reference to Northwest America. Faculty of Forestry Bulletin Number 7. University of British Columbia, Vancouver, BC.
Schmiegelow, F.K., C.S. Machtans, and S.J. Hannon. 1997. Are boreal birds resilient to forest fragmentation? An experimental study of short-term community responses. Ecology 78(6): 1914-32.
Schmiegelow, F.K.A., and M. Mönkkönen. 2002. Habitat loss and fragmentation in dynamic landscapes: Avian perspectives from the boreal forest. Ecological Applications 12: 375-89.

Schofield, W.B. 1976. Bryophytes of British Columbia III: Habitat and distributional information for selected mosses. Syesis 9: 317-54.
–. 1988. Bryogeography and the bryophytic characterization of biogeoclimatic zones of British Columbia, Canada. Canadian Journal of Botany 66: 2673-86.
–. 1994. Bryophytes of Mediterranean climates in British Columbia. Hikobia 11: 407-14.
Schowalter, T.D. 1995. Canopy arthropod communities in relation to forest age and alternative harvest practices in western Oregon. Forest Ecology and Management 78: 115-25.
Schowalter, T.D., S.G. Stafford, and R.L. Slagle. 1988. Arboreal arthropod community structure in an early successional coniferous forest ecosystem in western Oregon. Great Basin Naturalist 48: 327-33.
Schultz, C.B. 1998. Dispersal behaviour and its implications for reserve design in a rare Oregon butterfly. Conservation Biology 12: 284-92.
Scott, J.A. 1997. The Butterflies of North America. Stanford University Press, Stanford, CA.
Scott, J.M., B. Csuti, J.D. Jacobi, and J.E. Estes. 1987. Species richness: A geographic approach to protecting future biological diversity. BioScience 37: 782-88.
Scott, J.M., F.W. Davis, R.G. McGhie, R.G. Wright, C. Groves, and J. Estes. 2001a. Nature reserves: Do they capture the full range of America's biological diversity? Ecological Applications 11: 999-1007.
Scott, J.M., M. Murray, R.G. Wright, B. Csuti, P. Morgan, and R.L. Pressey. 2001b. Representation of natural vegetation in protected areas: Capturing the geographic range. Biodiversity and Conservation 10: 1297-1301.
Selva, S.B. 1994. Lichen diversity and stand continuity in the northern hardwoods and spruce-fir forests of northern New England and western New Brunswick. Bryologist 97: 424-29.
–. 2003. Using calicioid lichens and fungi to assess ecological continuity in the Acadian Forest Ecoregion of the Canadian Maritimes. Forestry Chronicle 79: 550-58.
Shindler, B.A., and L.A. Cramer. 1999. Shifting public values in forest management: Making sense of wicked problems. Western Journal of Applied Forestry 14(1): 11-17.
Shirazi, A.M., P.S. Muir, and B. McCune. 1996. Environmental factors influencing the distribution of lichens *Lobaria oregana* and *L. pulmonaria*. Bryologist 99: 12-18.
Sierra, R., F. Campos, and J. Chamberlin. 2002. Assessing biodiversity conservation priorities: Ecosystem risk and representativeness in continental Ecuador. Landscape and Urban Planning 59: 95-110.
Siira-Pietikainen, A., J. Haimi, and J. Siitonen. 2003. Short-term responses of soil macroarthropod community to clear felling and alternative forest regeneration methods. Forest Ecology and Management 172: 339-53.
Siitonen, P., A. Lehtinen, and M. Siitonen. 2005. Effects of forest edges on the distribution, abundance, and regional persistence of wood-rotting fungi. Conservation Biology 19: 250-60.
Sillett, S.C. 1994. Growth rates of two epiphytic cyanolichen species at the edge and in the interior of a 700-year-old Douglas-fir forest in the western Cascades of Oregon. Bryologist 97: 321-24.
–. 1995. Branch epiphyte assemblages in the forest interior and on the clearcut edge of a 700-year-old Douglas-fir canopy in western Oregon. Bryologist 98: 301-12.
Sillett, S.C., and B. McCune. 1998. Survival and growth of cyanolichen transplants in Douglas-fir forest canopies. Bryologist 101: 20-31.
Sillett, S.C., B. McCune, J.E. Peck, T.R. Rambo, and A. Ruchty. 2000. Dispersal limitations of epiphytic lichens result in species dependent on old-growth forests. Ecological Applications 10: 789-99.
Sillett, S.C., and P.N. Neitlich. 1996. Emerging themes in epiphyte research in westside forests with special reference to cyanolichens. Northwest Science 70 (special issue): 54-60.
Simberloff, D. 1986. Design of nature reserves. Pp. 315-37 *in* M.B. Usher, ed., Wildlife Conservation Evaluation. Chapman and Hall, London, UK.
–. 1998. Flagships, umbrellas, and keystones: Is single species management passé in the landscape era? Biological Conservation 83: 247-57.

Simberloff, D.S., and L.C. Abele. 1976. Island biogeography theory and conservation practice. Science 191: 285-86.

–. 1982. Refuge design and island biogeography theory: Effects of fragmentation. American Naturalist 120: 41-50.

Sippola, A-L., and P. Renvall. 1999. Wood-decomposing fungi and seed-tree cutting: A 40-year perspective. Forest Ecology and Management 115: 183-201.

Sippola, A-L., J. Siitonen, and P. Punttila. 2002. Beetle diversity in timberline forests: A comparison between old-growth and regeneration areas in Finnish Lapland. Annales Zoologici Fennici 39: 69-86.

Sisk, T.D., and C.R. Margules. 1993. Habitat edges and restoration: Methods for quantifying edge effects and predicting the results of restoration efforts. Pp. 57-69 *in* D.A. Saunders, R.J. Hobbs, and P. Erhlich, eds., Nature Conservation 3: Reconstruction of Fragmented Ecosystems. Surrey Beatty and Sons, Sydney, Australia.

Skalski, J.R., and D.S. Robson. 1992. Techniques for Wildlife Investigations: Design and Analysis of Capture Date. Academic Press, San Diego, CA.

Smith, D.M. 1962. The Practice of Silviculture. 7th ed. John Wiley and Sons, New York, NY.

Smith, J.E., R. Molina, M.M.P. Huso, D.L. Luoma, D. McKay, M.A. Castellano, T. Lebel, and Y. Valachovic. 2002. Species richness, abundance, and composition of hypogeous and epigeous ectomycorrhizal fungal sporocarps in young, rotation-age, and old-growth stands of Douglas-fir (*Pseudotsuga menziesii*) in the Cascade Range of Oregon, U.S.A. Canadian Journal of Botany 80: 186-204.

Söderström, L. 1987. Dispersal as a limiting factor for distribution among epixylic bryophytes. Pp. 475-84 *in* T. Pócs, T. Simon, Z. Tuba, and J. Podani, eds., Proceedings of the IAB Conference of Bryoecology. Symposia Biologica Hungarica 35. Akadémiai Kiadó, Budapest, Hungary.

–. 1988. Sequence of bryophytes and lichens in relation to substrate variables of decaying coniferous wood in northern Sweden. Nordic Journal of Botany 8: 89-97.

Soltis, P.S., and M.A. Gitzendanner. 1999. Molecular systematics and the conservation of rare species. Conservation Biology 13: 471-83.

Soulé, M.E., and M.A. Sanjayan. 1998. Conservation targets: Do they help? Science 279: 2060-61.

Speight, M.C.D. 1989. Saproxylic Invertebrates and Their Conservation. Council of Europe, Strasbourg, France.

Spence, J.R., R.C. Cartar, M. Reid, D.W. Langor, and W.J.A. Volney. 2003. Dynamics of Arthropod Assemblages in Forests Managed to Emulate Natural Disturbance (BUGS). Final project report, Sustainable Forest Management Network, Edmonton, AB.

Spence, J.R., D.W. Langor, J.K. Niemelä, H. Carcamo, and C.R. Currie. 1996. Northern forestry and carabids: The case for concern about old growth species. Annales Zoologici Fennici 33: 173-84.

Spies, T.A., J.F. Franklin, and T.B. Thomas. 1988. Coarse woody debris in Douglas-fir forests of western Oregon and Washington. Ecology 69: 1689-1702.

Spribille, T., G. Thor, F.L. Bunnell, T. Goward, and C. Björk. 2008. Lichens on dead wood: species-substrate relationships in the epiphytic lichen floras of the Pacific Northwest and Fennoscandia. Ecography 31: (in press).

Stadt, J.J., J. Schieck, and H.A. Stelfox. 2006. Alberta biodiversity monitoring program: Monitoring effectiveness of sustainable forest management planning. Environmental Monitoring and Assessment 121: 33-46.

Stanford, J.A., and J.V. Ward. 1988. The hyporheic habitat of river ecosystems. Nature 335: 64-66.

–. 1993. An ecosystem perspective of alluvial rivers: Connectivity and the hyporheic corridor. Journal of the North American Benthological Society 12: 48-60.

Stanger, N. 2004. Edge and Age Effects on Epiphytes in a Lowland Douglas-Fir Forest. Report prepared for Weyerhaeuser, Nanaimo, BC.

Stankey, G.H., and B.A. Shindler. 1997. Adaptive Management Areas: Achieving the Promise, Avoiding the Peril. USDA Forest Service, General Technical Report PNW-GTR-394, Portland, OR.

Stefaniak, O., and S. Seniczak. 1978. The effect of fungal diet on the development of *Oppia nitens* and on the microflora of its alimentary tract. Pedobiologia 21: 202-10.
Stem, C., R. Margoluis, N. Salafsky, and M. Brown. 2005. Monitoring and evaluation in conservation: A review of trends and approaches. Conservation Biology 19: 295-309.
Stener, B. 2001. Research results prove it's cheaper and better with wood. Forestry Chronicle 77: 31.
Stevenson, S.K. 1978. Distribution and abundance of arboreal lichens and their use as forage by black-tailed deer. MSc thesis, University of British Columbia, Vancouver, BC.
Stevenson, S.K., and J.A. Rochelle. 1984. Lichen litterfall: Its availability and utilization by black-tailed deer. Pp. 391-96 *in* W.R. Meehan, T.R. Merrell Jr., and T.A. Hanley, eds., Proceedings, Symposium on Fish and Wildlife Relationships in Old-Growth Forests, 12-15 April 1982, Juneau, AK. American Institute of Fishery Research Biologists, Bethesda, MD.
Stokland, J.N. 1997. Representativeness and efficiency of bird and insect conservation in Norwegian boreal forest reserves. Conservation Biology 11: 101-11.
Strayer, D., D.H. Pletscher, S.P. Hamburg, and S.C. Nodvin. 1986. The effects of forest disturbance on land gastropod communities in northern New England. Canadian Journal of Zoology 64: 2094-98.
Strittholt, J.R., and D.A. DellaSala. 2001. Importance of roadless areas in biodiversity conservation in forested ecosystems: Case study of the Klamath-Siskiyou ecoregion of the United States. Conservation Biology 15: 1742-54.
Sturtevant, B.R., J.A. Bissonette, and J.N. Long. 1996. Temporal and spatial dynamics of boreal forest structure in western Newfoundland: Silvicultural implications for marten habitat management. Forest Ecology and Management 87: 13-25.
Sultan, S.E., A.M. Wilczek, S.D. Hann, and B.J. Brosi. 1998. Contrasting ecological breadth of co-occurring annual *Polygonum* species. Journal of Ecology 86: 363-83.
Swanson, F.J., J.A. Jones, D.O. Wallin, and J.H. Cissel. 1994. Natural variability: Implications for ecosystem management. Pp. 80-94 *in* M.E. Jensen and P.S. Bourgeron, tech. eds., Ecosystem Management: Principles and Applications. Volume 2. USDA Forest Service, General Technical Report PNW-GTR-318, Portland, OR.
Symstad, A.J., J. Willson, and J.M.K. Knops. 1998. Species loss and ecosystem functioning: Effects of species identity and community composition. Oikos 81: 389-97.
Tansley, A.G. 1935. The use and abuse of vegetational concepts and terms. Ecology 16: 284-307.
Thomas, C.D. 1991. Spatial and temporal variability in a butterfly population. Oecologia 87: 577-80.
Thomas, C.D., J.A. Thomas, and M.S. Warren. 1992. Distributions of occupied and vacant butterfly habitats in fragmented landscapes. Oecologia 92: 563-67.
Thomas, J.W., ed. 1979. Wildlife Habitats in Managed Forests: The Blue Mountains of Oregon and Washington. USDA Forest Service Agricultural Handbook 553, Washington, DC.
Thomas, J.W., M.G. Raphael, R.G. Anthony [plus others]. 1993. Viability Assessments and Management Considerations for Species Associated with Late Successional and Old-Growth Forests of the Pacific Northwest. Portland, Oregon. USDA For.Serv. 523 pp.
Thompson, I.D. 2006. Monitoring of biodiversity indicators in boreal forests: a need for improved focus. Environmental Monitoring and Assessment 121: 263-73.
Thompson, I.D., and W.J. Curran. 1995. Habitat suitability for marten of second growth balsam fir forests in Newfoundland. Canadian Journal of Zoology 73: 2059-64.
Thompson, W.L. 2002. Towards reliable bird surveys: Accounting for individuals present but not detected. Auk 119: 18-25.
Thor, G. 1998. Red-listed lichens in Sweden: Habitats, threats, protection, and indicator value in boreal coniferous forests. Biodiversity and Conservation 7: 59-72.
Tilman, D., and A. Downing. 1994. Biodiversity and stability in grasslands. Nature 367: 363-65.
Trofymow, J.A., J. Addison, B.A. Blackwell, F. He, C.A. Preston, and V.G. Marshall. 2003. Attributes and indicators of old growth and successional Douglas-fir forests on Vancouver Island. Environmental Reviews 11: S187-204.

Trombulak, S.C., and C.A. Frissell. 2000. Review of ecological effects of roads on terrestrial and aquatic communities. Conservation Biology 14: 18-30.

Tscharntke, T. 1992. Fragmentation of *Phragmites* habitats, minimum viable population size, habitat suitability, and local extinction of moths, midges, flies, aphids, and birds. Conservation Biology 6: 530-36.

Turin, H., and P.J. Den Boer. 1988. Changes in the distribution of carabid beetles in the Netherlands since 1980. I. Isolation of habitats and long-term time trends in the occurrence of carabid species with different powers of dispersal (*Coleoptera, Carabidae*). Biological Conservation 44: 179-200.

UN (United Nations). 1992. Convention on Biological Diversity. New York, NY.

UNESCO. 1974. Task force on criteria and guidelines for the choice and establishment of biosphere reserves. Man and the Biosphere Report 22: 1-61.

Usher, M., J.P. Field, and S.E. Bedford. 1993. Biogeography and diversity of ground-dwelling arthropods in farm woodlands. Biodiversity Letters 1: 54-62.

Van der Wal, R., I.S. Pierce, and R.W. Brooker. 2005. Mosses and the struggle for light in a nitrogen-polluted world. Oecologia 142: 159-68.

Vanha-Majamaa, I., and J. Jalonen. 2001. Green tree retention in Fennoscandian forestry. Scandinavian Journal of Forest Research 3 (supplement): 79-90.

Veblen, T.T., and P.B. Alaback. 1996. A comparative review of forest dynamics and disturbance in the temperate rainforests of North and South America. Pp. 173-213 *in* R.G. Lawford, P.B. Alaback, and E. Fuentes, eds., High-Latitude Rainforests and Associated Ecosystems of the West Coast of the Americas. Springer-Verlag, New York, NY.

Vernier, P.R. 2004. Avian Indicator Modeling on Weyerhaeuser's Tenure on Vancouver Island, B.C. Report to the Sustainable Forest Management Network. Available at: http://biod.forestry.ubc.ca/sfm/tfl39/pdf/tfl39_modeling.pdf.

Vernier, P.R., and F.L. Bunnell. 2002. Habitat Associations of Northern Goshawk Nest Sites at Multiple Spatial Scales on Canfor's Forest Tenure on Vancouver Island. Report to Canadian Forest Products, Vancouver Island, BC.

Vernier, P.R., F.K.A. Schmiegelow, and S.G. Cumming. 2002. Modeling bird abundance from forest inventory data in the boreal mixedwood forest of Canada. Pp. 559-72 *in* J.M. Scott, P.J. Heglund, M. Morrison, M. Raphael, J. Haufler, and B. Wall, eds., Predicting Species Occurrences: Issues of Scale and Accuracy. Island Press, Covello, CA.

Villa-Castillo, J., and M.R. Wagner. 2002. Ground beetle (Coleoptera: Carabidae) species assemblage as an indicator of forest condition in northern Arizona ponderosa pine forests. Environmental Entomology 31(2): 242-52.

Vyse, A., C. Hollstedt, and D. Huggard, eds. 1998. Managing the Dry Douglas-Fir Forests of the Southern Interior: Workshop Proceedings, 29-30 April 1997, Kamloops, BC. Working Paper 34/1998. Research Branch, Ministry of Forests, Victoria, BC.

Wahlberg, N., A. Moilanen, and I. Hanski. 1996. Predicting the occurrence of endangered species in fragmented landscapes. Science 273: 1536-38.

Wall, D.H., and J.C. Moore. 1999. Interactions underground: Soil biodiversity, mutualism, and ecosystem processes. Bioscience 49: 109-17.

Wallenius, T. 2002. Forest age distribution and traces of past fires in a natural boreal landscape dominated by *Picea abies*. Silva Fennica 36: 201-11.

Walters, C.J. 1986. Adaptive Management of Renewable Resources. McMillan, New York, NY.

–. 1997. Challenges in adaptive management of riparian and coastal ecosystems. Conservation Ecology 1(2): 1. http://www.consecol.org/vol1/iss2/art1.

Wang, S. 2002. Wicked problems and metaforestry: Is the era of management over? Forestry Chronicle 78: 505-10.

Wästerlund, I., and T. Ingelög. 1981. Fruit body production of larger fungi in some young Swedish forests with special reference to logging waste. Forest Ecology and Management 3: 269-94.

Waterhouse, M.J., H.M. Armleder, and R.J. Dawson. 1991. Forage Litterfall in Douglas-Fir Forests in the Central Interior of British Columbia. Research Note No. 108. Ministry of Forests, Victoria, BC.

Weetman, G. 1998. A forest management perspective on sustained site productivity. Forestry Chronicle 74: 75-77.

Welch, N.E., and J.A. MacMahon. 2005. Identifying habitat variables important to the rare Columbia spotted frog in Utah (USA): An information-theoretic approach. Conservation Biology 19: 473-81.

Wells, R.W., F.L. Bunnell, D. Haag, and G. Sutherland. 2003. Evaluating ecological representation within different planning objectives for the central coast of British Columbia. Canadian Journal of Forest Research 33: 2141-50.

West, N.E. 1990. Structure and function of microphytic soil crusts in wildland ecosystems of arid to semi-arid regions. Advances in Ecological Research 20: 179-223.

–. 1994. Biodiversity and land use. Pp. 21-36 *in* W.W. Covington and L.F. DeBano, tech. co-ords., Sustainable Ecological Systems: Implementing an Ecological Approach to Land Management. USDA Forest Service, General Technical Report GTR-RM-247, Fort Collins, CO.

Wilcove, D.S., D. Rothstein, J. Dubow, A. Phillips, and E. Losos. 1998. Quantifying threats to imperiled species in the United States. BioScience 48: 607-15.

Wilhere, G.F. 2002. Simulations of snag dynamics in an industrial Douglas-fir forest. Forest Ecology and Management 174: 521-39.

Williamson, M. 1975. The design of wildlife preserves. Nature 256: 519.

Williston, P. 2000. Floristics and distribution patterns of lichens and bryophytes in microbiotic crusts of British Columbia's ponderosa pine forests. Pp. 769-78 *in* L.M. Darling, ed., At Risk: Proceedings of a Conference on Biology and Management of Species and habitats at Risk, 15-19 February 1999, Kamloops, BC. Ministry of Environment, Lands, and Parks, Victoria, BC.

Willott, S.J. 1999. The effects of selective logging on the distribution of moths in a Bornean rainforest. Philosophical Transactions of the Royal Society 354(1391): 1783-90.

Wilson, E.O. 1988. The current state of biodiversity. Pp. 3-18 *in* E.O. Wilson and F.M. Peter, eds., Biodiversity. National Academy Press, Washington, DC.

–. 1992. The Diversity of Life. Belknap Press, Cambridge, MA.

Wilson, E.O., and E.O. Willis. 1975. Applied biogeography. Pp. 522-34 *in* M.L. Cody and J.M. Diamond, eds., Ecology and Evolution of Communities. Harvard University Press, Cambridge, MA.

Wind, E. 2005. Aquatic-Breeding Amphibian Monitoring Program Pre-Treatment Wetland Buffer Study. Annual progress report, 2004, for Weyerhaeuser BC Coastal Group, Nanaimo, BC.

Wood, P. 1997. Biodiversity as the source of biological resources: A new look at biodiversity values. Environmental Values 6: 251-68.

–. 2000. Biodiversity and Democracy: Rethinking Nature and Society. UBC Press, Vancouver, BC.

Woodward, A., K.J. Jenkins, and E.G. Schreiner. 1999. The role of ecological theory in long-term ecological monitoring: Report on a workshop. Natural Areas Journal 19: 223-33.

WRI (World Resources Institute), IUCN (The World Conservation Union), UNEP (United Nations Environment Programme), FAO (Food and Agriculture Organization of the United Nations), and UNESCO (United Nations Education, Scientific, and Cultural Organization). 1992. Global Biodiversity Strategy: Guidelines for Action to Save, Study, and Use Earth's Biotic Wealth Sustainably and Equitably. WRI, Washington, DC.

Yin, X. 1999. The decay of forest woody debris: Numerical modeling and implications based on some 300 data cases from North America. Oecologia 121: 81-98.

Zackrisson, O., M.C. Nilsson, and D.A. Wardle. 1996. Key ecological function of charcoal from wildfire in the boreal forest. Oikos 77: 10-19.

Zartman, C.E. 2003. Habitat fragmentation impacts on epiphyllous bryophyte communities in central Amazonia. Ecology 84: 948-54.

Zielke, K., and W. Beese. 1999 (revised 2002). SPs for VR: Guidelines for Designing Variable Retention – Layout and Silvicultural Prescription. Weyerhaeuser BC Coastal Group, Nanaimo, BC.

List of Contributors

Bill Beese is Forest Ecologist for Western Forest Products. He is a Registered Professional Forester with a BSc in Forest Management (Southern Illinois) and MF in Forest Ecology (UBC). His career includes twenty-eight years in research on the BC coast, environmental consulting, and policy development for several forest companies. His research has focused on silvicultural systems, prescribed burning, forest regeneration and biodiversity, and includes coordination of multi-disciplinary research. His leadership helped accomplish phase-in of variable retention harvesting as part of the innovative program described in this book – a team effort that received the Ecological Society of America's Corporate Award for 2001. He has served on numerous research advisory and review committees and working groups on old-growth forests and ecosystem-based management. He was recently a member of an international science panel advising Forestry Tasmania on implementation of variable retention in old-growth eucalyptus forests.

Fred Bunnell is Professor Emeritus, Department of Forest Sciences, University of British Columbia where he has taught for thirty-five years. His higher education was in British Columbia (UBC), Switzerland (ETH, Zurich) and California (University of California, Berkeley). He has received provincial, national, and international awards for his diverse research interests, but over the past twenty-five years he has focused on relations between forestry and wildlife or biodiversity. He has held elected offices in provincial, national, and international scientific associations, served on over eighty committees dealing with resource management, and held commissions on resource management, which include serving as Independent Chair of the Scientific Panel for Sustainable Forest Practices in Clayoquot Sound. He has worked in the following countries: Finland, Japan, Malaysia, Norway, Panama, Sweden, Turkey, the United Kingdom, and the United States.

Glen Dunsworth is an ecological consultant providing services in forest resource management, conservation biology, and strategic planning. He has extensive experience in the BC coastal forest industry with Macmillan Bloedel and Weyerhaeuser, where he directed regeneration and biodiversity research and developed effective new strategic approaches to ecosystem management. He has facilitation, organizational, and project management experience in landscape and strategic planning.

The first graduate of the forestry program at the University of Alberta, he has an MSc in Forest Genetics and specializes in biodiversity, genetics, landscape ecology, and ecosystem-based management. Glen has authored over forty publications and has been associated with WWF Canada through their Pacific Region office. He has also provided ecological strategic planning advice to a number of coastal BC forest companies and conservation organizations. His leadership experience has helped him administer large, multi-disciplinary teams, and he is a member in good standing of the College of Applied Biology. His most recent experience has been in developing and implementing an adaptive management and monitoring program in support of Weyerhaeuser's Coast Forest Strategy, a major shift in forest management from a clear-cut to a variable retention approach.

David Huggard is a consultant in forest ecology and statistics. He holds an MSc in zoology and a PhD in forest ecology from the University of British Columbia. His research has included large mammal population ecology; effects of alternative forest harvesting and silviculture on carnivores, forest grouse, songbirds, small mammals, invertebrate groups, and habitat structures; landscape analysis; and forest projection modeling. He has contributed to large-scale monitoring programs for industrial, government, and academic clients in several jurisdictions in western Canada.

Laurie Kremsater is a project manager at the University of British Columbia and an independent consultant in the field of forest wildlife ecology. She has an MSc from UBC and is a Professional Biologist (RPBio) and Forester (RPF). She began her research on responses of black-tailed deer to forest edges and has since expanded her interests to include biodiversity more broadly (songbird, small mammal, and amphibian surveys) in many areas of BC. Her focus over the last decade has been on devising management strategies and monitoring programs to sustain biodiversity for forest companies and for government. She was a member of the Scientific Panel for Sustainable Forest Practices in Clayoquot Sound and of the Adaptive Management Working Group that helped establish and implement the program described in this book.

Jeff Sandford is an honours graduate in arboriculture with twenty-one years of experience in forestry research and project coordination with MacMillan Bloedel Limited, Weyerhaeuser, and Western Forest Products Inc. Jeff has strong skills in project management, database development, global positioning system (GPS) technology, geographic information systems (GIS), native plant ecology, and digital data collection, handling, and analysis. He has managed numerous silvicultural research projects involving planning, design, establishment, measurement, maintenance, analysis, and report writing. Jeff has managed the forest habitat structure component of the integrated adaptive management program described in this book for the last nine years. That work involved the development of a database management system, detailed methodology, software programs for data collection, mapping, implementation, contract management, and crew training. In the last four years, Jeff has been the Forest Investment Account program coordinator for Weyerhaeuser's BC Coastal Group and Western Forest Products Inc. This government-funded program was designed to aid in the development of a globally recognized, sustainably managed forest industry.

Index

Note: "(a)" after a page number indicates an appendix; "(f)" after a page number indicates a figure; "(m)" after a page number indicates a map; "(t)" after a page number indicates a table

Printed and bound in Canada by AGMV Marquis
Set in Stone by Artegraphica Design Co. Ltd.
Copy editor: Dallas Harrison
Proofreader and indexer: Dianne Tiefensee
Cartographer: Eric Leinberger

Québec, Canada
2009